Emil & Kathlee

Western History and Biography

Emil & Kathleen Sick Series in
Western History and Biography

With support from the Center for the Study of the Pacific Northwest at the University of Washington, the Sick Series in Western History and Biography features scholarly books on the peoples and issues that have defined and shaped the American West. Through intellectually challenging and engaging books of general interest, the series seeks to deepen and expand our understanding of the American West as a region and its role in the making of the United States and the modern world.

ATOMIC FRONTIER DAYS

Hanford and the American West

JOHN M. FINDLAY AND BRUCE HEVLY

Center for the Study of the Pacific Northwest

in association with

University of Washington Press | Seattle and London

CENTER FOR THE STUDY OF THE PACIFIC NORTHWEST
P.O. Box 353587, Seattle, WA 98195 U.S.A.

UNIVERSITY OF WASHINGTON PRESS
P.O. Box 50096, Seattle, WA 98145, U.S.A.
www.washington.edu/uwpress

LIBRARY OF CONGRESS
CATALOGING-IN-PUBLICATION DATA
Findlay, John M., 1955-
Atomic frontier days : Hanford and the American West /
John M. Findlay and Bruce Hevly.
  p. cm. — (Emil and Kathleen Sick series in Western history and biography)
Includes bibliographical references and index.
ISBN 978-0-295-99097-2 (pbk. : alk. paper)
1. Nuclear facilities—Waste disposal—Washington (State)—Hanford Site.
2. Radioactive waste disposal—Washington (State)—Hanford Site.
3. Nuclear reactors—Washington (State)—Hanford Site.
4. Hanford Site (Wash.) I. Hevly, Bruce William. II. Title.
TD898.12.W2F56 2011   623.4'51190979751—dc22   2011004808

Printed in the United States of America
Design by Ashley Saleeba / Composed in Minion Pro and Futura

Contents

Illustrations

Acknowledgments

We are grateful to many people and organizations for supporting this project. Our research benefited early on from funding from the U.S. Department of Energy (Cooperative Agreement no. DE-FC06-91-RL12260). We are indebted to Tom Bauman and especially Marji Parker of the Richland Operations Office for overseeing the grant and providing so much encouragement. The Cooperative Agreement helped us to produce a manuscript called "Nuclear Technologies and Nuclear Communities: A History of Hanford and the Tri-Cities" (1995, available in PRR). The project received financial support as well from the University of Washington Graduate School Research Fund, and from the Department of History, University of Washington, and in particular the Howard and Frances Keller Endowed Fund to support faculty research and the Lenore Hanauer Fund.

We received considerable support at the many libraries we visited, and extend our heartfelt thanks to the staffs at all of them: Special Collections, University of Washington Library, Seattle (where Carla Rickerson and Karyl Winn deserve special mention); the U.S. Department of Energy Public Reading Room in Richland, Washington (where Terri Traub proved as kind as she was helpful); the Richland Public Library, Richland, Washington; the Hagley Museum and Library, Wilmington, Delaware; the Penrose Library, Whitman College, Walla Walla, Washington; the Office of History and Heritage Resources, U.S. Department of Energy, Germantown, Maryland (where Roger Anders, Benjamin Franklin Cooling, and Terrence Fehner were most accommodating); the National Archives Center, Southeast Region, Morrow, Georgia (where Charlie Reeves guided us to wonderful materials); the National Archives Regional Facility, Seattle; and the National Archives in Washington, D.C. In searching for photographs, we benefited from the deep knowledge and keen interest of Dan Ostergaard, a photographic specialist who works for Lockheed Martin, a DOE subcontractor in the Tri-Cities, Washington. Gabriele Sperling of the Franklin County Historical Society and Museum, Pasco, Washington, and Corene Hulse of the East Benton County Historical Museum,

Kennewick, Washington, also helped us identify suitable photographs. Frederick Bird conscientiously prepared the maps for the book.

Over the years, many graduate students served as research assistants on this project. We thank Jennifer Alexander, Sharon Boswell, John Eby, Glen Furnas, Lorraine McConaghy, Chad Moody, Kathy Morse, Will Prust, Joe Roza, Robert Self, and Moran Tompkins. Matt Klingle shared his related research with us. We benefited from conversations with several former Hanford employees, including in particular the late Fred Clagett and the late Merle Harmon. Doug Tillson of Kennewick provided a guided tour of the hydroplane pits once home to the Atomic Cup during a Columbia Cup race. Over the years, numerous scholars listened to our ideas and offered thoughts or nuggets from their own research. Paul Forman of the Smithsonian's National Museum of American History was an informative companion for a tour of the B reactor early in our work. We are particularly grateful to colleagues Dan Pope and Michele Gerber, who commented helpfully on the entire manuscript at a late stage, and for their own scholarship on nuclear energy in Washington State. Finally, we appreciate the close reading by three anonymous readers for the University of Washington Press, the editing of Kerrie Maynes, and the patience and encouragement of Julidta Tarver and Marianne Keddington-Lang.

Hanford Timeline

JUNE 1942: President Franklin D. Roosevelt authorizes what becomes the Manhattan Engineer District, U.S. Army Corps of Engineers, to oversee wartime development of the atomic bomb, and commits the nation to a crash program to create the new weapon.

DECEMBER 1942: An experimental reactor or "pile" at the University of Chicago, under supervision of physicist Enrico Fermi, produces the world's first sustained, controlled chain reaction.

DECEMBER 1942: Dispatched to the American West to find prospective sites for the Manhattan Project's plutonium factory, Franklin T. Matthias identifies Hanford as the place best suited to the program's needs.

FEBRUARY 1943: The War Department finalizes plans to acquire the Hanford Site, and build and operate Hanford Engineer Works, and announces its intent to acquire the land.

MARCH 1943: Local residents and property owners receive letters from the federal government leading to the condemnation of private landholdings within a 670-square-mile reservation. Meanwhile, construction begins on the first buildings for Hanford.

SEPTEMBER 1944: The world's first production reactor, Hanford's B reactor is tested and begins making plutonium.

DECEMBER 1944: Plutonium processing begins in Hanford's 200 Area.

FEBRUARY 1945: With three new reactors operating and plutonium processing plants at work, Hanford makes the first shipment of plutonium to Los Alamos, New Mexico. The Hanford Camp shuts down as construction draws to a close and production operations accelerate.

JULY 1945: Hanford ships the one hundredth wartime batch of plutonium to Los Alamos. Moreover, the Manhattan Project detonates the world's first atomic bomb (the Trinity test) near Alamogordo, New Mexico, using Hanford plutonium for fuel.

AUGUST 1945: U.S. Army Air Corps planes drop the uranium bomb on Hiroshima (August 6) and the plutonium bomb on Nagasaki (August 9). The Manhattan Project, including the Hanford Site, is disclosed to the public for the first time, and World War II comes to an end.

DECEMBER 1945: DuPont announces it will leave the Hanford project.

SEPTEMBER 1946: The General Electric (GE) company replaces DuPont as the government's primary contractor at Hanford.

JANUARY 1947: The U.S. Atomic Energy Commission (AEC), a new civilian agency, takes over for the U.S. Army as the government bureau in charge of America's atomic-weapons complex, including the Hanford Site.

FEBRUARY 1947: The AEC sets out the goal of "normalizing" the towns it owns and operates as part of making atomic weapons.

AUGUST 1947: The AEC announces a large-scale expansion program for Hanford, confirming the site's continuing role as the nation's sole supplier of weapons-grade plutonium as the Cold War crystallizes. Hanford adds five new reactors by 1955, and revamps and expands the three reactors built during World War II.

NOVEMBER 1947: The *Tri-City Herald* begins publication.

JUNE 1948: The Richland Community Council holds its first meeting, in response to the AEC's initiative to "normalize" atomic communities.

AUGUST 1948: Richland holds its first Atomic Frontier Days celebration.

FEBRUARY 1949: The American Metal Trades Council of the American Federation of Labor becomes the collective bargaining agent for Hanford operations employees, and signs the first union contract with GE and the AEC in May 1949.

AUGUST 1949: The Soviet Union successfully detonates its first atomic bomb.

DECEMBER 1949: The infamous Green Run occurs at Hanford, during which the AEC intentionally releases radioactive iodine into the atmosphere as part of an experiment.

APRIL 1950: The U.S. Bureau of the Census reports that Richland has 21,809 people; Pasco 10,228; and Kennewick 10,106.

AUGUST 1951: The AEC releases the findings of its Panel on Community Operations, known as the Scurry Report, which maps out the steps to making Richland, Oak Ridge, and Los Alamos independent of the federal government.

OCTOBER 1951: Congressman Henry M. Jackson (D-WA) asserts before Congress that the United States needs to produce as many nuclear weapons as it can possibly afford to make, an argument that evolves into the contention that the United States cannot have enough nuclear weapons.

NOVEMBER 1952: The United States detonates the world's first hydrogen bomb.

AUGUST 1955: President Dwight D. Eisenhower signs legislation authorizing the sale of property and the transfer of municipal government to citizens in Richland and Oak Ridge.

JUNE 1956: Senator Henry M. Jackson introduces the first of several bills to Congress to authorize construction of the world's first dual-purpose reactor at Hanford, and at the same time pressures officials of the Federal Housing Administration to reduce the appraised value of homes to be sold in Richland.

JUNE 1957: The U.S. government begins to sell Richland homes and businesses to private owners.

JANUARY 1958: Construction begins on Hanford's N reactor, the ninth on the site, which will begin producing plutonium in December 1963.

DECEMBER 1958: Richland formally incorporates as its own municipality.

APRIL 1961: *Look* magazine and the National Municipal League proclaim Richland an All-America City.

FEBRUARY 1963: The Tri-City Nuclear Industrial Council is founded.

SEPTEMBER 1963: President John F. Kennedy visits Hanford to help break ground on construction of a nuclear power generating plant to be attached to the N reactor, realizing the dream of a dual-purpose reactor at the site.

JANUARY 1964: President Lyndon B. Johnson announces that the AEC will shut down three Hanford reactors, and the AEC announces a program of "segmentation" and "diversification" at Hanford—meaning that work at Hanford will be segmented among multiple contractors (rather than being concentrated in the hands of one firm such as GE) and that contractors will be required to invest in the diversification of the Tri-City economy.

OCTOBER 1964: The state of Washington produces a "master plan" for profitable participation in the nuclear age.

JANUARY 1967: The AEC chooses Hanford as the site of the Fast Flux Test Facility.

MARCH 1967: The AEC announces the creation of Hanford's Arid Lands Ecology Reserve.

JANUARY 1971: The administration of President Richard M. Nixon announces that Hanford's KE reactor will be shut down, marking the end of plutonium manufacture by reactors completed during World War II and the early Cold War. Only the dual-purpose N reactor will remain in operation—although Nixon proposes closing it down, too.

SEPTEMBER 1973: The Washington Public Power Supply System (WPPSS) begins construction on three nuclear power plants in the state—one at Satsop and two at Hanford. Eventually it will plan and start building an additional reactor at each site, for a total of five.

JANUARY 1975: The U.S. Energy Research and Development Administration (ERDA) and the U.S. Nuclear Regulatory Commission replace the AEC.

OCTOBER 1977: The U.S. Department of Energy succeeds the ERDA.

MARCH 1979: An accident occurs at the Three Mile Island nuclear power plant in Pennsylvania.

APRIL 1979: Washington governor Dixy Lee Ray identifies Hanford as a promising site for storing the nation's nuclear wastes.

NOVEMBER 1980: Washington voters pass a law, Initiative 393, banning the import of nuclear waste into the state. The measure is overturned by the courts beginning in June 1981.

JULY-AUGUST 1983: Financing for the WPPSS collapses, creating the largest default on municipal bonds in U.S. history.

NOVEMBER 1983: The PUREX processing plant is restarted for the first time in eleven years, marking the resumption of Hanford's production of weapons-grade plutonium as its contribution to the defense buildup of the administration of President Ronald Reagan.

SEPTEMBER 1984: The WPPSS no. 2 reactor at Hanford is dedicated—the only one of five planned plants to be completed and put into operation.

MARCH 1985: The Tri-City Nuclear Industrial Council changes its name to the Tri-City Industrial Development Council, deciding that focusing solely on "nuclear" industry no longer makes a lot of sense.

FEBRUARY 1986: The DOE releases 19,000 pages of documents detailing the history of the site, including the threat it posed to the surrounding environment and population. This is followed by releases of additional thousands of pages.

APRIL 1986: An accident occurs at the nuclear power plants at Chernobyl in the Ukraine.

JANUARY 1987: The N reactor is shut down, bringing to an end Hanford's mission of producing weapons-grade plutonium.

DECEMBER 1987: Nevada's Yucca Mountain site selected as the sole prospective national repository for high-level nuclear waste, ending consideration of Hanford for the same purpose and terminating the 1,200 jobs associated with the Basalt Waste Isolation Project.

JANUARY 1988: The Hanford Environmental Dose Reconstruction Project is launched.

JANUARY 1989: The Hanford Thyroid Disease Study begins.

MAY 1989: The U.S. Department of Energy, the U.S. Environmental Protection Agency, and the Washington State Department of Ecology sign the Hanford Federal Facility Agreement and Consent Order (also known as the Tri-Party Agreement), creating a legal and financial framework for long-term cleanup at the Hanford Site.

APRIL 1999: The DOE determines that the federal government shall retain control over more than 50,000 acres of Wahluke Slope, incorporating the land into the Saddle Mountain National Wildlife Refuge.

JUNE 2000: The administration of President William J. Clinton creates the Hanford Reach National Monument by executive order, thus protecting from further development more than fifty miles of the Columbia River adjacent to the Hanford Site. The Hanford Reach becomes the first national monument within the U.S. Fish and Wildlife Service system.

ATOMIC
FRONTIER
DAYS

INTRODUCTION

ELEVEN DAYS AFTER HIROSHIMA—ON AUGUST 17, 1945—Mrs. J. W. Nichols of Grants Pass, Oregon, wrote a letter to Colonel Franklin T. Matthias, the army officer who had overseen the construction and operation of the Manhattan Project's plutonium factory at Hanford in Washington State during World War II. She told Matthias about a ranch she owned in northeastern Oregon, property she was sure contained uranium ore that she would gladly sell to Hanford. "Thot [*sic*] maybe you folks who make the bombs would be interested in finding the ore close at hand. . . . I am a widow of a Vet. If interested please answer this." Mrs. Nichols signed her name and then added, "P.S. Your bombs are certainly wonderful."[1]

The letter from Grants Pass—consulted in a National Archives branch office in an Atlanta, Georgia, suburb—is the kind of thing that tends to get overlooked by those contemplating Americans' experience with atomic weapons. Historians have, for good reason, tended to put the bombs front and center, rather than tacking them on as an afterthought. Yet Mrs. Nichols's missive reminds us that for many people the bombs were *not* the most important thing about the Manhattan Project. Like many others,

Mrs. Nichols wanted to know whether the manufacture of nuclear weapons could have a direct effect on her life. She worried less about the extraordinary product that Hanford helped to make, and more about incorporating the facility into the ordinary economy, politics, and society of the American West and Pacific Northwest. In many ways, her concern about the relationship between atomic bombs and regions is ours, too.

Most other students of U.S. nuclear weapons programs have had other priorities. Those who have written about the Manhattan Project and its successors have tended to focus on the atomic bomb as a *national* story, a tale of the entire country's accomplishment that emphasizes the federal government. In their telling, the development of nuclear weapons begins in Washington, D.C., and then spills over into outlying areas. There is no doubt that through development of the bomb the federal government transformed the rest of the United States. But it is equally true that outlying areas of the country, including that around Hanford, exerted a surprising amount of influence on decisions made in Washington, D.C., regarding the nation's nuclear-weapons complex.

Our study of Hanford diverges from the standard accounts in another regard. Most students of America's atomic age, besides focusing on prominent federal officials, pay an inordinate amount of attention to such leading scientists as Albert Einstein, J. Robert Oppenheimer, Leo Szilard, and Edward Teller. These individuals make compelling figures around which stories can be organized, yet they played a very small role at Hanford, which was always more industrial than scientific in orientation. While the creation of nuclear weapons depended upon national political and scientific networks, it also would require the United States to become "one huge factory," as Danish physicist and Nobel Prize winner Niels Bohr had predicted to American scientists in 1939. We argue here that it is crucial to recognize Hanford's industrial character, partly because confusion persists about whether to assign it to a culture of production or one of research.[2] Most regions of the country underwent some industrialization through their participation in the production of nuclear weapons during World War II and the Cold War, but the lightly settled American West received much more than its share of such manufacturing. At such locales as Rocky Flats, Colorado; Pantex, Texas; and the Nevada Test Site, and especially at Hanford, the U.S. government and its private contractors performed the routine, often dirty work that is the hallmark of a large industrial process.

In offering an account of Hanford that prioritizes local and regional factors and that emphasizes the industrial nature of the site, this book adds to what might be thought of as a third generation of scholarship about America's nuclear weapons programs. A few words about the first and second generations of study are in order. Initial histories of the site were written or sponsored by the organizations responsible for building and operating it—the army, the DuPont company, and the Atomic Energy Commission (AEC). These accounts tended to have an eastern orientation. The official history of the Atomic Energy Commission, for instance, treats Hanford as a satellite of the nation's capital.[3] The leading figures in these early accounts tended to be political leaders, military officers, and prominent academic scientists. Events and workers at Hanford—a production center—seemed to be of less interest.

In early institutional histories of the U.S. atomic weapons program, moreover, the story was mostly one of triumph. While it is not terribly surprising that accounts sponsored by the army, DuPont, and the AEC tended to make the army, DuPont, and the AEC look good, there is more to the upbeat accounts. The earliest histories of nuclear weapons in the United States tended to emphasize the period of World War II, when numerous obstacles were overcome, new technologies were developed and tested, and America's atomic bombs sealed victory in the war against Japan, bringing an end to a prolonged global conflict. For many, World War II remains Hanford's historical heyday.

During the 1980s, however, a second generation of historical accounts of Hanford began to emerge, as did general accounts of the Manhattan Project and its aftermath. The nation's antinuclear movements contributed to this reassessment, and so did the federal government's decision to release formerly secret information about Hanford's operations, including data on the radioactive emissions that emerged as by-products of the nuclear process. Motivated by heightened concern that Hanford had negative effects on the environment and on public health, these new accounts differed substantially from their predecessors. For one thing, they focused attention very closely on the Hanford Site and on the downwind and downstream populations affected by the plutonium factory's wastes. No longer seen as being on the margins of a national project, Hanford played a central role in a different kind of story.

Second-generation authors focused on quite local—in fact, personal—stories of Hanford. They asked, especially, whom Hanford had injured.[4] During this second phase of study, scientists became historians in order to study the effects of Hanford emissions on downwind and downstream populations. So far their findings have not fully validated the popular perception that Hanford's operations increased the number of cases of disease, particularly of the thyroid, in surrounding areas. Plaintiffs and their attorneys became historians as well, arguing in court that Hanford caused illnesses including thyroid cancer. To date, their lawsuits have produced mixed results, with some "bellwether" verdicts decided in favor of plaintiffs and some in favor of the government and its Hanford contractors.[5]

Caught between first- and second-generation perspectives, the story of Hanford has become bifurcated. In earlier histories, Hanford and the Tri-Cities helped to solve the nation's problems. In later accounts, Hanford became primarily a source of problems. The first approach tended to emphasize the "bigger picture": Hanford helped to protect the nation during times of crisis, contributing to victory in World War II and the Cold War. In such accounts, little (if any) attention was paid to the site's effects on neighbors. The second approach tended to highlight the environmental and medical damage that Hanford caused to very specific places and individuals, without sufficiently explaining the context in which the damage was done. These accounts, for example, did little to evoke the broad political consensus that supported American participation in World War II and that perceived Stalin's Soviet Union as a threat. At places in some more recent accounts of Hanford, readers would have a hard time remembering that there had been a war on. In reviewing books about American downwinders, the journalist Thomas Powers explained the impact of this kind of polarization of attitudes in views of the past: "You can persuade some Americans that the downwinders were callously abused for no reason at all, and you can persuade some Americans that a resolute military posture and a willingness to confront Soviet expansionism won the Cold War at a modest cost, all things considered. But each of the two groups seems unable to grasp what the other is getting at."[6]

In this book, we want to help different groups see what the other is getting at. Hanford need not be cast as *either* a contributor to U.S. success on the global stage *or* a source of quite local afflictions in the Pacific Northwest. It was surely both at the same time. One task taken up by a third generation of historians is to tally up Hanford's costs and benefits, weighing its contributions against the problems it caused.[7] Yet Hanford cannot be fully understood in terms of ledgers of credits and debits. It was more complicated and more interesting than that. Moreover, previous historians have simply overlooked many aspects of the story, including ones that, if studied more closely, would have enabled them to better understand their specific topics of interest. The chapters that follow try to fill in some of the gaps and provide more historical context and, by so doing, encourage a less polarized understanding of the place and its significance. At the same time, they leave issues for others to study. We do little, for instance, that overlaps with the efforts of scientists, attorneys, judges, and juries to determine the extent to which Hanford put people's health at risk. Similarly, we have tried not to repeat the work on Hanford's environmental history that has been done by others.

Our approach is shaped in large part by our backgrounds. One of us (Bruce Hevly) works in the history of science and technology, and brings to this account a concern for the nature of the work undertaken at Hanford and the employees performing that work. Bruce also aims to explore the roles that the site played in larger scientific enterprises. Given Bruce's relative familiarity with physics, our chapters pay greater attention to Hanford's reactors than they do to its chemical-processing facilities. The other one of us (John Findlay) works in the realm of urban history, and brings to the study a concern for the development of communities surrounding Hanford. Both of us have long been concerned with the study of regions—the American West as well as the Pacific Northwest. More than anything else, perhaps, it is our concern with region that sets this history apart from others.

It is our view that a regional approach to Hanford offers a valuable *middle* way into the subject. We mean "middle" both in the sense of a level between the national and the very local or personal scales, and in the sense of a position between the triumphalist and accusatory accounts that have preceded ours. For us, "region" has meaning at more than one scale: as the grouping of towns that together became known as the Tri-Cities; as the Columbia Basin; as the state of Washington; as the Pacific Northwest; and as the American West. Considering Hanford's significance at these levels allows us to bring a different perspective to the plutonium factory. New insights can be gained by understanding Hanford within the context of the history of specific sections of the country. It turned out to be crucial, for example, that Hanford came to be located in a region where most inhabitants—such as Mrs. J. W. Nichols—believed that their economy and society were underdeveloped.

Yet our account is by no means strictly regional in approach. In fact, in trying to account for local, regional, and national perspectives, we have benefited from exploring relationships between places. For example, our understanding of Hanford and the Tri-Cities stems in large part from emphasizing connections between local communities, the Columbia Basin region, and the nation, particularly as they were mediated through elected officials of the state of Washington. While remaining cognizant of the site's mission nationally as well as its impact locally, we try to enhance understanding of how the national and the local interacted with one another. Situating Hanford within the regional contexts of the American West, the Pacific Northwest, the state of Washington, and the Columbia Basin provides new angles on both the national and the local stories.

In organizing our study of Hanford as a place, we have followed primarily a topical approach. Each of the book's chapters develops a particular theme—production, community, politics, environment—roughly from the 1940s to the 1990s. Because these themes are interrelated, readers will find some overlap among the chapters. But each is meant to have its own narrative trajectory, and to present arguments and related information in a logical sequence. Each builds in some measure upon what has come before, but to a certain extent each may be read as a stand-alone essay.

Chapter 1 offers an overview of Hanford's development that emphasizes the federal government's changing missions for the facility. It begins with the origins of the Hanford Engineer Works during World War II, then examines the site's rapid expansion during the early Cold War—the period in which Hanford became quite strictly defined as a production facility. Employees and neighbors liked to see themselves as dwelling on the cutting edge of science, but in fact Hanford's nuclear technology became obsolete rather quickly, and relatively little research and development was conducted on the site.[8] Nonetheless, Hanford served as a workhorse for the American nuclear weapons program during the Cold War. It produced 54.5 metric tons of weapons-grade plutonium over the years, about 60 percent of all weapons-grade plutonium made by the United States.[9] The industrial plant generated a great deal of pollution, most of which has so far been retained on, or in, the site's grounds. Hanford's employees now engaged in cleaning up the site are trying to contain those wastes, a task undertaken anew in the late 1980s, when plutonium production was finally halted altogether and the mission of cleanup was embraced. As the decades passed and the crises of World War Two and the Cold War wound down, the national government's tight control over Hanford eventually declined, yet during the 1990s the site was as dependent on federal initiative and funding as it had been during the Manhattan Project, although these efforts were now institutionalized in a more scattered group of entities, in contrast to the big, single-mission agencies that had dominated the three decades from World War II on.

During the twentieth century, mobilization for war in the American West led to the substantial in-migration of people, often through their connections to defense-related industry, and the great majority of this influx was channeled into towns. Chapter 2 focuses on the changing populations around Hanford by explaining the creation and transformation of nearby communities from the time of the Manhattan Project into the 1990s. It particularly highlights the evolution of Richland, which was taken over, reshaped, and operated by the federal government through its corporate contractors between 1943 and 1958, when, somewhat reluctantly, Richland attained its independence from federal control. Inhabitants of Richland had little doubt about either their place in western American history or their role in helping to usher in the promising new atomic age. Both perceptions heightened their own sense of importance, and shaped community identity; both encouraged them to frame their experiences in explicitly regional terms by likening themselves to earlier generations of pioneers. Thus they came to call their town the "Atomic City of the West." Chapter 2 compares and contrasts Richland with its neighbors, Pasco and Kennewick. The three towns, which came to be known as the Tri-Cities, had much in common—most of all their great dependence on Hanford and the related federal spending. At the same time, across the decades their respective populations remained quite differentiated, due in large part to decisions made by the federal government about Hanford in the two particularly intense bursts of mobilization for World War Two and the early Cold War during the 1940s and 1950s.

Chapter 3 addresses the politics of Hanford. As Americans and Westerners became more critical of nuclear technologies, especially after 1980, there emerged a tendency to view Hanford as something that the federal government simply imposed on the region without the consent of locals. To a certain extent this perception was accurate. During World War II and the Cold War, the U.S. Army and the Atomic Energy Commission built and operated Hanford in a shroud of secrecy. Nonetheless, during the 1950s and 1960s people in the Tri-Cities and around the state of Washington became increasingly effective at using political channels to influence Hanford's development for their own economic benefit. Because they regarded the plutonium plant as a tool with which to encourage economic development, they lobbied to secure more spending and new missions for Hanford. They also toiled to gain access to adjacent lands that had been set aside as a buffer zone around the plutonium factory; used political channels to negotiate a better deal when purchasing Richland homes and businesses from the government; and persuaded the AEC to support the development of a dual-purpose reactor capable of manufacturing not only weapons-grade plutonium but also kilowatts for the regional power grid. In all these instances, it was Westerners who imposed their agendas on Hanford, rather than vice versa. Although the president and the AEC had explicitly promised that Hanford would not become some sort of New Deal–era public works project, by the 1970s the AEC had shut down eight of the site's nine reactors, and only 25 percent of its budget went for programs related to national security. Through the political efforts of Tri-City and Washington State interests, the site became part of America's welfare state.

Chapter 4 situates Hanford within the environmental and economic contexts of the Columbia River Basin. Few historical accounts make much of the fact that the plutonium-making complex was superimposed on another massive federal engineering program—the Columbia Basin Project launched by the New Deal administration of President Franklin D. Roosevelt. Yet the two projects proved to be allies as well as rivals in transforming one corner of the American West. Both Grand Coulee Dam and the relatively free-flowing Columbia River were essential to the successful operation of Hanford. At the same time, however, the region's farmers, fishers, and merchants toiled to minimize Hanford's ability to interfere with other economic opportunities in the Columbia Basin. Between the 1930s and 1960s, both the federal government and regional interests generally advocated full-scale development of the Columbia Basin, for purposes of both national security and economic gain. After the mid-1960s, however, the development of the area became much more contested in the wake of the growing interest in preserving the environment. Students of the American West have begun examining how defense industries and military bases were integral to the region's economic development during the mid-twentieth century. They have paid less attention, however, to those later decades, when environmental concerns increasingly challenged the emphasis on economic advancement.[10] Many people around Hanford and the Tri-Cities initially resisted the call for heightened ecological protections, but in the end Hanford had to submit to the new day. The nuclear future envisioned by so many local people—consisting of weapons production, power generation, atomic research, and so on—did not come to pass in the twentieth century, at least not as predicted. The region's development took three surprising turns. First, electricity demand and economic development were not tied together as many had expected that they would be. Second, the national-security agencies that had done so much to pollute the environs suddenly proclaimed themselves to be forceful advocates for environmental protection. Third, contrary to initial expectations, the task of dealing with Hanford wastes made the Tri-Cities of the 1990s as prosperous as they had ever been, as cleanup became the basis for further economic development.

These thematic elements applied to the histories of Hanford and the Tri-Cities also provide contexts for the development of nuclear reactor technology on the site. The wartime construction of graphite-moderated, water-cooled nuclear piles established the ethos of production (as opposed to research) that came to characterize Hanford. Solving the problems that arose in the processes of extending the first reactors' working lives and producing copies of them to meet Cold War demands for plutonium stocks was part and parcel of putting the Tri-Cities on a permanent footing as an urban industrial center. Hanford's ninth reactor, the dual-purpose N reactor, reflected the deeper history of economic development in the Columbia River valley and helped forge new political relationships with the rest of the state. Finally, the Fast Flux Test Facility, which grew out of a project to develop a breeder reactor that could produce its own fuel, reflected a new and distinctive environmental sensibility in the Tri-Cities.

The new appreciation for Hanford's natural setting marked a contrast to earlier attitudes toward these arid lands. Like the plutonium factory itself, the Columbia Basin has been easy to overlook, literally. The area around the present-day Tri-Cities is the low point of a plateau defined in large part by the looping course of the Columbia River, which comes from Canada, enters the United States in northeastern Washington, and then tracks mainly to the west before flowing southward along the foothills of the Cascades and then east and south again to join the Yakima and the Snake rivers. Overlooking the rivers are sere ridges visible to the northwest and the northeast. The area around Pasco, including the Ringold Formation comprising the white bluffs for which one of the towns erased by the Manhattan Project was named, donated the soil blown to the east-northeast to pile up in the fertile Palouse hills on the Idaho border. Farm families there received a downwind legacy long before the Manhattan Project came to the Columbia.

According to geologists, "the Columbia Plateau is . . . most exciting . . . for the country has been shaped by geologic processes that have acted on a gigantic scale," a landscape shaped by lava coming from the west and titanic floods coming from the northeast at the end of the last ice age.[11] By contrast, Anglo-American newcomers during the nineteenth century viewed the lands as dreary and unwelcoming. In 1885 the settler Guy Waring, newly arrived from Boston with a wife and family in tow, saw the country defined by the Columbia's Big Bend as a tiresome obstacle to cross over on the way to the Okanogan valley. The hostile territory was bordered by the Columbia River to the west and north, and the tracks of the Northern Pacific that angled from Spokane to Ainsworth, a hell-roaring construction town near the later site of Pasco. Leaving the railroad at Sprague, Waring's party cut across the sandy, treeless no-man's land, contending with deep gullies scoured out by the ancient floods, each requiring wagons and gear to be painfully lowered down one side and hauled up the other. The largest of these was the Grand Coulee, which would later provide a reservoir for the Columbia's waters impounded behind the Grand Coulee Dam. For Waring the basin was something to get across with the least effort or, preferably, to go around.[12]

During the late nineteenth and early twentieth centuries, plans were made to provide irrigation water to the area and to reclaim it for agriculture, making use of the soils in place rather than just letting them blow into the Palouse. Farms and ranches started to appear in the vicinity of the Big Bend of the Columbia. One irrigation colony sponsored by Cornelius H. Hanford, a Seattle lawyer and judge, provided a name for the vicinity. Government support for homesteading and irrigation assisted in the development of resources, culminating during the 1930s with a commitment by the United States to build a series of dams and irrigation canals in the Columbia Basin. The news promised some degree of investment in the local economy. However, lands to the west of the river, in the area that became the Hanford Site, were not included in initial plans for irrigation. In fact, during the Great Depression the farm population of the region diminished as families left to look for opportunity elsewhere. The census of 1940 counted roughly 7,425 people in the towns of White Bluffs,

Hanford, Richland, Kennewick, and Pasco. The region was empty enough to seem attractive, a couple of years later, to army officers looking for an isolated locale for manufacturing plutonium. At the same time, there was enough infrastructure in place, in the form of towns, railroads, and electricity-producing dams, to make a massive construction project feasible. Almost instantaneously, this patch of the American West would be underdeveloped no more.

One

PLUTONIUM, PRODUCTION, AND POLLUTION

Hanford's Career as Federal Enclave

WHEN THE ARMY CONCEIVED, PLANNED, AND BUILT A PLUTO-
nium plant in south-central Washington between 1942 and 1945, it called
the place the Hanford Engineer Works. The name is telling, because it
reminds us of both Hanford's specific role in the federal weapons program
and the confidence invested in the place. The term "engineer" connoted
that Hanford was not a facility devoted to research and development, as
were the Metallurgical Laboratory at the University of Chicago and "Site
Y," the Los Alamos laboratory run by the University of California in New
Mexico. It belonged to the realm not of physicists and academic chemists
but of engineers and technicians. It was created to manufacture a product.
In American usage, the term "works" refers to a "factory, plant, or similar building or complex of buildings where a specific type of business or
industry is carried on."[1] And, indeed, between 1943 and 1971 the Hanford
Engineer Works (shortened during the Cold War to Hanford Works) functioned as a factory and generated a very specific kind of commodity.

Today, the same place is known much more ambiguously as the Hanford Site. It no longer seems to be a place that "works." Since about 1971 the
federal facility has lacked a clear-cut purpose—particularly, a well-defined

production mission. Because the site's main functions—the electricity-producing reactor, the environmental reserves, and the waste management programs—come wrapped in controversy, Hanford is also a "site" of concerns. The place has become notorious for the immense and problematic amount of radioactive and chemical wastes it possesses, and for the sharp division of opinion about its future. In this context, the term "site" serves as euphemism; calling the place the Hanford Site is one way to avoid making explicit the fact that for most people the place has become a *problem*. During the 1940s and 1950s, the name Hanford Works suggested that it was a place that *solved problems*. A lot has changed over the decades.

This chapter examines how Hanford changed over time in conjunction with the changing needs of the U.S. government. The reservation came into being during World War II as part of the Manhattan Project, the army's secret program to make atomic bombs. Reflecting the urgency of wartime, Hanford emerged in great haste. Once atomic bombs helped bring World War II to a close in August 1945, things slowed down considerably for a little more than a year. By the end of 1946, however, Hanford was returning to an emergency footing as the nation geared up for the Cold War. Another crash construction program expanded the plant and its support facilities far beyond their original size. From the late 1940s to the mid-1960s, Hanford focused on manufacturing plutonium to meet the nearly insatiable appetites of the national security state. Its strict orientation toward full-scale production set it apart from other nuclear weapons sites in the U.S. After 1964, however, this mission changed. Enough plutonium had been produced that it was no longer necessary to keep Hanford going full-bore. A number of other missions were proposed for the site, including generating electricity, conducting research, and serving as the nation's repository for high-level radioactive wastes. None of these truly panned out, and by the 1990s the site's mission had become cleaning up and managing the wastes produced during World War II and the Cold War.

Hanford and the Manhattan Project, 1942–45

For a brief time, at the very beginning of this story, the plutonium plant that became Hanford existed primarily as a secret in the minds of Easterners. For about two months, in late 1942 and early 1943, a handful of army officers based in Washington, D.C., DuPont executives from Wilmington, Delaware, and university physicists in Chicago had a free hand to script developments at Hanford while the inhabitants of the region remained unaware of the unfolding drama. During these precious weeks, the Pacific Northwest seemed ready to meet every need that the government's atomic-weapons program had.

The Manhattan Project's investment in the region emerged as scientists raced to figure out how best to build an atomic bomb, and as a result of two important shifts in the Anglo-American nuclear weapons program. While the creation of tremendously powerful explosives had been a recognized technical possibility since the discovery of nuclear fission in the last days of 1938 and the realization of its mechanism in early 1939, the Roosevelt administration became fully committed to a crash program only in the fall of 1941. The scientific consensus at the beginning of the war in Europe was that a scarce isotope of the element uranium, uranium-235, was desta-bilized enough by a subatomic particle, the neutron, that it would shake itself apart and release more neutrons as a consequence, along with other forms of energy. In principle, it was possible that this phenomenon could take the form of a chain reaction in which the fission of one atom released neutrons causing the fission of others, in a rapid reaction to be manifest as a powerful explosion. By 1942 an American political consensus had been created along these lines: it was believed that German scientists were try-ing to create nuclear weapons to add to Hitler's arsenal, necessitating a competing effort on the part of Allied scientists and American industry. Because the project would require an enormous construction program to build and operate facilities, responsibility was given to the United States Army Corps of Engineers, whose efforts in this area were directed by a hard-charging commanding officer, General Leslie Groves, who made the Manhattan Project a top priority for manpower and materiel, and who as manager largely supplanted government science advisors.[2]

Because scientists did not know of a single best method for making the new weapon, and because the United States could not afford to choose the wrong one, Groves committed the Manhattan Project to explore different methods simultaneously. When scientists told him that either uranium-235 or another substance, plutonium-239, could serve as the fissionable mate-rial in an atomic explosion, Groves determined to secure a supply of both. In September 1942 he acquired 59,000 acres in Tennessee for the Oak Ridge facility, devoted to isolating enough uranium-235 to fuel atomic weapons. (The fissionable isotope uranium-235 comprises less than 1 percent of natu-rally occurring uranium, most of which is uranium-238. Most work at Oak

Ridge was devoted to isotope separation, that is, distinguishing between the chemically identical U-235 and U-238 atoms based on the tiny difference in their masses.) In November, seeking a site for the laboratory where bombs would be designed and assembled, Groves arranged to purchase a boys' boarding school at Los Alamos, New Mexico. Then he turned his attention to the matter of manufacturing plutonium-239.

Groves's initial inclination was to put the needed reactors and processing plants at Oak Ridge.[3] In the middle of deliberations, however, on December 2, 1942, plutonium's promise was affirmed when scientists led by Enrico Fermi at the Metallurgical Laboratory at the University of Chicago achieved the world's first controlled chain reaction. The experiment demonstrated how a "pile," or atomic reactor, could produce the fissionable material. The army now had a new problem. Whereas Fermi's experimental pile fit into a squash court at an urban university, the enormous production reactors needed to make fuel for bombs would each be 500 million times as powerful as the Chicago pile. A "scaling-up" of such unprecedented dimensions, to be conducted without extensive knowledge of the dangers of radiation, required isolation from major population centers. Oak Ridge, it turned out, would not work; it lay too close to Knoxville, Tennessee, a city of about 125,000 people. Groves and E. I. du Pont de Nemours and Company, the contractor chosen by the army to build and operate the plutonium plant, wanted a more spacious, isolated place for the untried and dirty work of making plutonium. Moreover, Groves decided that adding another mission at Oak Ridge would stretch that site's resources, particularly its overtaxed construction crews, beyond their limits.[4] A third locale was needed, in all likelihood to be situated in the West.

In mid-December 1942, DuPont executives met in Wilmington with Chicago scientists and Lt. Col. Franklin T. Matthias to establish the site-selection criteria. By this time Groves had appointed Matthias to supervise development of the plutonium facility for the Army Corps of Engineers. Keeping notes of the meeting in an unauthorized diary, Matthias recorded a consensus that the site ought to be an uninhabited and roadless rectangle measuring twenty by twenty-eight miles, or roughly 560 square miles. Moreover, no towns with a population of more than a thousand were to be located on an additional 1,300 square miles or so of land surrounding the rectangle. The plant demanded plenty of other resources as well. It needed a river to run through it—more specifically, "'relatively pure' and 'relatively low temperature water' flowing through the reservation at a volume of 'not less than 25,000 gallons per minute.'" The plant also required access to at least 100,000 kilowatts of electricity to pump river water through the reactors, run the machinery, and provide power for the construction camp and operating village. Ideally, the source of electricity would be near the plant to avoid building lengthy transmission lines. DuPont officials added that the perfect site would have solid, level ground, a mild climate, and plenty of gravel available locally.[5]

Not many places in the country could meet these specifications. By trying to put a substantial distance between the plutonium plant and the local population, the specifications demonstrated that the Manhattan Project made safety and secrecy primary criteria. In retrospect, the plutonium plant could have proven less threatening to its neighbors if the site-selection criteria had paid greater consideration to the winds that might carry atmospheric emissions beyond the site and to the conditions affecting underground waste deposits. But the Manhattan Project had neither enough time nor a sufficient number of suitable alternative sites to take meteorological and geological concerns into serious consideration; moreover, it probably did not have adequate expertise to make such considerations decisive factors for site selection. (Instead, meteorology became an operational concern, to be dealt with in practice as the plant operated.) As it was, the Manhattan Project was already pushing beyond the limits of scientific knowledge. Matthias remembered returning from the meeting in Wilmington and telling Groves, "Well, I think I have to read some Buck Rogers to get properly oriented."[6]

Like the rest of the army, Matthias had no time to do background research. Instead Groves dispatched him, along with Gilbert Church and A. E. S. Hall from DuPont, to find a site. They had instructions that "an investigation of the Grand Coulee area" ought to come first. Locations along the Columbia River appeared attractive not only because of the heavy flow of cold, clear water but also because of the abundance of hydroelectricity generated by recently completed dams. On December 22, 1942, Matthias, Church, and Hall flew over north-central Oregon and south-central Washington. Matthias held out little hope for the "productive wheat land" along Oregon's Deschutes River, but the arid territory in the vicinity of the Washington towns of White Bluffs, Hanford, and Richland looked "far more promising." He and his companions agreed after driving around the area that "the site was so good that there couldn't be a better one in the country. It looked perfect in almost every respect." Matthias called Groves from Portland, Oregon, to say that he saw no reason to look elsewhere, but Groves sent the party on to California to scout out lands along the Colorado River. Matthias, Church, and Hall found nothing matching what they had already seen, and by the last day of 1942 Matthias was back in the nation's capital to recommend that the army acquire Hanford. By the time Matthias took Groves on a personal tour of the Hanford Site on January 16, 1943, the army had already started proceedings to acquire the land. By war's end the Hanford Engineer Works of the Manhattan Engineer District of the U.S. Army Corps of Engineers would occupy 670 square miles, roughly 429,000 acres, of Benton, Franklin, and Grant counties in Washington State.[7]

In February 1943, when the U.S. government announced that it was acquiring the land that would become Hanford, local people finally learned something about the nation's intentions for them and their environs. They were none too happy about the news. One of the reasons Matthias had favored the Hanford Site was that he felt that an army occupation of it would disrupt relatively few lives, but this strategy provided no comfort to those who were evicted. One-third of the territory belonged to other government agencies, but the rest belonged to private owners. The army

Indigenous lands?

1.1 Franklin T. Matthias was the thirty-four-year-old officer appointed by General Leslie Groves to supervise construction and operation of the Hanford Engineer Works on behalf of the U.S. Army Corps of Engineers. Matthias helped to identify the site in December 1942 and stayed with the project through the end of the war. He served as intermediary between Groves, the army, other Manhattan Project sites, DuPont, the workers imported to build and run Hanford, and the numerous civilians of the Pacific Northwest who took an interest in the secret project. Photograph courtesy of U.S. Department of Energy.

condemned about 3,000 tracts of land held by roughly 2,000 individuals and in many cases gave occupants only thirty days to leave.[8]

Naturally, the government takeover of the land became a matter of contention. In discussing the matter, the army and DuPont tended to emphasize the aridity and infertility of the land and the poverty of local rural society, particularly since 1929. Their evaluation justified relatively low appraisals for the condemned properties. Private owners, of course, saw things differently. They felt that they had already accomplished what Hanford would later claim to have done: make productive a barren landscape. Few accepted the idea of being displaced, and some suggested that the army leave the irrigated farms around Richland alone and move its operations farther north. Once the government takeover became inevitable, most still felt that they deserved both more time to depart and more money in recompense. They were having the best growing year in a long time, and there were high expectations for new sources of irrigation from the Columbia

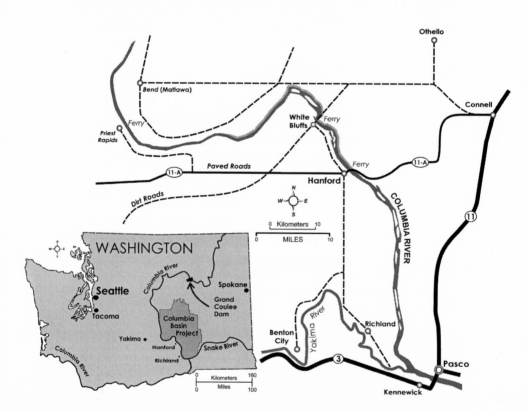

1.2 Map of Hanford in 1940. Map by Frederick Bird.

Basin Project of the Bureau of Reclamation. Taking all this into account, displaced land owners expected greater compensation for their land than the government had offered. Many went to court. Numerous evictees won higher prices for their land, either from sympathetic local juries or from federal officials who eventually decided to settle at higher figures.[9]

The plutonium-producing facility of the Manhattan Project now existed in a place other than the imagination. Virtually none of the local residents would learn what the army was doing at Hanford until August 6, 1945, but it was no longer a secret that the government was doing something big there. While local property owners demonstrated that they would not simply acquiesce to federal initiatives, throughout World War II the military retained absolute control over Hanford. In short order, the army separated local people from their homes and lands. Only a handful of residents, who went to work on the project, stayed behind in their housing around Richland, and they now paid rent to DuPont. Even the dead were evacuated: the government exhumed 177 burials from the White Bluffs cemetery and reinterred them in the nearby town of Prosser. It also promised the Wanapum tribe that it would "treat Indian graves . . . with reverence and dispose of the bodies in some reasonable way."[10]

Of all the groups, only a handful of Native Americans retained some access to the Hanford Site. The Wanapum traditionally fished for salmon at Priest Rapids on the Columbia for two to three weeks every fall, camping on land that had become the Hanford reservation. Matthias tried to persuade the Indians to accept some other arrangement—a cash payment in lieu of fishing; camping and fishing further downstream; letting the army catch and deliver fish to them—but the Wanapum insisted on their rights. In the end Matthias agreed to provide a truck and driver to transport the fishers to their camps each day during the season. No overnight stays were permitted, however, and the army supervised the Indians while they were on the military reservation.[11]

Visitors and new arrivals from the East found a dusty, wind-swept site, emptied of agriculture and most structures, and traversed by the Wanapum on their supervised trips to fish. It was an artificially desolate landscape. As one of the scientific settlers described it, "They [army contractors] bulldozed the place flat, got rid of whatever top soil there was and brought in silt from the Yakima River flats. They put six inches of this stuff over the town. . . . When the wind blew you wouldn't be able to see across the street." Another eminent visitor from the project's Chicago laboratory called it "the flattest, most lonesome territory I have ever seen." The land-scape had been made more dusty and lonesome than necessary by federal intervention.[12]

In vacating the site, the Manhattan Engineer District established pri-orities. It wanted to begin construction of three production reactors, two chemical separations plants, and assorted other facilities, clustered mainly on the northern part of the site. It evacuated the villages of White Bluffs and Hanford immediately and began erecting the huge temporary con-struction town, called Hanford Camp, where the town of Hanford had been. At the southern end of the site, the district simultaneously began to remake Richland into the residential village and administrative head-quarters for the plant, carefully situating it more than twenty miles south of the industrial core of the site. Because the completion of Richland was less urgent, most of its former residents had until November 15, 1943, to vacate their homes.[13]

The project's success rested in the hands of a pair of organizations, the Army Corps of Engineers and DuPont. The Army Corps served as the military's construction arm, and its Manhattan Engineer District over-saw development of the entire atomic-weapons complex during the war. General Groves was in charge of the Manhattan Engineer District, and he appointed Matthias to supervise the work at Hanford. Matthias, who stayed on the job through the end of the war, had the unenviable job of serving as on-site taskmaster as well as intermediary among Groves, DuPont, Manhattan Project scientists, laborers, and the diverse people of the Pacific Northwest. He had to placate the local interests who both feared the enormous impact of Hanford and hoped to capture some of the fed-eral expenditures ($390 million at the time; more than $6 billion in 2010 dollars) for themselves. He worked as a troubleshooter for DuPont and a broker between the army, suppliers, the company, and organized labor. Above all he had to follow his superior's unrelenting, urgent commands. Groves "had a reputation for being an old sourpuss and a bastard," Mat-thias recalled in 1993, "which I guess he was."[14]

The government oversaw and financed the show at Hanford, but it was hardly prepared to perform the actual work. The Manhattan Project was at bottom an effort to build and operate an immense industrial complex — one that, by war's end, would rival in size and cost the entire U.S. automo-bile industry.[15] The government, of course, had neither the experience nor the resources to accomplish such a feat, and so it turned to an assortment of contractors. The University of Chicago hosted the Metallurgical Labora-tory (Met Lab), and the University of California oversaw Los Alamos. At both Oak Ridge and Hanford, the Manhattan Engineer District entered into agreements with big construction and manufacturing firms.

DuPont, a major producer of chemicals and explosives, was well-suited to Hanford's unprecedented production processes. The company had an extensive pool of skilled manpower, allowing it to divert some key per-sonnel from other divisions despite overall wartime demands. DuPont drew on skilled machinists from firearms companies, which historically had provided a home for exceptionally skilled craftsmen who were able to set up specialty apparatuses for the precisely controlled production of metal parts. Along with scientists and engineers, DuPont recognized that some problems would be solved by the application of tradecraft. The com-pany brought machinist E. E. Swensson from the Remington Arms plant in Bridgeport, Connecticut. He arrived five days before the first uranium came to the site in October 1943, and began to organize the machinists who would engage the problem of fabricating fuel elements for the reactors.[16] Moreover, DuPont had a tradition of building its own factories, which made it a logical candidate to construct as well as operate the Hanford plant. Finally, it was a huge organization with considerable strengths that was already deeply involved in the war effort; by 1945 it would build fifty-four war-related plants at thirty-two locations.[17]

General Groves pressed hard to get DuPont to agree to the job, but the company needed to be convinced, since it was already overcommit-ted. Groves appealed to the company directors' patriotism, explaining the importance of beating the Nazis to the atomic bomb and predicting that the new weapon would save thousands of lives. His pitch proved decisive for getting DuPont to commit to the project, but he had to overcome other reservations on DuPont's part as well. The company regretted the reputa-tion it had acquired as a "merchant of death" when critics denounced the profits it had made during World War I. To forestall such charges in the future, DuPont insisted on a contract that limited its profit to one dollar over costs. The company worried as well that it would not, during the mid-dle of the war, be able to hire all the trained personnel it needed to accom-plish its mission, so it insisted on corporate (rather than government or university) pay scales for employees. Finally, DuPont had serious doubts about whether plutonium-239 could be produced—and safely. After gain-ing some reassurance from Fermi's experimental success on December 2,

1942, DuPont insisted on another clause in its contract that specified that
the government would reimburse the company for all losses incurred as a
result of its work at Hanford, including any legal liabilities in the future.
(Such provisions became standard for other nuclear contractors.)[18] In the
end DuPont managed to get many of its terms met.

The choice of DuPont was often second-guessed during the war. Both
General Groves and Colonel Matthias clashed with the company and
lamented its seemingly over-cautious approach to things. Scientists at the
Met Lab were even more critical, complaining that the contractor was too
"conservative," too slow, too obsessed with safety. At one point, wanting
to accelerate the rate of progress, the scientists proposed that they should
take over the design, construction, and operation of Hanford's reactors.
DuPont, for its part, did not trust the Chicago scientists and determined
to allow none of them to visit Hanford unless invited. Tensions ran deep.
Yet by the end of the war DuPont's virtues had become apparent. The company's industrial experience, engineering orientation, and painstaking
approach staved off chaos and prevented severe disruptions to production
schedules. After the war Hanford received high marks for DuPont's safety
programs, especially when compared to those at Manhattan Project sites
overseen by universities.[19]

A major producer of explosives, DuPont understood the need for a safe
work environment and for generous open spaces around munitions plants
in particular, for example, allowing safe storage of smokeless gunpowder
in quantities of 100 to 150 tons. These earlier corporate experiences provided a template for safety plans at Hanford, with "particular reference to
spacing of units and distances to inhabited area," standards that came from
army ordnance manuals that DuPont engineers had helped to write just
before the war.[20] As Groves had understood from the start, it takes one
successful industrial organization to create another.

Much of the tension and disagreement surrounding Hanford stemmed
from the urgency of the Manhattan Project. Americans raced to beat Germany to the first atomic bomb; once the German threat diminished, based
on intelligence gathered by the fall of 1944, they hurried to complete, test,
and deploy the bomb against Japan before the war ended. The construction
and start-up of Hanford proceeded at a pace that would have been incon-
ceivable outside of wartime, and the haste led to numerous frustrations.
Designs for the first reactors were changed routinely by Met Lab scientists,
even though construction by DuPont was underway and finished work

had to be torn apart and redone. Matthias noted that building crews took it upon themselves to slow the work down when they grasped that they were taking certain tasks "out of scheduled order." In several instances officials omitted certain safeguards and accepted inferior materials to save time. Late in 1943, for instance, Matthias instructed DuPont to install certain water-treatment equipment on the second reactor but not the first in order to reduce the amount of time needed to manufacture the initial batch of plutonium-239. The colonel prioritized Hanford's goals baldly: "Our first requirement is the early production of some material, and . . . our second requirement is a large quantity of material."[21] Questions of public health and environmental safety—both near- and long-term—took a backseat to production.

Before reactors could operate they needed to be built, and they would not get built until Hanford had acquired enough labor for the job. The army and DuPont found that recruiting and retaining an adequate number of trained workers was nearly impossible. Other employers and the armed services had had more than a year's head start in mobilizing for war; the isolation and weather of the Hanford locale remained unappealing to many; the War Manpower Commission did not prove very helpful; and living conditions remained primitive. There was never enough labor to go around, and turnover among the workers was high. Early in the project Matthias reported a monthly employee turnover rate of 10 percent, and throughout 1943 he delayed construction on the plant so that labor could be concentrated on building the barracks and amenities needed to keep workers from leaving the job. Even the accounting supervisor reported in 1943 that among his office personnel, "some have been able to stick it out only a few days before resigning."[22]

By one estimate Hanford had no more than 50 to 70 percent of the labor force it needed through the spring of 1944. Finally, in June 1944—when project employment peaked at around 45,000—Matthias could say for the first time that he had "almost enough laborers and craftsmen to accomplish the work scheduled." Yet in the same month the labor turnover rate stood at 21 percent. Groves insisted on conducting exit interviews to discover why workers were leaving and then trying to fix the causes of their dissatisfaction at Hanford. The army and DuPont implemented programs to improve morale, provide entertainment, and alleviate on-the-job aggravations. Yet they could not address every grievance. One complaint from some departing workers, which stemmed from the secrecy of the Manhat-

1.3 During World War II the army and DuPont imported construction workers from all around the United States to build the Hanford facility. The workforce reached a peak of 45,000 people in mid-1944. Whites, African Americans, and Mexican Americans were all recruited to help build Hanford, but they lived in segregated housing. Here laborers stand in line at the branch of the Seattle First National Bank that opened to serve the Hanford Camp. Photograph courtesy of the U.S. Department of Energy.

tan Project, was that they felt they were not making a substantial contribution to the war effort.[23]

In some contrast to the patriotic civilians who left to make a greater impact elsewhere were the pipe fitters who arrived in late 1944. Hanford had always had a hard time recruiting certain kinds of skilled construction workers, and it turned to the armed services on occasion to make up the deficit. In some cases it arranged with local draft boards to keep skilled workers from being called up to military service, but in the summer of 1944 not even that expedient had provided enough workers to complete the extensive plumbing required for the first reactor. So the Manhattan Engineer District furloughed trained pipe fitters from the ranks of enlisted men, enrolled them in reserve units, and hired them as civilians at Hanford, with a promise to return them to the army if their work proved unsatisfactory. "This will give adequate control" over these employees, Matthias noted in his diary on September 4, 1944—when unionized pipe fitters stood on the verge of a walkout. The furloughed men may not have relished the army's control over them, but, having been closer to the front lines, they apparently preferred civilian work to active duty.[24]

Hanford left few stones unturned in its effort to find workers. In 1940 people of color had amounted to 2.2 percent of Washington's population. Against the preferences of the almost exclusively white local population, DuPont hired thousands of African Americans and Mexican Americans, many of whom came from Texas, Louisiana, Oklahoma, and Arkansas. People of color accounted for more than 16 percent of Hanford's construction labor force, including about 5,400 blacks by mid-1944. Citing the need to expedite the work, officials segregated African Americans in their own quarters at Hanford Camp, thus defying the guidelines of the U.S. Fair Employment Practices Commission. "This was no place to risk racial conflict," one worker recalled. Thinking along similar lines, Matthias only reluctantly agreed to take Mexican American recruits from Texas in early 1944. Believing that whites, blacks, and Chicanos would refuse to live among each other, he had not wanted to spare the time and labor to erect still another set of segregated facilities for Mexican Americans. He relented only because the number of other available workers was so small, and only after he determined that he could house Chicano recruits off the Hanford Site, near Pasco. In this, Matthias followed the army's more general institutional attitudes. The army took a survey of white enlisted men in 1942, revealing "a strong prejudice against sharing recreation, theater or post exchange facilities with Negroes." Based on this, the army's adjutant general opined, "The Army is not a sociological laboratory," a view echoed by army chief of staff George Marshall, who said, "Experiments within the Army in the solution of social problems are fraught with danger to efficiency, discipline, and morale."[25]

Women comprised another substantial segment of the Hanford population, as much as 13 percent of the workforce during construction. Most women performed duties in the mess halls and barracks of the Hanford camp, or labored as clerks, secretaries, nurses, and teachers elsewhere on the project. One army engineering officer commented on the presence of eight "women 'chemists'" working in the vicinity of the D reactor in March 1945: "They are doing satisfactory work, and fill the role for which they were hired; namely, following of carefully outlined procedures or carefully given instructions in the performance of routine analysis. In general, they are by no means capable of carrying out tests or analysis without prior instructions."[26] Accomplished women scientists such as physicist Leona Marshall Libby, who came to Richland from Enrico Fermi's Chicago research group, were exceptional; the massive project required large numbers of less-skilled female workers, whose contributions were vital but who, as with non-white workers, were also grudgingly employed.

As many as 14,000 women, not all of whom were employed on the project, resided on the site during mid-July 1944. One of these was Nell Lewis MacGregor from Puget Sound country, whose reluctance to accept employment in the desert-like area around Pasco was overcome by the offer of an irresistibly high salary. Among other things, McGregor worked as house mother to a barrack full of nurses. She commented tellingly on the overall quality of the workforce in a 1969 memoir: "One can't imagine a place with more strange characters than Hanford had. With so many men in service or frozen to jobs, the recruiters had taken people who ordinarily would not have been considered. We had morons picking up papers on the grounds and physically abnormal freaks doing work within their scope." MacGregor's uncharitable assessment bespoke the very limited choices facing Hanford. Manhattan Project chemist Glenn T. Seaborg found Richland's jail "full of the motliest-looking aggregation of tough-looking characters" imaginable; the DuPont supervisor explained that the site was scraping the bottom of the barrel for construction labor. More than half of Hanford's male employees during the war were over the maximum draft age of thirty-seven years; of those in the age bracket of eighteen to twenty-six years, 75 percent were unfit for military service.[27]

Not only did the Manhattan Project have to get workers wherever it could find them but, because there were too few workers during most months of construction, it also required longer hours from those it employed. Groves and Matthias instituted first a five-and-a-half-day and then a six-day workweek, and also insisted upon ten-hour days. Fortunately for the project, many employees wanted the chance to earn the extra income, and did not grumble about the overtime. At the same time, the army and DuPont hoped to keep the workers content enough to remain on the job by building morale and addressing workers' complaints.[28]

One way to prevent laborers from departing was to keep them from learning about certain workplace hazards at Hanford. As the time for the start-up of the first reactor neared in August 1944, DuPont executives in

Wilmington ordered the managers at Hanford to practice a "complete evacuation" of the areas around the piles, the processing plants, and the construction barracks. Groves and Matthias—no doubt construing DuPont's
plan as further evidence of its excessive caution—simply refused to allow
such an exercise and authorized instead only a very limited evacuation of
the reactor complexes. Matthias explained in his diary that a wholesale
drill "might be disastrous to the project as it might cause a large number
of people to leave if their fears for safety were increased. It would also be
sufficiently upsetting so that we could expect a serious effect both on security and manpower facilities to finish the job." For the sake of efficiency
and secrecy, the army wanted not to alert workers to some of the risks
presented by Hanford.[29]

Given the army's approach, it proved very difficult for unions to inter-
vene effectively on workers' behalf. The army and DuPont regarded organized labor as a threat to both efficiency and secrecy. In their haste to finish
the job they had no patience for rules that prevented, say, employees from
one craft going to work in another craft in short supply, for jurisdictional
disputes between unions, or for "agitators" and strikers who slowed the
project down. They also appeared to assume that the tremendous urgency
of the Manhattan Project, and the fact that the contractor did not profit
directly, placed them on a kind of moral high ground in dealing with
unions. It was also true that wartime Hanford presented problems that
defied the peacetime conventions of labor relations. For example, it created unprecedented jobs for which no union had trained members, and
its extreme secrecy prevented union agents from visiting the work site
and workers from describing their jobs to union agents. More often than
not, while the army and DuPont tried to respond to workers' complaints
because they could not afford to lose employees, organized labor deferred
to management during wartime.[30]

Among the workers attracted to Hanford was James W. Parker, a sev-
enteen-year-old from Idaho who accompanied his parents to seek work
at Hanford in 1943 while he waited to become eligible for induction into
the U.S. Army Air Corps. The son of an electrician, Parker hoped to pitch
in at the Hanford Works as a summer job. Arriving at the construction
camp, he recalled, "The first impression was of blown sand and rows
of poles. . . . six inch insulated hot water pipes everywhere, suspended
like oversized lines from poles along the street," evidence of the community central heating system in place. Other striking sights for Parker

were Hanford's restrooms, which were segregated by race. Too young to
be hired for heavy construction at the main site, Parker worked first as
an electrician's helper at the Pasco airport, and then as a laborer for a
housing contractor in Richland, where he helped in the assembly of prefabricated homes being put together at a frantic pace by gangs of men
and women who followed after trucks that dropped loads of building
materials on one lot after another. Housing was scarce; Parker's family
lived in a trailer they had brought with them, and he and a buddy slept
outside. The trailer camp had a wash house with showers, where Parker
found other workers using the changing-room benches as beds. The family escaped the cramped life of trailer-dwellers when Parker's father was
taken on as a production employee and the family was allowed to rent a
house in Richland.[31]

Having a family connection became characteristic of life at Hanford
and contributed to the close-knit character of the communities surrounding the reservation and the interpenetration of workplace and living space. Parker was eventually taken on as an electrician's assistant for
work on the site. After moving with his crew from one building project
to the next, and having lied about his age, he received special clearance to
work inside the B reactor building, a large, square, concrete building near
the Columbia's banks with a control room behind a window to one side.
The immense volume comprising most of the building's interior space
was dominated by the "pile"—a huge cubical stack of graphite blocks
pierced through all three faces. The pile's front presented the ends of an
array of aluminum pipes, called process tubes, which would be loaded
with uranium fuel elements and would carry cooling water around the
hot fuel while the reactor operated. Seeing the pile under construction,
Parker "was almost staggered by the spectacle. . . . It was immense. Workers were crawling around on it like ants. The tubes were being cleaned by
blowing folded-up sanitary napkins through them with compressed air.
The napkins came sailing out in a blizzard of cotton. The room resembled
a snow ball fight in a school yard." Informed by a co-worker that "he
had firm information that the whole plant was designed to make plas-
tic tableware for the post-war world," Parker and his crew began running cables for lighting and for motors to run the reactor's controls and
safety systems.[32] Parker's work had brought him to the heart of Hanford's
construction: the reactor, the successful operation of which would allow
Hanford's staff to complete its mission.

The reactor complexes located along the Columbia River were not the only sites of activity on the federal reserve. Employees at Hanford were distributed across the reservation in several clusters, each assigned to a particular district, which in turn was designated by a number and devoted to a particular part of the production process. Supervisors and administrators, employed by either the U.S. Army or its primary contractors, worked out of offices in Richland, also known as the 700 Area (the residential and commercial neighborhoods of Richland were the 1100 Area). These offices were deliberately sited at the very southern tip of the reservation—close

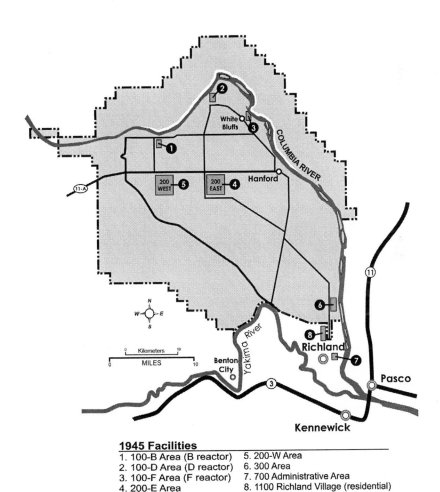

1945 Facilities

1. 100-B Area (B reactor)	5. 200-W Area
2. 100-D Area (D reactor)	6. 300 Area
3. 100-F Area (F reactor)	7. 700 Administrative Area
4. 200-E Area	8. 1100 Richland Village (residential)

1.4 Map of Hanford in 1945. Map by Frederick Bird.

enough to be managed as part of the Hanford Engineer Works, yet as far as possible from the site of the reactors, which were regarded as the part of the project that posed the greatest danger to public health. Just to the north of Richland, along the western bank of the Columbia River, the army and DuPont built and operated the fuel fabrication facilities, known as the 300 Area. This complex received uranium mined as ore in Africa, Canada, Colorado, and Utah and then refined into metallic blocks at facilities in Missouri, New Jersey, and Ohio. At Hanford the metal was shaped into short cylinders called "slugs." The slugs were then covered in aluminum jackets, or "canned." These "uranium elements" were then moved to the atomic reactors at the far northern end of the site that were collectively designated the 100 Area.

The army and DuPont built and operated three reactors during the war, designated by the letters B, D, and F (also called 100-B, 100-D, and 100-F, to identify them with the 100 Area). Operators inserted the uranium slugs into tubes in the graphite cores of the reactors, where they were bombarded with neutrons. The activity created a chain reaction that converted the uranium-238 into plutonium-239, an isotope suitable for detonation in a bomb. The immersion of uranium in a bath of neutrons, in the presence of a substance that would moderate the neutron's velocities and in the absence of materials that absorbed neutrons and removed them from the reaction, had two consequences: atoms of uranium-235 would undergo fission and release neutrons, keeping a self-sustaining chain reaction in motion, and atoms of uranium-238 would absorb neutrons and change their atomic identity, ending up as plutonium-239. The moderator principle went back to some of the work done by Fermi before he left fascist Italy for the United States. He had then pursued fission processes at Columbia University, and in particular the questions of whether a chain reaction was possible and what the most suitable moderators might be. At Chicago, Fermi and his collaborators had translated his insights into the geometry of a pile: graphite blocks, some containing lumps of uranium, stacked to create a matrix of uranium amid a graphite medium. Control rods of neutron-absorbing material were run in from the side; as these were withdrawn, the reaction became self-sustaining and its intensity increased. Safety rods were positioned to drop down from above, to stop the reaction in case it began to run too hot. At Hanford the production piles were built with process tubes running from the front so that new fuel elements could be loaded in front, with used fuel being shoved out the back as a result.

1.5 General Groves insisted on programs to maintain morale at Hanford to reduce the high levels of turnover among construction workers. The army and DuPont imported entertainers, deployed propaganda, and organized holiday celebrations to strengthen workers' commitment to the project. Occasionally Groves himself visited to exhort the workforce in person, as in this picture. Photograph courtesy of the U.S. Department of Energy.

Connected to each process tube was an end cap with a plumbing pigtail, which carried cooling water that was forced through the pipe. Conceived and built by academic scientists as an experiment to prove the basic principles, redesigned by Fermi's staff and by DuPont's production experts for practical use, and built by the work gangs housed in trailer parks and at the Hanford Camp, the graphite-moderated pile went from a concept to a working reality.[33]

The manufacture of plutonium created heat and waste products. Water drawn from the Columbia River was pumped through the reactors to cool them; in the process, the cooling water became radioactive and tinged with chemical effluents. The most intense radioactivity decayed away before the cooling water was released back into the Columbia, but some chemical and radioactive contaminants were carried along when the water was returned to the river. The fuel elements, consisting now of plutonium as well as uranium and weighing just under ten tons altogether, were pushed out of the reactors to make room for more uranium slugs. The irradiated slugs then cooled in tanks of water until they were less radioactive and therefore less dangerous to process further, even though the spent reactor fuel was moved from the reactors in shielded railcars and handled by remote control throughout the rest of its processing.

From the 100 Area, workers transported the plutonium-tinged fuel elements south to the 200 Area, near the middle of the Hanford reservation. There, moved by remote control and manipulated by workers from behind shielded windows, they made their way through reprocessing plants that stripped away the unwanted by-products produced in the reactors, and purified the plutonium for use in weapons. The chemical processing required to separate plutonium from the radioactive stew of used reactor fuel was a major challenge for DuPont but (in contrast to operating reactors) was within the realm of DuPont's experience. The T and B reprocessing plants, completed during 1944 and 1945, relied on bismuth phosphate batch processing to accomplish these tasks. Reprocessing the plutonium generated the greatest amount of hazardous pollution, especially before new processes, technologies, and facilities shrank the stream of wastes after World War II. By contrast to the 100 Area, which distributed most of its pollution to the Columbia River, most of the radioactive and chemical wastes from the 200 Area went into the ground and the atmosphere. Meanwhile, the plutonium yielded by the reprocessing plants was concentrated further "by chemical treatment and heating," then packaged in a can for shipment to Los Alamos, where it would be treated further, transformed into plutonium metal, and manufactured into components of atomic bombs.[34]

The delivery of plutonium to Los Alamos marked a crucial turning point. In late 1944 and early 1945, Hanford's reactors, reprocessing plants, and fabrication facilities were completed. The site's construction phase ended, and its operations phase began. In their wartime haste, some people, including Manhattan Project physicists, felt that construction had taken too long. In retrospect, however, it is remarkable just how quickly the Hanford Engineer Works—and especially its three reactors and two reprocessing plants—had been finished. Work on the world's first production pile, Hanford's B reactor, had begun on June 7, 1943, and reached completion less than fifteen months later on September 15, 1944. By February 10, 1945, two additional reactors and one more processing plant were completed, and by March the remainder of the industrial, administrative, and residential facilities at Hanford had been finished.[35] (All three wartime piles ran at their designed power levels for the first time on March 18, 1945.)[36] DuPont laid off the great majority of construction-related employees, and by February 23, 1945 the army had evacuated the construction camp. (See Appendix 1 for the start and completion dates of wartime construction.)

While one frenzied phase came to an end, however, there was no letup, and General Groves permitted no complacency. For example, one week before completion of the B reactor, he had Matthias and DuPont make plans to ensure that the anticipated surrender of Nazi Germany did not distract workers. Matthias oversaw the preparation of a propaganda campaign to keep Hanford's employees mobilized. Whether or not the workers were influenced by posters, rallies, and speeches designed to inspire their dedication to the ongoing project, Hanford drove ahead to the finish of the war, keeping up a delivery schedule of plutonium shipped to Los Alamos. There was an overwhelming determination to build and deploy atomic bombs before the war ended.[37]

As Hanford underwent the transition from construction to operations in 1944–45, the remainder of the Manhattan Project waited impatiently. The scientists at Los Alamos were responsible for designing, assembling, and testing the first nuclear weapons, but their efforts could go only so far without the quantities of plutonium-239 that only Hanford could provide. This was all the more true because reactor-produced plutonium carried a contaminant that would cause a premature detonation, resulting in a fizzle, if used in a gun-type bomb, the original configuration that the designers had settled on and that proved suitable only for weapons fueled with uranium-235. Research and design of a second ignition method—implosion—and a pre-deployment test would be crowded into the first seven months of 1945.[38]

Groves placed intense pressure on Matthias to deliver quickly, and Matthias in turn pushed DuPont. The B and D reactors were both in operation by the end of 1944, and the plutonium-239 that they produced began making its way through the processing plants in the last days of 1944. Hanford yielded its first "units" of fissionable material on February 2, 1945. The next day, Matthias drove them to Portland, carried them aboard a southbound train, and in Los Angeles handed them over to an "agent from Los Alamos." Shipments of plutonium became frequent and regular, leaving Hanford every five days. By May 1945, to save time, they went by truck rather than train, and by the end of July Groves and Matthias pulled out all stops and had the plutonium delivered by airplane. DuPont delivered its hundredth batch of plutonium refined from spent reactor fuel to the army on July 4, 1945. The plant's rapid pace generated the fissionable material necessary for perfecting the design of the plutonium bomb, which was tested at White Sands, New Mexico, on July 16, 1945, and used on Nagasaki on August 9, where it killed as many as 70,000 people.[39]

On August 6, the day the first, uranium-fueled bomb was dropped on Hiroshima, President Truman released a statement prepared by the Manhattan Project telling the American public about the project in general terms, which was followed a week later by the more substantial Smyth Report *Atomic Energy for Military Purposes*. People around the Pacific Northwest finally learned what the federal government had been doing at Hanford. This moment ended more than two years of work by the army to muzzle the media. From the start it had been clear that the Manhattan Project had had a tremendous effect on life in south-central Washington, and reporters had flocked to the story. A fairly well-informed article in a Lewiston, Idaho, newspaper in April 1943 had provoked Matthias to despair that "trying to restrict publicity on this project is like keeping water in a sieve." Yet continued secrecy had been crucial, so Matthias had doggedly lobbied editors to get them to withhold from publication stories about Hanford that the army had not approved. He had gotten many to cooperate by promising them special treatment once the real story broke. So on August 6, 1945, when the Manhattan Project became public knowledge, the army paid its debt by holding press conferences and providing carefully managed tours of Richland. Matthias himself traveled to Walla Walla, Spokane, and The Dalles to tell communities about Hanford.[40] The federal site had taken its first step toward becoming less isolated from the surrounding region.

The army had finally allowed the public to peek over Hanford's wall of security, but it never wished to take the wall down. The site continued to produce plutonium after the war, so it could not simply be thrown open to the public. Virtually everything about the plutonium-manufacturing process remained secret, with the army trying (apparently unsuccessfully) to keep spies from conveying information about Hanford to agents of foreign powers.[41] One aspect of such secrecy was to prevent people from learning just how much pollution had been created in the manufacture of plutonium. The army calculated that if the enemies of the United States learned how much waste Hanford had generated, they could determine how much plutonium had been produced and therefore how many atomic bombs the nation possessed. The army also hoped to prevent Hanford's employees and neighbors from becoming alarmed about plant operations. When President Truman and the army revealed what Hanford had been doing during the war, employees flooded the site's medical department with questions about the health risks associated with the work. Based on screening of the workers on the site, DuPont reported, "It is reassuring that thus far it has been possible to tell these persons that no one received any injury from the special hazards associated with operations here." Throughout the Manhattan Project, health physicists and physicians had built on prewar knowledge of radiation hygiene and had monitored workers' exposure to radioactivity, perhaps more carefully than their exposure to chemical contaminants.[42]

As it turned out, there was something for Hanford's workers and neighbors to be alarmed about. During the planning and design of the plutonium plant, both the army and DuPont expressed concern about dangers to the environment. DuPont realized right away, in November of 1942, that producing even very small quantities of plutonium would generate a great deal of waste: "For one ton of [uranium] from a pile worked up to provide one gram of X-10 and one gram of fission materials, 8,000 gallons of hot material must be handled." Much of that material would be poured into waste tanks and trenches on the Hanford Site. Additional pollution would, by design, enter waterways and the atmosphere. One reason for locating reactors along the Columbia was to permit the river to carry away and dilute radioactivity from the piles. But the greatest immediate threat came from the gases released from the processing plants to the atmosphere.

Even before condemnations got underway, in January 1943 scientists hastily studied weather patterns to make sure that prevailing winds would disperse waste products. They came away relatively confident that the winds would dilute and drive away dangerous gaseous emissions.[43] Events would not have turned out differently even if the scientists' initial study had produced a less favorable weather forecast. While the stagnant air over the Columbia Basin would have to be dealt with in practice by designing an exhaust stack for the processing facilities, modeling the local wind conditions, and setting up appropriate operating rules, the Hanford Site's strengths were overwhelming, and Groves insisted on decisive action in acquiring land and beginning construction in 1943. In a February 2 memo, Lt. Col. Kenneth D. Nichols, one of Groves's subordinates, wrote, "Decision relative to acquiring the site is held up pending results of the meteorology study being made by Dr. Compton's group. Upon the completion of this study, DuPont will make its recommendation and the site will be acquired."[44] In other words, the army seemed determined to occupy the site regardless of the scientists' conclusions about wind patterns. There wasn't much in the way of an alternative, and the momentum to move forward with the production of atomic bombs was overwhelming.

Conducted under wartime pressure, the scientists' initial study of weather patterns proved rushed and ill informed. Investigations over the next several months revealed that winds could not be relied upon to disperse atmospheric emissions. By this time, of course, it was much too late for the army to find another site, so plant operators decided that releases of emissions would be timed as much as possible to coincide with favorable winds and, during the war at least, to avoid releases during rainy weather.[45] They seemed confident that they could exert strong control over the production schedule. Once more, however, they were mistaken. In late 1944 and early 1945 the pressure to deliver plutonium to Los Alamos became overwhelming. Operators at Hanford did not let irradiated fuel rods (the slugs pushed out of reactors into water tanks) cool as long as they were supposed to—increasing the amount of radiation that would be released in the processing phase—and they began making atmospheric releases at times when the winds were not altogether favorable for dispersing the pollution. The result was an irregular flow of radioactive emissions into the atmosphere. Doses of radiation from this airborne pollution were in all likelihood absorbed by people working at Hanford, residing in the Tri-Cities, and living tens or hundreds of miles downwind. All along, the army suspected that plutonium production could present some public-health risks, but it went ahead anyway, confident that the benefits of accelerated manufacture outweighed the costs. Moreover, it remained optimistic that wartime exigencies would be temporary and that necessary corrections would occur afterward.[46]

When Hanford was initially conceived, scientists knew that its industrial wastes would present health risks, but they were not sure exactly how. At first they worried about the effects of radioactive xenon gas; only after the beginning of plant operations, in late 1944, did officials begin to real-

1.6 and 1.7 The army and DuPont selected the Hanford Site relatively quickly, believing that emissions from operations could be managed well there. Closer study of the environment, however, identified some of the risks associated with plutonium production. Even before starting up the manufacturing process in 1944, for instance, atmospheric tests revealed that winds would not necessarily disperse gaseous wastes. The photograph in fig. 1.6 (from 1943–44) shows how one wind pattern rather than carrying wastes aloft and diluting them in the atmosphere would deposit pollutants quickly back on the site. Over the years operators and scientists continued to study environmental conditions at Hanford to understand the impact of site wastes on the environs. The investigations produced, among other things, fig. 1.7, which illustrates five different prevailing wind patterns that affected atmospheric releases. Both images courtesy of the U.S. Department of Energy.

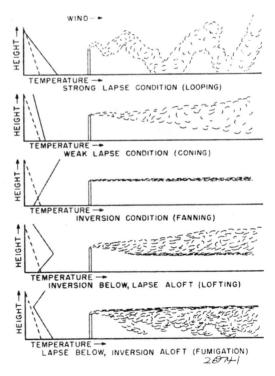

ize that radioactive iodine (I-131) presented more of a health threat. The army and DuPont thus began monitoring iodine releases, but the demand for plutonium production made it impossible to control them effectively. Moreover, scientists did not yet understand the full dangers of radioactive iodine. They worried especially about concentrations of it in the atmosphere, but, as it turned out, the isotope presented an even greater threat to human health by accumulating on vegetation that could be consumed by animals (especially dairy cows) and humans. The connection between radioactive iodine and the contamination of milk was not made until the 1950s, long after Hanford's wastes had been ingested by animals and humans (especially small children) downwind from the plant. Gaseous emissions from Hanford peaked in 1945; of the 739,000 curies of radiation from iodine-131 released between 1944 and 1972, three quarters were emitted in 1945. The relaxed pace of production in 1946 and, more importantly, the installation of filters on the stacks of the processing plants reduced the hazardous atmospheric emissions.[47] (A measure allowing comparison of the effects of different radioactive substances, a curie is roughly equal to the radioactivity released by 1 gram of radium, or 37 billion atoms dissolving per second. The curie might be distinguished from units such as the roentgen, the rad, and the rem, measures of dosage taken up by living things exposed to radioactive substances.)

With the beginning of plant operation in 1944, Hanford also began discharging radioactive waste products into the Columbia. River water was pumped through the reactors to cool them, and on its way through the plant the water received radioactive isotopes, mainly via exposure to the piles' neutron flux. The wastewater was then cooled in holding tanks adjacent to the reactors, where the shortest-lived radioactive materials decayed, before being pumped back into the river. The Columbia diluted the radioactive material, especially during the seasons of heaviest runoff, and the long voyage to the ocean (gradually delayed even more by the completion of dams on the stream) allowed most radioactive wastes to decay before they reached the river mouth. Nonetheless, people who drank, swam in, or boated on the water, or who had eaten fish, shellfish, or waterfowl that had been exposed to the Hanford waste products (including as far away as Willapa Bay on the Pacific Coast), absorbed at least minor doses of radiation. These releases began in 1944–45 but did not peak until the two decades between 1950 and 1970.[48]

By 1945, then, the plutonium plant at Hanford had begun to exert a considerable influence over the surrounding people and environs. Much about the exact nature of that influence remained unknown to the local population, of course, but there could be no doubt that the mid-Columbia region and the Pacific Northwest had entered into a new relationship with the federal government. Just how long that relationship would last was a question that neither the federal government nor the region was prepared to answer.

Cold War Hanford and the Culture of Production

At the end of World War II, Lt. Gen. Frederick Clarke replaced Matthias as the supervisor at Hanford. Upon his arrival, Clarke later recalled, the Manhattan Project had "just enough material for one nuclear bomb." The plutonium factory continued to operate over the ensuing months, and by 1947, when the army was replaced by the Atomic Energy Commission (AEC), a new civilian agency in charge of Hanford, there was, according to Clarke, enough plutonium for several bombs, a number "in the double digits." So Hanford did not close down after the war, but neither did it keep up the frantic pace of production (and pollution) that had characterized late 1944 and most of 1945. The changes were immediate: in September of 1945 DuPont slashed the Hanford workweek from forty-eight to forty hours; by December it had shut down one reactor and begun to run the other two at reduced power. By January 1946, with the pace of production slowed, the atmospheric releases of radiation had diminished substantially.[49] The number of employees dwindled. In September 1945 Hanford had about 10,000 contractor employees; fifteen months later the total had fallen to 5,000. Richland's population fell from 15,400 in March of 1945 to around 13,000 for the first few months of 1946, with women making up more than half the residents.[50]

Between the end of World War II and the crystallization of the Cold War, Manhattan Project sites existed in a kind of limbo. Hanford and Richland stagnated as they waited for key decisions to be made about their future. Would the nation continue to want an atomic-weapons program? If so, would the United States rely on plutonium-239 or uranium-235 as the fissionable material for future bombs? Would Hanford continue to produce fissionable material, or would some other production facility be required? Would the government continue to own and operate Richland, or would it close down the town? These and other questions made it impossible during most of 1946 to know what lay in store for Hanford, all the more so because Hanford remained something of a stepchild within the atomic complex. "Only the facilities for producing uranium-235 at Oak Ridge, Tennessee had been transformed from a temporary, emergency effort during wartime to a stable industrial operation," noted an AEC production history in 1963. Hanford's postwar stability was an open question.[51]

Then the answers came, shaped largely by the onset of the Cold War. As conflict intensified between the Soviet Union and the United States, the federal government committed the nation to a much expanded atomic-weapons program. It also determined that plutonium-239 provided the best primary fuel for atomic bombs. As the nation's sole sup-

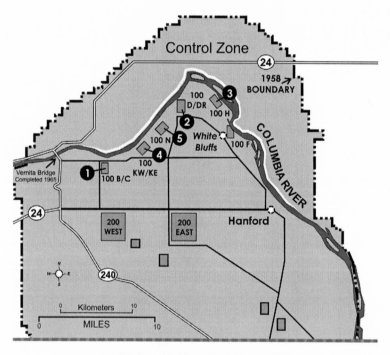

MAJOR FACILITIES ADDED SINCE 1945

REACTORS	200 WEST AREA	200 EAST AREA
1. C Reactor	Plutonium Finishing Plant	B Plant
2. DR Reactor	Reduction-Oxidation (REDOX) Plan	Plutonium-Uranium
3. H Reactor	T Plant	Extraction (PUREX) Plant
4. KE & KW Reactors	U Plant	
5. N Reactor		

1.8 Map of Hanford in 1966. Map by Frederick Bird.

plier of the fissionable material, Hanford would be the workhorse of the American nuclear industry during the Cold War. The site produced about seventy-five tons of weapons-grade and reactor-grade plutonium between 1944 and 1988—approximately one-fourth of the world's supply. It also produced a tremendous amount of waste, so much that during the 1990s Hanford became identified as the most toxic place in the United States. The level of contamination was surpassed only at the site's Cold War counterpart in the Soviet Union, Chelyabinsk-65, known as "the most polluted spot on Earth."[52]

Under the auspices of the Manhattan Project, Hanford had been something of a wartime experiment. During the early Cold War it became a larger, more permanent feature on the American landscape. By 1955 it possessed eight healthy production reactors, five built after the war, all of them substantially more powerful than the initial three had been (even the original three wartime reactors were run at higher power levels after the war). It had added two new reprocessing facilities as well as a plutonium finishing plant, and a sizeable urban area and administrative bureaucracy had grown up around the nuclear facility.[53] These additions were accompanied by new patterns of pollution. The amount of atmospheric emissions declined markedly between 1945 and 1955, but the plutonium plant now released more radioactivity than ever into the Columbia River and deposited more long-lasting chemical and radioactive wastes into the surrounding soils.[54]

As in the early 1940s, decisions about Hanford during the early Cold War stemmed almost entirely from national security considerations. The immediate pressures of 1947–52—intensified when the Russians carried out their first successful nuclear explosion, which was detected by American authorities in September 1949, and when war broke out in Korea in June 1950—required not only that operations at Hanford continue without interruption but also that they expand. U.S. military planners determined that the nation simply could not produce plutonium quickly enough to meet the urgent demand. David Lilienthal, chairman of the newly created Atomic Energy Commission, explained the imperative to arm "this country atomically, to erect a giant deterrent to aggression in the world." That deterrent would succeed only if it was massive enough to forestall the Soviet Union while permitting the U.S. armed services to keep their size at a level substantially below that of World War II. The nation needed to understand, particularly after September 1949, that "numbers are crucial in providing a deterrent to aggression: that high level of production and quality could provide what a dwindling monopoly [of control over atomic weapons] has lost us."[55]

Insistence on heightened production began early on. In 1947 the army and navy agreed to a postwar industrial mobilization plan that stipulated that, even during peacetime, America must have the industrial capacity to immediately produce munitions at or above the peak level of production that had been reached during World War II. Yet this was too delimiting for the Atomic Energy Commission, which required an even higher level of output. In contrast to the military, the AEC's operations manager explained to the armed services, the AEC was "operating at full capacity for the purpose of producing weapons," even in peacetime.[56] The legislation creating the AEC required it to recommend to the president, in consultation with the Joint Chiefs of Staff, a production schedule for fissionable material. The commission quickly adopted a standard phrase used to set yearly production goals for its plants: "We recommend that you approve the production . . . of plutonium and uranium 235 in the maximum amounts attainable consistent with safety and good operating practice." This policy simultaneously discouraged all but the most essential research during this period, so that the greatest amount of plutonium could be used in the fabrication of atomic weapons. A 1949 plan to use some space in Hanford's reactors for materials tests to gather data for future reactor designs provoked a reaction from the AEC general counsel, who warned against violating plans calling for the maximum practicable output of plutonium.[57]

Even when the AEC was able to exceed its planned output of new weapons, it appeared to fall behind, because the dangers presented by communism grew more threatening in the eyes of military planners. "The accumulated total number of bombs expected to be on hand January 1, 1950, will be ahead of the scheduled requirements for that date," the AEC chairman and the secretary of defense reported in April of 1949 (prior to Americans learning in September of that year that the Soviets had detonated their own atomic bomb). "However, from an analysis based on the Eniwetok tests, and from a continuing consideration of the critical international situation, the Joint Chiefs of Staff express a strong view that the current objective . . . is inadequate." Taking into account these concerns, President Truman was asked to approve an increase in production goals and the acquisition of plutonium and fissionable uranium in the greatest quantities possible.[58] During the early years of the Cold War there really was no production target for making plutonium; the nation simply required all that could be manufactured, and then it required some more.

Given the urgency of the situation, Hanford emerged as the obvious place to increase the output of plutonium. Nonetheless, the decision to expand operations there was complicated. One problem with the site was its proximity to the Soviet Union. By late 1947 Hanford was "theoretically vulnerable to Soviet long-range bombing," and by 1950 its vulnerability was no longer theoretical. Most of the western and northern states stood well within range of bombers from Russia, so some advisors recommended developing new reactors somewhere other than Hanford. The site's relatively dated reactors presented another challenge. Its piles had been state of the art in 1943–44, but by 1947 they were recognized as an increasingly obsolete technology, and AEC experts worried that their operating life might be nearing an end. The technological shortcomings of the site became even more pronounced with the development of fusion weapons between 1949 and 1952. The new weapons required not only plutonium but also tritium, which Hanford reactors could make, but not especially well. Hanford produced some tritium and hoped to be awarded the mission of supplying the substance for production of hydrogen bombs. Instead, to meet the need for tritium, add productive capacity, avoid interruption to Hanford's production schedule, and address the matter of the Pacific Northwest's proximity to bombers from the Soviet Union, the AEC built another generation of production reactors at Savannah River, South Carolina, between 1950 and 1956 (with operations beginning in late 1952).[59]

Over the long term, Savannah River presented a serious challenge to Hanford. It not only broke Hanford's U.S. monopoly on plutonium production but also possessed a brighter future because its multipurpose reactors reflected newer design ideas. Moreover, the AEC succeeded in developing the Savannah River site without a government-owned and -operated town, so its political and economic overhead there was lower. At the same time, however, the AEC determined that while Savannah River might become the production center of the future, nothing could really replace Hanford in the near term—and the near term was all-important at the start of the Cold War. The commission decided that replicating Hanford's productive capacities in some other state would prove too expensive and time-consuming. It even ignored advice from the military about locations on the Hanford Site least vulnerable to Soviet attack, because following it would have cost too much time and money.[60] America at the dawn of the Cold

War needed to multiply its productive capacity quickly. Unable to wait for Savannah River's facilities to come on line, the United States committed itself anew to the plutonium plant along the Columbia River.

After a brief sense of postwar relaxation between August 1945 and November 1946, then, Hanford again took on the characteristics of an emergency project. Although the rapid growth and change of wartime were the norm rather than the exception at Hanford between 1942 and 1955, after 1946 the government began to operate the facility on terms different from before. First, it had to find a new contractor. DuPont announced that it would leave the Manhattan Project upon the expiration of its contract in 1946. Despite the government's pressure, DuPont corporate leaders resisted pleas to remain in control at Hanford after the war, preferring to undertake research and development on nuclear power with government funding, "to retain and utilize the Company's present knowledge of the field so that prompt advantage could be taken of developments of possible commercial interest when they occur." This did not include continuing the management of the Hanford production site.[61] On September 1, 1946, General Electric (GE) replaced DuPont as the prime contractor, beginning two decades of management at Hanford.[62] Around the same time, August 1946, Congress authorized the creation of the Atomic Energy Commission, which took over at Hanford on January 1, 1947, and would stay for twenty-seven years.[63]

Both of these changes were significant. The political struggle over the control of nuclear technology after World War II vested responsibility in the civilian commissioners of the AEC, deliberately a step removed from the Executive Branch and the War Department. In turn, the AEC stayed at arm's length from its contractors in the government-owned, contractor-operated facilities the commission managed, thus institutionalizing the wartime, ad hoc division of responsibility. In the words of a 1947 AEC report assessing its relationship to contractors such as GE, "nongovernmental institutions and operations" were like the "remote-control techniques developed for handling radioactive materials. . . . [The] AEC works at its job with artificial hands" via contracts, which translate "will" into "motion."[64] Those dependent on Hanford in Richland, Pasco, and Kennewick, though, were no one's first concern. They would learn to stand up for themselves.

The transition to GE and the AEC did not always proceed smoothly. The new organizations took over at the very moment an "almost frantic urgency" set in at the start of the Cold War, and neither possessed the kind of experience that DuPont and the army had accumulated by the end of the war. So mistakes were made and could no longer be entirely shielded behind a wall of secrecy. As the task of building more weapons for the purpose of deterring the Soviet Union loomed ever larger, General Electric received immediate criticism because it gave too much priority to "building more permanent housing units and storage tanks for radioactive waste, instead of focusing on . . . production and separation facilities." General Electric's problems with construction delays and cost overruns became the focus of congressional investigations and journalistic criticism, as did the AEC's peculiar management of its company towns.[65] Hanford may not have been more mistake-prone than before, but its mistakes certainly became more public.

The Atomic Energy Commission and General Electric eventually steered the site successfully through the years of rapid growth. Between 1947 and 1953 the AEC decided to add five reactors at Hanford, which were completed between 1949 and 1955. At the same time the commission boosted the output of the original three reactors several times. Two additional reprocessing facilities, the Reduction Oxidation Solvent Extraction, or REDOX (1952), and Plutonium Uranium Extraction, or PUREX (1956), plants eventually supplanted the B and T plants built during the war. The initial burst of new building, between 1947 and 1949, was the largest federal peacetime construction project ever. In 1946 Hanford had 4,479 operations employees and 141 construction employees. Within two years those figures jumped to 8,628 and 14,671, respectively. Between 1950 and 1955 the average number of operations employees at Hanford stood at 8,770, and the average number of construction workers hovered around 5,555.[66]

Richland grew from 14,000 people in early 1947 to almost 22,000 by 1950.[67] The housing shortage became acute by the spring of 1947. The lack of homes inhibited production, and the AEC expressed alarm. "Competent workmen . . . cannot be obtained in sufficient numbers for lack of housing on the project," reported J. E. Travis of the AEC. General Electric reported a shortage of 212 houses, and Travis requested authorization to build another 500 units in anticipation of the city's growth. Three weeks later he asked for authorization to construct still another 1,000 homes.[68] In addition, GE and the AEC built another construction camp between the village and the plant, called North Richland, which had a population of 13,000 in 1948.[69] Additional workers spilled over into Kennewick and Pasco. The second great Hanford boom was underway.

As the Atomic Energy Commission expanded operations, it gave clearer definition to its facilities. The commission ratified the army's decision to divide America's nuclear program between research and design centers, which would be run by academic contractors, and centers for production of basic materials, which would be industrial in nature. Hanford became more and more clearly identified as an industrial site — "the production front," as one AEC spokesman put it.[70] Hanford's distinctive contribution to the Cold War was not research or strategy or training or testing or assembly but production — the manufacture of fissionable material for bombs as quickly and as cost-efficiently as possible. The special mission became apparent in the way that the AEC defined tasks for its sites. In 1949, when the commission authorized a new plutonium fabrication complex at Hanford, chairman David Lilienthal explained the addition: "The activities for which this facility was designed had been carried on at Los Alamos, largely by scientists. For security reasons, and because it was unsound to continue operating a production line with scientists, a second facility of different design was desirable, which could be operated by production men."[71]

Lilienthal's remarks merely confirmed the thinking that had already determined what kind of reactors would be added at Hanford. At the start of the Cold War, the urgency of renewed wartime mobilization meant that there was no time to experiment with anything too different from technologies that had been proven during World War II. So, while the AEC planned and built a new and untested kind of production reactor in South Carolina, in Washington between 1947 and 1955 it basically constructed variations on what had been fashioned during World War II. For instance, the designs of DR reactor and H reactor (which began operation in October 1949) incorporated only minor changes based on lessons learned from operation of the first three reactors since 1944. "Because of the time element involved, however, major design changes which would have interfered with the expeditious prosecution of the work were not undertaken."[72] Similarly, the "jumbo" reactors completed in 1955, KE and KW, were mostly bigger — not technologically different — versions of their predecessors.

While Hanford focused on maximizing the output of plutonium, research on new reactors went elsewhere. The AEC decided to build a separate Materials Testing Reactor in Idaho, at what would become the

Idaho National Engineering Laboratory. Meanwhile, GE expressed a strong preference for a single contract covering its operation of both the Knolls Atomic Power Laboratory in New York State and Hanford. The AEC spurned the proposal because it did not conform to the commission's preferred functional division between research and production. Knolls was concerned with research, and the AEC believed that GE should operate it without expecting a profit. Hanford was concerned with production and manufacturing and it was there, in the AEC's view, that GE should expect to receive "appropriate profits and incentives." The AEC separated the contract with GE into two parts, on the grounds that reactor research and plutonium production were "essentially different."[73] This bifurcation reflected the federal organization chart rather than that of GE: within the AEC, Hanford fell within the Division of Production, as opposed to the Division of Reactor Development.

By defining Cold War Hanford's purpose primarily in terms of production, and by determining that reactor technology dating from World War II would serve best for meeting the nation's immediate needs, the AEC profoundly shaped the workforce, culture, and long-term prospects of the site. At the end of World War II, the government's wartime coordinator of research and development had called science "the endless frontier."[74] More than ever, America needed to relentlessly colonize this territory of radar and computers, of guided missiles and nuclear weapons. In the eyes of many, the boundaries of this frontier were pushed back most successfully by scientists at sites such as the AEC's Los Alamos Laboratory, a facility managed by a university rather than an industrial contractor.

In contrast, Hanford no longer seemed to be on the cutting edge of technology. Its first three reactors were graphite moderated, water-cooled uranium piles with pass-through coolant, a model that by the end of World War II was already viewed as a dead end in nuclear reactor technology. Yet Hanford not only held on to its wartime reactor model, it added five more based on the same fundamental design. The United States relied heavily on the production of plutonium from Hanford's increasingly obsolete reactors throughout the 1960s, because they were cost-effective—in terms of time as well as money—and reliable. The U.S. strategy of a high-technology defense mandated that Hanford stick to a relatively low-technology method of producing plutonium. The site did not generate new scientific insights; rather, its operation depended on the industrial application of dated scientific ideas and technical processes.

1.9 Between 1947 and 1955 the Atomic Energy Commission added five new production reactors at Hanford to increase the output of plutonium for the Cold War. Rather than adopting newer reactor designs, the commission relied on the same basic model that had been used during World War II. Thus the first eight Hanford reactors consisted of water-cooled graphite. In this 1952 photograph, workers are laying the blocks of graphite for the C reactor. On the two side walls can be seen the water pipes and process tubes that would penetrate the pile. Photograph courtesy of U.S. Department of Energy.

The AEC decided that Hanford was not meant to stand at the forefront of nuclear science. It never became a national laboratory, as did Oak Ridge and Los Alamos, and Hanford did not require many highly specialized researchers and scientists in its workforce. As early as 1947 the AEC had determined that Hanford needed "men of broad training and experience along engineering and scientific lines," not highly educated specialists. Each new reactor required a few scientists during start-up, but industrial operators generally played the leading role.[75] In 1947 the AEC and GE together employed 91 chemists, 28 physicists, and 9 metallurgists at Hanford. At the same time, 104 chemical engineers, 65 mechanical engineers, 52 electrical engineers, and 36 civil engineers were on the payroll.[76] It was a professional workplace dominated by engineers rather than scientists.

Among those scientists on the site, the more industrially-oriented disciplines (chemistry and metallurgy, for example) prevailed.

Many experts in the new technology endorsed this approach. The journal *Nucleonics*, which sought to encourage development of an industry based on nuclear technology, editorialized in 1948 that scientists were not essential for most of Hanford's purposes: "Experience . . . has shown that capable chemical, electrical, electronic, and mechanical engineers can acquire the small amount of nuclear physics necessary . . . for competence in reactor design."[77] Chemical engineers were considered the appropriate professionals for managing a chemical process operation such as the production of plutonium. In another article, the chairman of Brookhaven's Reactor Science and Engineering Department assured *Nucleonics* readers that "the production of plutonium in a reactor[,] while involving a nuclear transformation, usually thought to be a subtle process of nuclear physics, has the atmosphere of a water treatment plant, or an oil refinery."[78] Hanford created a culture of production, one that secured its relevance but limited its scope and eventually led to its obsolescence.

Hanford reflected the army's and the AEC's division of labor between academic and industrial organizations. Hanford was a factory, not a university. This difference may help explain a recollection by Herbert M. Parker, health physicist at Hanford, of how different Manhattan Project sites contemplated the dropping of atomic bombs at the end of World War II. Other sites with higher populations of academic scientists were more receptive to a petition circulated by scientist Leo Szilard opposing the use of the bomb on populated areas. At industrial Hanford, no such doubts surfaced: "The attitude . . . was more to the point. . . . The belief was that we were here to make an explosive atomic device, and we got about the job."[79]

Despite the fact that it was associated with the mysterious power of the atom, after 1945 Hanford ran not on Nobel-level scientific genius but on another kind of intelligence. This knowledge was technological rather than scientific, and had as much to do with hands-on skill as with theory. Like most technologies, this one advanced through secrets learned during judicious banging on the pipes. "Anyone who has visited these plants knows that they are extremely difficult situations, they are steamfitters and plumbers nightmares [*sic*]," one congressman explained to his colleagues on the Joint Committee on Atomic Energy in 1949. He described the new REDOX reprocessing plant as "the damndest thing you ever saw in your life, with pipes and wires and things running all through the thing."[80] It was the realm not of ethereal intelligences, of Einsteins and Oppenheimers (at least according to the public's perception of Einstein and Oppenheimer), but of craftsmen—of plumbers, steamfitters, boilermakers and machinists. "You haven't got the foggiest notion of what it is going to cost you," the congressional visitor continued his report, "except generally that there is going to be half a million tons of pipe and valve fittings and a lot of other stuff and eventually you may have to rip out half of it and replace it with something else because it won't work."[81]

This essential character of Hanford's technology, and of its workforce, can be understood by considering the operation of those complicated plumbing systems: the site's reactors. Hanford's piles consisted of large stacks of graphite bricks with holes drilled through them to admit a battalion of water pipes. The pipes held the small slugs of uranium that fueled the reactor, and they also delivered cooling water that carried off the heat of the reaction. While the aluminum walls of the fuel elements held the fission fragments produced by uranium fission and only a small part of the radiation was released into the cooling water supply, the radiation's heat passed directly out and raised the water's temperature.[82] Much of the expense of constructing and operating the reactors, and much of the difficulty of maintaining them, came from pumping a reliable supply of treated cooling water from the Columbia River through the plumbing system and determining the most effective pressure and flow rate.

Both to keep the reactor's heat within manageable limits and to allow for the easy passage of fuel elements into and out of the reactors, it was vital to keep the pipes clear. Even though the river water was regularly treated, sediments built up in the pipes, so operators developed a simple method to scour out the plumbing. Borrowing a rifle from the army, the Hanford staff sawed off the barrel and welded a flange that bolted over the end of a partially blocked reactor pipe. The pipe could then be cleared by firing a blank charge into it, allowing the expanding gases from the charge to blast out the sediments.[83] Using a .30-caliber rifle to make adjustments on an advanced piece of nuclear equipment is not Nobel Prize–winning stuff, but it is a reminder that Hanford's workings were as much a matter of crowbars and sledge hammers as slide rules and electronics.

During the early years of the Cold War, Hanford's workforce and its method of operation reflected the site's orientation as a production-oriented, cost-conscious facility. The graphite reactors were bulky but had proven reliable and efficient. For all the talk about more exotic moderators, such as beryllium and heavy water, only graphite piles were in practical service, and they worked economically. In plutonium production, $3.50 worth of graphite did the duty of $55.00 worth of heavy water.[84] Actual operating practice at Hanford, which involved manipulating the piles' control rods by hand, reflected a similarly industrial approach to the manufacture of plutonium. To keep the self-sustaining reaction running at peak efficiency throughout the pile, an operator adjusted control rods to the best position for each part of the large reactor. As one chronicler explained in 1957, "A good operator can beat a machine in maintaining the trim of a pile, particularly a big one like [Hanford's], where the neutron flux may behave quite unpredictably at times, sometimes making the operator think he is operating several small reactors in parallel from the same set of controls."[85] The piles were run not by theory or automatic controls or feedback circuits but by hands-on knowledge.

Historians of technology argue about whether analytical precision and historical accuracy require a clear distinction between science and technology—both as activities and as bodies of knowledge.[86] Post–World War II, science-based technologies are especially prone to being treated as scientific artifacts, and so the features best understood by treating them as technologies (with economic, industrial, and material development factors) are easily overlooked.[87] In the case of Hanford, it is crucial to remember that such phenomena as neutron fluxes and graphite creep, while esoteric, were treated as elements of a production process characterized by an industrially-oriented workforce of craftsmen and engineers. Postwar developments in reactor operation, developed on the site, kept the Hanford piles built according to wartime designs working as reliable production equipment well after the war's end. On the one hand, this secured the site's future as a plutonium factory for two decades after the war. On the other, though, it largely curtailed prospects for research and development projects beyond those directly related to plant operations (including environmental safety).

The maintenance and expansion of graphite reactors and separation facilities were successful because of improvements developed by workers and engineers concerned with their own idiosyncratic technologies. This was especially important after 1946, when plutonium production increased markedly. Briefing AEC Chairman Lilienthal for a visit to Hanford in the late 1940s, the manager of Hanford Operations Office, Fred Schlemmer, urged him to praise employees for their problem-solving ability. Thanks to these talents, Schlemmer wrote, "threatened shutdown of certain production areas did not materialize due to "on the job" improvements made by regular operating personnel. The plants were built for a single wartime job, and the modifications made by plant personnel were made without production loss. The accomplishment is one of the greatest hurdles taken since the Commission took over."[88] Schlemmer identified the two main themes in the postwar phase of Hanford's production plants: the steady adaptation of Manhattan Project–era reactors (rather than the use of new reactor types) to meet the skyrocketing demand for plutonium, and an overriding concern with avoiding production losses.

As Hanford steadily surmounted technological challenges and endured the lengthy construction boom of the late 1940s and early 1950s, its mass-production mission apparently became fully internalized. General Electric's monthly reports to the AEC emphasized the growing efficiency of the plant. For example, in June 1949 the contractor told of the quantities of metal produced, the improved quality (or concentration) of the final product, the higher operating efficiency and expanded nominal power of the reactors, a "new high record yield" of acceptable slugs canned, new construction completed on the site, and, to ensure that such gains continued, a "9-Point Job Improvement Program." This production mentality permeated even the public-affairs office, which reported that in December 1950, "39 news releases were written and distributed, and a total of 274 column inches was obtained in Pacific Northwest newspapers." An enlarged plant, coupled with heightened efficiency throughout the site—fuels processing, production reactors, chemical separations, plutonium fabrication, and public relations—resulted in dramatic productive increases and per-unit cost reductions for decades after the war.[89]

As Hanford's productivity increased, there was a growing belief that the manufacture of plutonium was—or was becoming—routine. One manifestation of this awareness appeared at Hanford in the person of the life insurance salesman. Since 1943 the army had prevented underwriters from learning about the hazards of manufacturing plutonium, but in January 1947 the AEC permitted inspectors to tour the plant so they could set rates for life insurance policies. The gap between Hanford and the rest of America shrank a bit.[90]

Another factor in the normalization of Hanford was the unionization of its workforce, a process that had begun in the building trades. During and after the war, the government had worked with organized construction labor. The AEC recognized in 1947, as it negotiated in Spokane for a "construction agreement with the AFL Building and Construction Trades Unions," that it had to conform to precedents and expectations in the "highly unionized West Coast and particularly the Pacific Northwest." The resulting contract produced "very definitely a union shop."[91] The need for secrecy, however, prevented some parts of the agreement from being explicit. Labor may not have received, for example, a satisfactory explanation of why the AEC refused to reimburse employees for travel time to and from the job site. The union had noted that some employees working on the most remote parts of the reservation received a higher wage than others of similar rank and assumed that GE was paying a travel allowance. In fact, GE was paying skilled (but not unskilled) construction laborers a bonus "for hazards"; their job sites on the most remote part of the reservation were right next to the production reactors. "While this could not be explained to the unions directly," an AEC party to the negotiations reported, "it was pointed out that the work for certain craftsmen, in what happened to be a more remote area, received a differential for very much the same reasons that carpenters receive a differential for handling creosote."[92]

While unions had played a role in construction at Hanford since its beginning, they did not gain a foothold among operating employees until early 1949. Organized labor had initially agreed with the secretary of war that "no recognized unions existed at Hanford Works," but on September 27, 1948, the AEC "announced it would not now object to recognition of unions." The American Metal Trades Council of the American Federation of Labor petitioned for an National Labor Relations Board–sponsored election, and on February 8–9, 1949, won certification as the collective bargaining agent for numerous plant employees. Thereafter the Hanford Atomic Metals Trade Council (HAMTC) negotiated contracts directly with GE. The Atomic Energy Commission, however, reserved the final say on many terms of any agreement, and also insisted that the unions and management respect its security requirements.[93] Hanford did not exactly typify American industrial relations during the postwar period, but it was steadily acquiring most of the trappings of a conventional manufacturing site.

Throughout the 1950s, public portrayals increasingly discouraged the idea that Hanford was a truly exceptional kind of factory by likening plutonium production to the basic manufacture of chemicals. D. S. Lewis, writing for *Electrical Engineering* in 1957, explained that Hanford's "work force is comparable to that of any normal chemical plant." In the same year

H. E. Hanthorn of GE claimed similarly that the separation "processes conducted in the canyon buildings do not differ in principle from those of the normal chemical industry"—except for the "massive quantities of radioactive material" requiring remote handling. Both authors found Hanford comparable to what seemed standard elsewhere in the American industry. The only major difference, according to a *New York Times Magazine* story in 1950, was that "it is ten times as safe to work at Hanford as in an average chemical plant."[94]

The issue of safety, like virtually everything else about the site, came framed in terms of production. In order to maintain a high level of output and minimize public criticism of Hanford operations, the AEC toiled to keep employees relatively content. So it reassured them about the safety of working at Hanford. These reassurances were not entirely propaganda. During World War II and after, Hanford earned high marks for its attention to workplace safety. A specialist in industrial safety reported in 1948: "Every operation in this plant impressed me as having been thoroughly engineered and every safeguard included in the design. The plant is a fine example of what can be accomplished when safety and common sense are built into a new operation."[95]

Although safety remained an important issue for Hanford managers, and a key to sustaining employee morale, it did not override the more critical concern with production. There were ways to make the manufacture of plutonium safer but the AEC did not embrace them—partly because its budget did not permit it to. Moreover, the AEC weighed potential improvements to safety against schedules of production, and safety often came in second. For example, the commission's Advisory Committee on Reactor Safeguards expressed concern throughout this period about the lack of containment shells around Hanford's reactors. The design of the reactors, and the fact that they operated at an exceptionally high level of power, made them particularly worrisome to the expert committee. Without some kind of enclosure around the piles, the committee feared, a severe accident at the plant would be much more likely to threaten downwind populations as far away as Spokane. Yet when Edward J. Bloch, the AEC director of production, contrasted the benefits of installing new safeguards with the prospective costs in 1958, he opposed spending the time and money to make the plants safer. "The contribution to national security through gains in plutonium-production," he wrote, "appear to outweigh the . . . consequences . . . [of] the unlikely event of a major reactor accident."[96]

The commission's approach in this instance reflected two related tendencies in its thinking. First, it focused much more on the possibility of a catastrophic release resulting by accident than on the hazards presented by "routine" operations. Second, it gave more attention to the safety of workers on site than to off-site residents. Postwar Hanford released radiation and other wastes into the environment on a regular basis as part of its normal production schedule. These releases, which were monitored, were generally regarded as a part of doing business. After reducing the amount of atmospheric emissions since the end of the war, site managers expressed confidence that its ongoing releases did not present a danger. A 1948 report summarized this thinking: "Due to the extraordinary record established by the [Health Physics] group, cumulative poisoning by radioactive materials has not yet been observed. . . . In most though not all places the risks now are well understood, discipline is extremely good, and the caliber of craftsmen and laborers is far better than the industrial average. The danger has, therefore, become that of accidents rather than of chronic exposure as in poisoning from lead." The same report admitted that the on-site disposition of radioactive and toxic wastes created during the manufacturing process remained an urgent, unattended problem.[97] But site managers, overall, felt confident about their efforts to ensure safety.

In fact, the production of plutonium at Hanford generated a substantial amount of pollution, both on and off the site. For example, there occurred in December of 1949 the infamous episode known as the "Green Run." In this experiment, run by the Department of Defense, Hanford processed "green" spent reactor fuel (that is, material that had not been allowed to "cool" for more than two weeks after leaving the production reactors) and intentionally released highly radioactive emissions to the atmosphere. The Green Run released approximately 11,000 curies of iodine-131, a figure that accounts for less than 2 percent of the plant's cumulative emissions of radioactive iodine. The point of the project, at least in part, was for the United States to test instruments that tracked gaseous emissions so that the country would be better prepared to measure the wastes generated by weapons programs in the Soviet Union and thereby monitor the Russians' nuclear capabilities. The experiment began despite iffy weather, and then the weather turned even worse. Meteorological conditions prevented the dilution of the radioactive clouds released by Hanford, and both the site and the surrounding lands were exposed to a concentrated dose of airborne radioactivity.[98] A one-time occurrence, the Green Run showed that Hanford continued to emit wastes into the atmosphere.

While gaseous releases from Hanford diminished over the early Cold War years, releases to the Columbia River increased substantially as the number of reactors grew from three to eight and as the power levels increased. The plant pumped river water through the reactors to cool them down, and in the process some of the water was contaminated with radioactivity. The water was allowed to stand for a while before returning to the Columbia to mitigate the problem, but radioactive isotopes (primarily including sodium-24, phosphorus-32, zinc-65, arsenic-76, and neptunium-239) made it to the river and, eventually, the Pacific Ocean. The city of Richland began drawing drinking water from the Columbia in 1963, a change that increased townspeople's exposure to radioactivity; residents in Pasco and Kennewick were already drinking river water. Hanford authorities tried to monitor the local population discreetly throughout the postwar years, bringing health physicists to local schools and undertaking a program of autopsies on longtime workers to check for medical conditions associated with radiation exposure. Some workers on the site placed confidence in the fact that they were protected by "a pretty darn good dosimetry program," that is, routine evaluation of their exposure to radioactive emissions, which gave them the sense that radiological danger was a known, and minor, concern.[99]

Hanford released radioactivity intentionally and routinely during its operations, from the start of production in 1944 until the reactors and processing plants closed down in 1986–88. In monitoring the releases, officials consistently interpreted the data to suggest that the emissions did not present a danger to human health. Again, they remained more worried about a major accident than about the emission of wastes from normal operations. From time to time the AEC and GE issued reassuring information about the pollution produced by Hanford operations. For example, a 1955 article in *Popular Mechanics* claimed that the amount of waste "that now goes out over the southeastern Washington countryside would have to be increased in concentration by thousands of times before any effect could be detected on the people and animals who live there."[100]

Whether or not the AEC accurately assessed the impact of Hanford pollution on people and the surrounding ecosystem, what is clear is that the commission generally avoided going public with specific or disconcerting information about plant emissions. Security was one reason. If the United

1.10 During Hanford's peak years of production, Herbert M. Parker served as the leading health physicist on site and was responsible in large part for evaluating the risks presented by site wastes to workers and nearby communities. In the photograph in fig. 1.10, Parker poses with two recent graduates of Hanford's technical training program. During the later part of 1954 Parker assumed a leading role in assessing the effects of a release of radioactive ruthenium to the Hanford Site and nearby towns. Photograph courtesy of U.S. Department of Energy.

States could keep track of the Soviet Union's nuclear weapons program by measuring releases of radioactivity into the environment, then the Soviet Union could do the same thing. The AEC refused to detail its wastes because it did not want outsiders to get information about how much plutonium Hanford produced. But there was another motive for remaining silent: Hanford's sensitivity to how the public perceived operations. In particular, plant managers hoped to avoid creating any alarm that might threaten productivity—perhaps by provoking outside inspection and criticism or by endangering employee morale. When confronted with situations that raised the possibility of injury to the public, Hanford managers generally tried to downplay the extent of risk.

The handling of the accidental release of an unusually high amount of ruthenium-106 in 1954 typified Hanford's approach to radioactive pollution. As the hot particles fell to the ground, they contaminated "large areas of the reservation," wrote health physicist Herbert M. Parker on August

19, in excess of "levels that can be considered unequivocally safe." Managers responded by declaring parts of the site off-limits, posting signs that read "Contaminated—Keep Out." Because the acreage of affected land was more extensive than ever before, Parker worried, "The [public] relations problem will be a severe one, and educational efforts to assure ready compliance without excitation of undue alarm will be needed." He intended to handle the problem "without public release of information," but admitted that "the program is so large that publicity may be unavoidable. A suitable press story will be prepared so that outside questions could be answered, if necessary."[101] There was a striking ambivalence to Parker's response. On the one hand, the contamination had occurred at a level beyond that which he considered "unequivocally safe"; on the other hand, he seemed concerned less with minimizing people's exposure to the radioactive isotope than with minimizing the possibility of "undue alarm." No public outcry was heard.

As Parker worked through the problem over the next month, he tried to calculate the possible effect of the contamination on people and livestock. He was not entirely sure just how dangerous this emission was, but he generally assumed the best possible picture. Ruthenium-106 was detected in cattle droppings on the Wahluke Slope, downwind from Hanford, for example, but it was somewhat casually assumed to represent no "great threat." Parker speculated that "a secondary hazard" to humans might result from consuming the kidney and liver of contaminated cattle, but public-relations concerns prevented the AEC from finding out for sure. "We were interested in obtaining such organs from local stock, but could not do it without risk of exciting too much comment. As kidney and liver are a low percentage of the normal diet, it seems safe to assume that the hazard to man would be insignificant."[102] Parker may have been correct in his assessment of the risks presented by the release of ruthenium-106, but he seemed less concerned with attaining scientific certainty than with avoiding public concern and all that such concern might entail for Hanford. Secrecy trumped certainty.

Although the AEC kept people in the dark about the effects on the environment of plutonium production, during the 1950s the public's involvement with Hanford increased. During the 1940s it had been standard for federal agencies to make decisions and take actions with virtually no warning to or consultation with the local population. Events at the reservation were driven primarily by the agendas of U.S. policy as determined in the nation's capital. In the realm of public health and environmental impacts, the government's quest to maximize production and secrecy continued to shape Hanford's relationship with its neighbors into the 1950s and 1960s. In other realms, however, the site's managers became steadily more responsive to the interests and desires of the people who lived nearby. The change resulted in part from the transfer of America's nuclear weapons program from military to civilian control. After 1946 Congress pressured the AEC to "normalize" some of the conditions at the sites it managed. So, for example, the commission decided that the inhabitants of Oak Ridge, Los Alamos, and Richland should "enjoy those facilities, services, and activities which are properly a part of American community life."[103] This meant, among other things, that the residents of Richland ought to get the opportunity to purchase property, form a municipal government, run their own affairs, and develop an economy apart from Hanford.

As Richland attained a degree of independence during the 1950s—people bought their own homes and Richland was incorporated as a city—its population showed increasing initiative, and its actions directly affected the work at Hanford. A major concern in Richland, as well as in Kennewick and Pasco, was that the local economy revolved too much around plutonium production. As residents took more control of the town and invested in it, they wanted to diversify the economic base so that the Tri-Cities would not have to rely exclusively on U.S. defense spending. One obvious solution was for Hanford to branch out into applications of the peaceful atom. During the mid-1950s and early 1960s, the Tri-Cities spearheaded a campaign in Congress to create a ninth, dual-purpose reactor at Hanford, one that could both make weapons-grade plutonium for nuclear bombs and generate kilowatts of electricity for the Pacific Northwest. The way that the Tri-Cities pursued this goal illustrated another major move toward "normalcy" at Hanford. They enlisted the support of powerful U.S. senators and representatives, notably Washington's Senator Henry M. Jackson, who put their case before both Congress and the AEC and finally won approval for a dual-purpose reactor in a series of legislative steps between 1958 and 1962. More and more, Hanford was seen as a statewide interest, not just a localized one; like Seattle's Boeing company, which produced for both civilian and military clients, Hanford represented a modern manufacturing establishment, diversifying the economy of a state otherwise dependent on extraction of natural resources via logging, fishing, and farming.

The N reactor began producing plutonium in 1964. Two years later, when it began generating kilowatts for regional electricity markets, it was responsible for 35 percent of the country's "nuclear-generated power on line."[104] In most important respects, of course, Hanford remained entangled within the security needs of the AEC, but by the 1960s the public had demonstrated its commitment to having some say about the affairs of the reservation as well. The site was becoming less of a federal enclave, but it was not clear that it could successfully be weaned from the U.S. government. Private, municipal, and state interests tried to increase their involvement with Hanford, but few of their ventures panned out. Developments of the 1980s and 1990s reaffirmed that the federal government would continue to play the dominant role.

In Search of a New Mission: The Federal Enclave after 1964

On September 26, 1963, President John F. Kennedy visited Hanford to dedicate the N reactor and attend the groundbreaking ceremony for construction of the steam-generating works that would enable it to produce electrical power. He took the opportunity to celebrate the site's diversification into civilian uses of atomic power while also noting its contributions to the nation's "military strength." A dual-purpose reactor, he added, represented a form of utilitarian conservation that would assist the United States in using its "resources to the fullest." Hanford, he concluded, made America prosperous as well as powerful. With the N reactor, he proclaimed, "this country will be richer and our children will enjoy a higher standard of living."[105]

The president's speech came at the site's high point as a production facility for weapons-grade plutonium. The three reactors built during World War II continued to operate, albeit at levels far above those for which they had been designed in the early 1940s. The site had added five more production reactors during the decade after 1945. Based on the original models dating from 1942–44, these had been designed to provide greater levels of output and then had been improved to attain still higher production. The AEC was also completing construction on the N reactor, which had a dual mission of producing plutonium as well as kilowatts. In addition to these nine reactors, of course, Hanford possessed an assortment of other facilities devoted to the manufacture of plutonium. The processing plants, which refined the material from the reactors into fuel usable in nuclear weapons, remained the most substantial. All parts of the Hanford plant continued to disseminate gaseous, liquid, and solid wastes to the surrounding environment. The amount of radiation leaving the site by air or water declined gradually and unevenly, but Hanford was not systematically treating the majority of its on-site wastes. It remained focused on production.

Between 1943 and 1963 the Hanford Works had been the scene of the most active devotion to industrial production of a crucial strategic commodity. As the United States had reacted or overreacted to one Cold War crisis after another, it had looked to Hanford to accelerate its output of fissionable material for atomic weapons. This was true even given the AEC's hesitation to keep Hanford in business after the war. Workers responded, using federal investment to generate increasing amounts of plutonium at a steadily decreasing expense. In January 1964, all of this changed when President Lyndon B. Johnson announced cutbacks in the nation's manufacture of plutonium. That moment initiated a reversal of Hanford's fortunes: the government began to shut the factory down. By 1971 the AEC had closed eight of the nine reactors, and by 1972 the site had ceased shipments of plutonium. Only the N reactor continued to operate—generating primarily electricity rather than fuel for nuclear weapons.

It was a shock to people in Washington State and, particularly, to the population of the Tri-Cities. Used to a steady diet of rapid construction and increased production "under the urgent demands of wartime schedules," the families at Hanford found it difficult to adjust to this sudden reversal.[106] The changes were not all that surprising to top officials in Washington, D.C. While AEC managers had raced to add productive capacity during the early 1950s, they had realized that one day, not too far into the future, they would no longer need all the industrial facilities they were assembling. Plutonium-239 has a half-life of approximately 24,000 years, which means that it does not decay appreciably once it has been stockpiled. In 1952 commissioner Gordon Dean had correctly predicted that the AEC would have a surplus of plutonium by the mid-1960s.[107] The realization of this prediction had an immediate effect on different AEC sites. At Savannah River in 1953, the AEC put a new plutonium fabrication facility on standby and, after starting to build and operate five reactors, decided not to proceed with a sixth one that had been planned. But in the long run Savannah River would survive the downsizing with its mission intact. Its newer reactors produced not only plutonium but also tritium (an essential ingredient for hydrogen bombs) and other materials. The South Carolina

site would not be hit so hard by the cutbacks of the 1960s. While Hanford
lost eight out of nine reactors, Savannah River lost only two out of five.[108]

When the federal government decided to reduce its production of plutonium, choosing which reactors to shut down and when to do it seemed
like simple decisions. Hanford's reactors were the oldest and would cost
the most to keep up, and their output was not needed any longer. But
Hanford's fate became entangled with several other national agendas, and
abandoning the reactors proved to be not as simple as just declaring them
to be excess to the nation's requirements. For example, President Johnson had timed the announcement of cutbacks in early 1964 to coincide
with a diplomatic overture to the Soviet Union. Seeking to advance nuclear
disarmament on terms favorable to the United States, Johnson presented
a plan whereby the United States offered to permit the Soviet Union to
verify reactor shutdowns, hoping (in vain, as it turned out) that the Russians would do the same for the United States.[109] In other words, the timing
of shutdowns was tied to nuclear diplomacy. Even in its decline, Hanford
remained a creature of the Cold War.

It also remained a creature of the AEC, an agency that brought its own
agenda to changes at the plutonium plant. To implement cutbacks at its
plutonium facilities, the simplest and most cost-effective course of action
for the commission would probably have been to close down one of its
two production sites altogether—with Hanford being the obvious choice.
But the AEC did not wish to lose either site entirely, or the organizational
resources that accompanied it. The commission worried in part about the
negative economic impact that total shutdown would have on surrounding
communities, but it was even more concerned about maintaining control
over and appropriations for both sites, and about being able to resume a
higher level of production if needed in the future. So it kept both sites
operational even as it reduced their output. One 1965 AEC memorandum
defined the problem as one of "excess capacity, rather than plant close-
out." Federal managers accepted the idea of reducing or stopping production, but they did not intend to abandon Hanford. Playing its cards close to
the vest, the AEC strung out the shutdowns at Hanford over several years
without giving much warning of bad news to workers and neighbors.[110]

One effect may have been that workers did always not know what was
coming, so they may have been less inclined to worry and leave their jobs.
Atomic Energy Commission secrecy and delay may also have given communities and political leaders time to make arrangements to find other

work for people at Hanford, so that neither the AEC nor the local economy lost as much in the way of "human resources" as they might have. The cutbacks at Hanford were severe, but they were never nearly as harsh as they could have been—apparently because the AEC wanted otherwise. Between 1964 and 1976, employment at Hanford fell by only 5.2 percent, from 9,539 to 9,030.[111] Given that Hanford had lost eight out of nine reactors and virtually ceased the production of fissionable material during this period, its resilience was remarkable.

What was left for those nine thousand employees to do? Not very much, said some former employees, who remembered fondly the intensely productive period of 1944–63.[112] Even if their perception represents an exaggeration, one thing is clear—fewer and fewer people worked at defense-related jobs. Until 1964 almost all federal expenditures at Hanford supported weapons work. By 1967 expenditures on military and on non-military purposes were about even. By 1975 military purposes accounted for only 25 percent of federal spending at Hanford. At the same time, the plant's release of radioactivity to the environment, and particularly to the Columbia River, declined sharply as reactors were taken out of production between 1964 and 1971.[113] Because Hanford's function changed, so did the impact on its environs.

Cutbacks posed the question of what the plant would do in the future. One answer that seems obvious today—managing and cleaning up a tremendous amount of waste—did not seem for many to be a realistic response to Hanford's plight in the late 1960s and early 1970s. Instead, local communities prodded the Atomic Energy Commission to support a new mission that perpetuated the site's orientation to *production*, all the while hoping to keep in place the facilities and personnel that would permit—if needed—a resumption of the old mission. The search for another purpose entailed a balancing act between the agendas of the AEC and its successor agencies, the Energy Research and Development Administration and the Department of Energy, and the increasingly influential voices of local and regional interests. In marked contrast to the situation of the 1940s and most of the 1950s, federal officials concerned with national security no longer dominated decision making about Hanford. They now had to consult others who claimed a stake in the plant. Two sets of interest groups in particular demanded responses from federal managers during the 1960s and 1970s: the first represented nearby communities' concern about Hanford as the mainstay of the local economy; the second included proponents as well as critics of nuclear energy from around the nation and the world. Local, state, and regional interests gained more say than before.

During the 1950s, including in particular the campaign for the N reactor and the transfer of real estate from the government to the inhabitants of Richland, residents of the Tri-Cities began to assert themselves in the political arena in matters concerning Hanford. During the 1960s, as the AEC began cutting back at its production sites, the communities rallied anew to protect their economic interests. Boosters enlisted Washington State's two powerful U.S. senators, Henry M. Jackson and Warren G. Magnuson, to lobby on their behalf in Congress and before the AEC, and as a result received more support from the federal government during the 1960s and 1970s than the government had expected to give. In this period, Hanford became less a fixture of the national security state and more a project of the welfare state, although defense readiness was one underlying rationale behind the new investments. The site's altered status illuminated class tensions, as residents of the Tri-Cities debated how demobilization had produced winners and losers.

Between the early 1960s and the early 1980s, it seemed that Hanford might replace plutonium production with a range of nonmilitary applications of nuclear energy. The N reactor's initial generation of kilowatts for regional power consumption in 1966 marked the beginning of a new career for Hanford in producing electricity. Eager to put atoms to work for peace as well as war, an assortment of federal officials, scientists, utilities, and corporations had been pushing for nuclear power plants since the early Cold War period. With the emergence of the "energy crisis" of the early 1970s, this movement gained momentum. Numerous people argued that atomic energy would provide an inexpensive and clean way of solving the nation's apparent shortage of electricity. And the communities around Hanford stepped forward to offer themselves as hosts of electricity-generating reactors. The main impetus for this initiative came not from the AEC but from a consortium of regional utilities affiliated as the Washington Public Power Supply System (WPPSS), and from the Bonneville Power Administration, whose forecasts of energy shortages encouraged widespread acceptance of nuclear power plants as a solution to the need for more kilowatts. The Washington Public Power Supply System started building three new power reactors at Hanford during the 1970s, an enormous construction project that reinvigorated the Tri-City economy.

The Atomic Energy Commission warmed only gradually to this change in Hanford's mission. In 1965, when most plutonium-producing reactors continued to operate, it had discouraged the notion of a "commercial nuclear industrial park" because it did not want civilian activities to interfere unduly with the site's military-oriented focus. Within a few years and after more reactor closures, however, it changed its perspective. In 1968 AEC chairman Glenn T. Seaborg visited Richland on the occasion of its twenty-fifth birthday and spoke enthusiastically about a "nuclear-powered industrial complex," or "nuplex," that would feature a variety of atomic reactors and other facilities for generating kilowatts, conducting research, advancing nuclear medicine, and producing other civilian benefits.[114] Seaborg's vision resembled those of many others who imagined a bright future for the peaceful atom and believed that the Tri-Cities, because they possessed a suitable workforce and experience with atomic energy, would become a capital of the nuclear future. This prospect was most closely associated with plans for a research facility organized around a fast-flux test reactor. The plan, which grew out of design studies begun in the late 1950s, would give Hanford's staff a role in the ongoing development of nuclear technology rather than leave it yoked to the existing production reactors. In the guise of a prospective site for large-scale development of civilian nuclear projects, Hanford became part of a larger national and international discourse over the benefits and costs of the atom.

During the 1970s the communities around Hanford must have felt that the conversation surrounding nuclear energy favored them. The apparent shortage of electricity and the apparent virtues of power reactors worked in their favor. By the beginning of the next decade, however, both public sentiment and economic reality had turned against them. The accident at the Three Mile Island reactor in Pennsylvania in 1979 energized a growing distrust of nuclear power. Meanwhile, closer to home, the Washington Public Power Supply System found its projects called into question by skyrocketing costs, lengthy construction delays, and shrinking demand for new kilowatts. In 1982 the WPPSS defaulted on its bonds. It completed only one of its promised reactors, in 1984, and mothballed the two others being built at Hanford and another pair underway in western Washington. The peaceful atom proved unable to prop up the Northwest's nuclear economy.

The travails of nuclear power and the WPPSS struck a serious blow to the economy and psyche of the communities around Hanford, but its effect was cushioned by the resumption of the site's Cold War mission under

President Ronald Reagan. Between 1972 and 1982 the N reactor had run "on a cycle that produces plutonium for energy research rather than weapons." But as Reagan expanded America's nuclear arsenal, the N reactor reverted to producing "bomb-grade plutonium." In 1983 the Department of Energy also reactivated the Plutonium Uranium Extraction Plant, which had been shut down for a decade.[115] After all the efforts to diversify the Tri-City economy, by the mid-1980s it remained nearly as dependent on federal defense spending as it had been between 1943 and 1963. Thousands of people had left the region after the collapse of the WPPSS, yet employment by the DOE and its contractors remained steady at roughly 13,000.[116]

The renewed production of plutonium at Hanford did not last. By the late 1980s the Soviet Union and its empire in eastern Europe had begun to implode, causing many to question whether America needed such a stockpile of nuclear weapons. Yet even before the demise of the Cold War, Hanford's days as a bomb factory had ended. Late in 1986 the DOE shut down the N reactor, partially in response to that year's Chernobyl nuclear disaster in the Ukraine. The N reactor shared some design features with Chernobyl's doomed facilities. After lengthy study of the plant's flaws and the projected costs of upgrading the N reactor, the DOE decided in 1987 not to restart it. The PUREX plant was shut down in 1988, not to be restarted, and in 1989 the Plutonium Finishing Plant followed suit. Hanford's mission of producing weapons-grade plutonium had come to an end. The closure of the N reactor was predicted to entail the loss of 6,400 jobs directly (and perhaps another 7,800 indirectly).[117]

The loss of Hanford's Cold War mission brought the site and the surrounding communities to a critical turning point. Leaders in the Tri-Cities had developed a strong orientation to production reactors—those that made plutonium for nuclear weapons, that generated electricity for civilians, and that contributed to business and research. By the late 1980s there was no more demand for Hanford's weapons-grade plutonium or for new, electricity-generating reactors in the United States. And proposals for others kind of reactors never seemed to go anywhere. Hanford's Fast Flux Test Facility (FFTF) had been built for the DOE's breeder reactor program, which Congress had killed in 1983. For ten years, the DOE and Tri-City leaders had tried to find a mission for the FFTF, in particular the production of specialty isotopes for medical and industrial purposes, even inviting proposals from other countries and from private industry. But nothing satisfactory emerged, and in early 1993 the DOE placed the reactor on cold standby. In a letter explaining this decision, the secretary of the Department of Energy explained that the DOE would not need the FFTF to produce plutonium-238 for battery packs for NASA, because it had "now concluded negotiations to purchase plutonium-238 from the Russians at a cost far less than production in FFTF."[118] It was an extreme case of post–Cold War irony—Hanford's last federally supported reactor being put out of commission because its longtime rival, the Soviet Union's nuclear complex, was having its own going-out-of-business sale.

If Hanford could not have any new operating reactors, then its other possible mission was as waste manager. In 1982 the U.S. Congress, in the Nuclear Waste Policy Act, decided to select two sites, one east and one west of the Mississippi River, as repositories for 77,000 tons of the nation's high-level radioactive wastes. The DOE proceeded to consider ten sites around the country, but through 1984 it seemed to favor Hanford so much that it had spent $300 million "studying Hanford's basalt" formations as a potential underground storage facility "while virtually ignoring most of the other sites." The agency's interest in Hanford stemmed in part from the perception that people in Washington (or at least in the Tri-Cities) were more receptive to their state becoming a waste-storage site than were people in most other states. The DOE was wrong about Washington but right about the Tri-Cities. A 1986 poll conducted by the *Tacoma News Tribune* found that the state as a whole opposed putting a national waste repository at Hanford, while within the Tri-Cities supporters of the idea outnumbered its opponents.[119]

Washingtonians, particularly those west of the Cascade range, no longer saw Hanford as economically significant. Many probably knew little about Hanford at all. It is unlikely that the opposition of Washingtonians to the waste repository made much difference, in any case. In 1987 Congress decided that further study should focus elsewhere. The DOE terminated Hanford's Basalt Waste Isolation Project (costing reportedly 1,200 more jobs) and focused on Yucca Mountain, Nevada (over the strong objections of that state's governor) as its new, preferred repository site for high-level nuclear waste. Many in the Tri-Cities wondered whether there would ever be another federal mission at Hanford. They also wondered how the local economy would survive.

As the world cheered the end of the Cold War, those who managed and worked for nuclear facilities around the globe wondered what the future held. In the former Soviet Union and the United States, towns built and

1.11 During the early 1980s the U.S. Department of Energy examined the Hanford Site as a possible repository for nuclear wastes from other states. Eventually, the department focused its attention on other places in the West. Hanford had no need to import wastes, though, to make them integral to its activity. By the 1980s it had accumulated plenty of its own wastes over decades of production of weapons-grade plutonium. Distributed across the site in tanks, burials, and barrels (as in this 1984 photo of low-level radioactive waste storage), the chemical and radioactive by-products of plutonium production made Hanford, by some accounts, the most polluted spot in the Western Hemisphere. Photograph by Greg Gilbert / Seattle Times.

subsidized by their national defense establishments now looked for ways to transfer their skills and resources to civilian, peacetime, global markets. At the DOE's two weapons labs, Los Alamos and Lawrence-Livermore in northern California, where coherent research-and-development infrastructures were in place, the transformation would prove difficult enough. It would be even more challenging at a production facility such as Hanford—which heretofore had never been designated a national laboratory, failed to develop close ties to a research university, and carried the enormous burden of having to manage nuclear wastes.[120] The Tri-Cities thus remained unattractive to many businesses that might otherwise have considered investing in the area.

At the same time, Hanford's reputation was blackened by new revelations. Early in 1986, the year of the Chernobyl disaster and of the closure of the N reactor, the Department of Energy released 19,000 pages of documents detailing some of the story of the site's radioactive emissions. It was just the first in a series of releases of previously classified data describing the regional impact of Hanford operations. Formal studies began of the impact of the site's radioactive emissions on surrounding populations. By 1990 the DOE reached a historic turning point when it conceded that Hanford's emissions during the 1940s and 1950s had been high enough to cause cancer and other illnesses among residents of the Pacific Northwest. The data also indicated that the army and the AEC had known about the emissions—and sometimes even planned them, as in the case of the Green Run in 1949—yet had never informed the public. U.S. Department of Energy leaders, particularly concerned about the use of prisoners as research subjects in health physics research, called for a new spirit of openness and confession. In January of 2000, the DOE further conceded that workers at America's nuclear weapons facilities had been "exposed to radiation and chemicals that produced cancer and early death."[121] This new information provoked much criticism of Hanford, encouraged the efforts of watchdog organizations, and facilitated the filing of lawsuits against the government and its Hanford contractors.

Even longtime defenders of Hanford and the government changed their tune. The *Tri-City Herald*, for so long the leading mouthpiece of Hanford boosters, reflected the shift. When the DOE released the initial 19,000 pages of declassified documents in February 1986, *Herald* editors simply did not accept the idea that Hanford had caused "significant offsite environmental contamination or health hazards." Within weeks, however, having had time to study the documents, the newspaper reversed course, determining that "Hanford's early years of operation were dirtier than most of us imagined." In April the Soviet nuclear power reactor at Chernobyl released contaminants over a wide area after an explosive fire. By October the *Herald* admitted that "there's no question any more that serious harm [to offsite populations] may have resulted from Hanford activities of the 1940s and 1950s."[122] All of this soul-searching, when combined with a host of other events and trends, ultimately led the paper to conclude that Hanford had outlived its original purpose.

In late 1986 and early 1987, even as the *Herald* digested the meaning of newly declassified documents, it urged the DOE to provide some sort of "replacement reactor" so that Hanford could continue its mission of production, ideally on behalf of the nation's military establishment. But by 1990, after months of intense scrutiny and criticism, it finally grasped that such hopes were badly misplaced. "Huge community resources and tremendous amounts of dwindling political currency have been expended to preserve a defense mission for Hanford," the *Herald* editorialized. "It isn't working and likely won't."[123]

Hanford and the Tri-Cities were not alone in making this reconsideration. Department of Energy production-oriented sites at Fernald, Ohio; Pantex, Texas; and Rocky Flats, Colorado went through a similar experience during the late 1980s and early 1990s. It was a time when the entire nuclear weapons program of the DOE was under attack, both because of previous practices and because of the ongoing threat that nuclear and other wastes presented to workers and neighbors of the department's industrial facilities. The DOE responded by embracing the mission of cleanup, a bureaucratic conversion that was heralded as a "changing culture" within the department. Perhaps that characterization was accurate, but it also seemed clear that if the department did not address its environmental and public-health problems, its ability to perform its more fundamental mission of producing nuclear weapons would be severely compromised. The DOE became more ecologically minded in order to survive, and along the way perhaps its "culture" changed. At Hanford the new era dawned in May of 1989 in the form of the Hanford Federal Facility Agreement and Consent Order (or Tri-Party Agreement), signed by the DOE, the Environmental Protection Agency (EPA), and the Washington State Department of Ecology. The deal promised that the Hanford Site would be cleaned up within thirty years at a cost of $57 billion. (Both the time and money needed to complete the job would increase dramatically.) Moreover, the agreement made the public more of a partner than before in the management of Hanford by requiring that a wide range of stakeholders be consulted regularly.[124]

Implementing the Tri-Party Agreement was problematic from the start. First of all, the local population had long resisted the idea of environmental cleanup as Hanford's main mission. It would take some time before the Tri-Cities offered widespread support for the new cause, but the mammoth sums now being spent at the site certainly boosted the local economy and encouraged among local residents a greater appreciation for environmentalism. More importantly, success at managing the wastes was elusive, and controversy ensued as cleanup deadlines repeatedly had to be postponed. At least one investigation of the effectiveness of the effort, five years after it had begun, found significant waste of money but little technical and environmental progress.[125] The delays, frustrations, and cost overruns worried those committed to cleanup, who feared that "taxpayers will become impatient with the pace and cost of the cleanup and give up." The remediation effort also produced local economic anxiety by bringing in new contractors and placing a premium on certain skills that were not in abundant supply in the local labor force.[126]

Yet, despite its problems, the agreement at least promised a long-lasting, productive activity with which the Tri-Cities could recapture economic vitality. It also required that highly complicated scientific and engineering problems be solved, thereby offering Hanford a chance to remain on a cutting edge. Finally, the agreement provided a framework in which the differences between the federal government, the Tri-Cities, nearby Indian tribes, and the other peoples of the Northwest could continue to be resolved in relatively open fashion. Roughly fifty years after the Hanford Engineer Works had come into being in an aura of near-total military secrecy, the past, present, and future of the place now known as the Hanford Site had become matters of sustained public discussion. At the same time, one basic thing remained largely unchanged: at the end of the twentieth century, Hanford was about as much of a federal enclave as it had been during the 1940s and 1950s. Its politics were considerably different, but the key decisions about the place and about the funding keeping it and the Tri-Cities economy alive continued to originate in Washington, D.C.

Two

THE ATOMIC CITY
OF THE WEST

Richland and the Tri-Cities

THE EXPERIENCE OF RICHLAND, WASHINGTON, HAS BEEN QUITE unusual, as towns go, but its story is not so odd as to have been unimaginable. Perivale St. Andrews, a setting imagined in George Bernard Shaw's 1909 play *Major Barbara*, anticipated some of Richland's exceptional ways. Perivale St. Andrews is the company town of Undershaft and Lazarus, a firm that manufactures arms and explosives. In contrast to the chaos and destruction implied by the weaponry it produces, the town is "beautifully situated and beautiful in itself." It is "perfect! wonderful! real!" and compels most observers to revise their unfavorable opinion of Andrew Undershaft, company owner and arms merchant to the world.

Among the most remarkable features of Perivale St. Andrews is the manner in which it looks out for its residents, the employees of Undershaft and Lazarus and their families. There are a nursing home, libraries, good schools, a "William Morris Labor Church," and "a ball room and banqueting chamber in the Town Hall." Employees participate in "the insurance fund, the pension fund, the building fund, the various applications of co-operation." The town takes care of its residents and permits no poverty. Undershaft's son Stephen worries about one problem, though: "This provision for every want of your workmen may sap their independence and weaken their sense of responsibility. . . . Are you sure so much pampering is really good for the men's character?" Andrew Undershaft replies that no such concern is justified. By meeting the residents' every need, the town prevents most social problems. Moreover, Perivale St. Andrews does not allow for too much complacency: "Our characters are safe here. A sufficient dose of anxiety is always provided by the fact that we may be blown to smithereens at any moment."[1]

Like Shaw's invention, Richland was a company town whose residents made explosives. Its population consisted primarily of the families of employees who worked at Hanford producing plutonium for atomic bombs. Designed rather carefully for its mission, Richland also possessed amenities contrived to attract and retain workers. Unlike Perivale St. Andrews, Richland was managed not only by a private company but also, indirectly, by a public agency. The U.S. government and its assorted corporate contractors held powerful sway over the lives of residents. Their influence was especially strong before 1960, yet the decisions they made—from the composition of the population to the arrangement of buildings—profoundly shaped the town and its neighbors well into the 1990s.

Richland did not live up to all of the expectations of its managers, but it nonetheless became a model community. This success resulted from initial scripting. The army and its contractors, in deciding who would and who would not live in Richland, had created a town of predominantly white, middle-class, well-educated families. To many Americans in the period after 1940, such a select population seemed exemplary. So did the community's planned, suburb-style layout and surfeit of cultural amenities. Richland became a kind of atomic-age utopia. Not long after the Hanford Works' existence became public knowledge, a guidebook for American automobile tourists recorded the federal enclave's boasts: full employment, no crime, empty jails, outstanding schools, and no one dependent on state relief payments.[2]

Like Perivale St. Andrews, that "perfect triumph of modern industry," Richland thrived by serving the dictates of economic production. Hanford managers viewed the town as a tool for increasing the output of plutonium, so they did not mind inefficiencies so long as the manufacturing went smoothly. During the late 1940s and early 1950s, however, some began to ask whether Richland was un-American. As a government-owned town at a time when the United States was increasingly hostile toward communism and most other "applications of co-operation," Richland became increasingly untenable. Slowly but surely, the Atomic Energy Commission (AEC) divested itself of the community, turning ownership and government over to its residents—perhaps even safeguarding their independence and sense of responsibility. This change only added to Richland's exemplary reputation: in April 1961 *Look* magazine named it an All-America City.

The residents of Richland didn't worry too much about the town undermining their character. Like the inhabitants of Perivale St. Andrews, the population drew strength from the perceived risks that came with the place where they lived. Being so closely associated with nuclear weapons presented a certain danger, and the residents of Richland sometimes claimed credit for exposing themselves to it. But they were likelier to say that their character had been shaped in a more classically American fashion—through the rigors of the frontier. Townspeople hardly expected to get "blown to smithereens at any moment"; after all, most possessed an incomplete knowledge of the health risks associated with Hanford and, in general, tended to downplay those risks anyway. Instead, they embraced an identity as frontier Americans, portraying themselves as modern pioneers who had moved west to fulfill a national mission and who made sacrifices for their country in doing so. The idea of the frontier had several connotations, each useful as the people of Richland slowly but surely took possession of their city.

Imagining themselves as pioneers served as one way for the people of Richland to create a sense of autonomy from the control that the U.S. government and its contractors exerted over the town. Seeing the town as a frontier was part of interpreting their own presence there as a matter of migration, a consequence of free choice. In fact, the town was very much the creation of the U.S. military and the demands of mobilization during World War II and the Cold War. Regardless of how residents imagined themselves, Richland changed primarily to suit the changing needs of the American security state. Even so, it would be wrong to portray federal managers' control of Richland as monolithic. The public- and private-sector entities in charge of the town seldom saw eye to eye, and they also found themselves coping with conditions that limited their power. Richland was a relatively new town undergoing fairly constant change; unlike other parts of America's secret nuclear weapons program, it remained open to inspection by an inquisitive public, and townspeople gradually asserted their own interests and autonomy.

As a result, development in Richland did not always proceed according to a single script. In contrast to Perivale St. Andrews, which is frozen in Shaw's third act, the Atomic City of the West evolved unpredictably over the course of time. Similarly, Richland was not isolated but, rather, evolved through interactions with its neighbors. Along with Pasco and Kennewick, it came to be known as one of the Tri-Cities clustered at the confluence of the Columbia, Yakima, and Snake rivers. Because of the communities' mutual dependence on Hanford, Richland shared financial interests as well as an atomic identity with the other two towns. Numerous socio-economic differences between the communities, however, placed Richland's distinctiveness in sharp relief. As a built environment, the town of Richland long remained what the army and the AEC had made it during World War II and the Cold War. But the federal government eventually let go of the town and ultimately abandoned Hanford as a plutonium-producing facility, changing its purpose and its identity.

Company towns have been common in American history, particularly in the timber and mining economies of the Northwest. Workers in these communities also lived in company-owned housing, purchased food and medical care from their bosses with their labor, and undertook dangerous work within the context of communities structured around the hazards of the job. Even in the region's largest cities, a single employer could dominate and define community life. The tension between free-market ideals and company socialism was present in Richland, just as it was in Henry Ford's worker communities. Residents who depended on Hanford and identified with its mission, workers proud of their skills and accomplishments, might be willing to accept a certain amount of pollution as part of the cost of doing business. They might even dismiss hazards as nonexistent or claim that they were not a threat to those who knew what they were doing. What distinguished Richland from other company towns was the requirement that workers had to keep much of what they did secret and that contractors ran the town with federal backing.

Also, the Tri-Cities comprised not a single town but a group of distinctive and at times conflicting communities. Overall, though, the Tri-Cities, for all their differences, together reflected the fact that the Hanford Works extended its identity as a federal reservation beyond the site proper and on into the surrounding communities. Federal lands reserved for nuclear operations, like those defined by treaty or by policy for other uses—for Native nations, as wilderness, for mining, or for impounding water behind dams—at least partly impose an identity on those who live on or near them. To at least some extent, the purpose for which the land was used and the identity of the local inhabitants came to be mutually defined. Outside criticism, directed at the symbols produced by the place, only strengthened the symbols' value as markers distinguishing community members from others. Such communities became sites for the negotiation of support between patrons and clients on the basis of mutual need and recognition, even if residents and federal agents sometimes seemed to speak different languages.[3]

Hanford's Bedroom Community:
The Origins of Richland, 1943–48

In 1940 Pasco and Kennewick, with populations of 3,913 and 1,918, respectively, were the largest towns in the region. Three years later, when the Manhattan Project arrived, the army found a predominantly rural society. The precarious economy revolved around ranching and farming, with much land remaining undeveloped and only small parcels given over to towns. The army turned the three communities into an industrial center, and laid a foundation for urban growth. As of July 1, 2009, Kennewick, Pasco, and Richland contained almost 174,000 people, within a metropolitan statistical area of more than 245,000.

In their search for a production site, representatives from the army and DuPont were looking for a place remote from big towns. Having found such a place, they proceeded to build a city there. In the process, the Army Corps of Engineers was responsible for eliminating some towns and creating others around Hanford, yet it would be a mistake to say that the corps had really thought much about communities.[4] Rather, it took an instrumentalist view of towns such as Richland, Oak Ridge, and Los Alamos— they existed to facilitate production. Towns were a cog within the larger machinery of bomb making. The army would do what was necessary to reach its demanding production goal: enforce security rules, regulate the population, maintain morale. Yet the army's control over towns was always limited, due in large part to its uncertainty about what the towns required and to the opportunities such uncertainty created for others to assert their own vision of what the towns should be. Much of a town's operation depended on the contractor, placing the army one step removed, and this distance increased with time. (In other aspects of a town's life, such as security and law enforcement, the army took more direct control.) As a result, the army's authority in Manhattan Project towns was substantial, but it never matched its aspirations or its rhetoric.[5]

In the beginning, when Colonel Franklin T. Matthias arrived at Hanford early in 1943, among his responsibilities was to provide for four separate urban realms for staging, construction, general support, and operations. First, he needed an entrepôt through which resources and employees would pass on their way to the project. Second, he needed a construction camp, a temporary community, for housing and feeding the workers who built the plant. Third, he needed a network of off-reservation communities to accommodate the diverse, fluctuating populations associated with wartime Hanford, especially after the closure of the construction camp. Fourth, he needed a more permanent town on the reservation where operations personnel and their families would live and from which managers would administer the project.

Matthias's initial requirement was a staging ground—a place through which personnel, supplies, and information could be funneled from the outside world to the reservation and a place where people could live and work until their own communities had been established on Hanford ground. Because of its rail connections and size, Pasco performed this function. Pasco was also where thousands of newcomers disembarked from the trains that brought them to work on the project. In addition, it served as Matthias's headquarters until completion of offices in the recently condemned village of Richland.

The moment of arrival could be traumatic. Francis McHale, who came by rail from Pennsylvania to set up police and fire departments for DuPont, recalled his first moments in Pasco: "The wind was blowing like hell, and if a train were going back east right away I would have been on it." (The howling winds to which McHale referred earned the nickname "termination winds" for encouraging workers to quit the project.) Many new recruits arrived in Pasco by train during the early morning hours; concerned about the perpetual labor shortage at Hanford, General Leslie Groves insisted that they be housed and fed right away so that they would not simply turn around and leave before they could go to work. The army had to toil around the clock to counter Pasco's shortcomings. Nell Lewis MacGregor, a fifty-seven-year-old widow from Seattle, initially declined offers of employment at Hanford because she regarded the climate around Pasco as unbearable. Only after months of intense recruiting and the offer

2.1 Numerous conditions at the Hanford Engineer Works made it difficult to retain labor. This photo shows a dust storm at Hanford Camp, the construction town where workers building the plant were housed in barracks and trailers. Dust storms have been part of the weather patterns in the vicinity at least since the first fur traders arrived to record them during the 1820s and 1830s. The extensive earth-moving and building associated with creating the Hanford Engineer Works during World War II surely made the storms worse. Photograph courtesy of the U.S. Department of Energy.

of an "incredible" salary did MacGregor accept a job.[6] New employees were advised to bring a Thermos bottle and a padlock, along with the essential tools for their craft.

Second, Matthias needed a construction camp to house workers while they built the plant. Although some laborers lived off site, the army crowded most of the workers into living quarters at the Hanford Camp. DuPont selected the now-evacuated Hanford townsite as the location for the construction camp because it lay close to the plant area, had some buildings and utilities worth reusing, and was served by a branch line of the Milwaukee Road railway. Beginning with this infrastructure, DuPont erected a short-lived town complete with housing, mess halls, recreation facilities, and social services. Residents lived in tents between May and October 1943 while more durable housing was built. At its highest occupancy in mid-1944, the camp had 131 barracks for 24,892 men and another 64 barracks for 4,357 women, 880 smaller hutments to house 9,834 men, and 3,639 trailer lots. Hanford Camp housed a maximum of about 48,000 people; just over 45,000 were employed on the project at its peak, and the rest were workers' family members, who were permitted to live only in trailers.[7] As the reactors and separation plants reached completion and began operation late in 1944 and early in 1945, construction workers were laid off and the camp's population dwindled. When workers vacated the last barracks on February 23, 1945, one departing employee observed the arrival of some new tenants: "The Army Engineers turned into [the camp] an abandoned flock of goats they ferried from the other side of the river. Whether this was an act of mercy or for experimental purposes, we didn't know."[8]

The abandonment of Hanford Camp marked a transition in the larger project. In the same way that Pasco's role as staging ground lasted only until equivalent facilities could be erected elsewhere, so the camp endured only until construction on the main parts of the plutonium plant had been completed and the focus turned to operations. What emerged to succeed these temporary realms were two permanent urban arrangements. One was the assortment of towns surrounding the federal reservation, including Pasco and Kennewick, and towns along the lower Yakima River, such as Prosser and Sunnyside. This third urban realm absorbed those employees and related activities that did not go to the town of Richland, the fourth settlement. Located about twenty miles by road from the chemical separation plants and twenty-five miles from the Hanford reactors, Richland was where project administrators worked and operating personnel lived. From the start, Richland differed by design from such outlying towns as Pasco and Kennewick. It was part of the reservation taken over by the army and managed by DuPont during World War II, while the other towns lay largely outside the control of the Manhattan Project. Moreover, the army and DuPont built Richland to serve certain groups of people, while relegating other groups to towns off the reservation. In time Richland would become one of the Tri-Cities, along with Pasco and Kennewick, but during the early 1940s it set out along a different path. (For estimated population growth of nearby towns, see Appendix 2.)

One way to illustrate the divergence between Richland and the other towns is to note how places like Pasco and Kennewick inherited the functions and character of the construction camp while Richland stood against virtually everything the camp represented. From the start of planning Richland was identified as "the Village" or "Richland village," the name suggesting permanence and cohesion.[9] The name Hanford Camp, by contrast, stated explicitly that it had been thrown together hastily and was not meant to last. Hanford Camp housed primarily laboring people without families, while Richland housed middle-class operators, engineers, and administrators, along with their families. Hanford Camp seemed violent and exciting, as Ainsworth had been in the early railroad days; like Ainsworth's successor community, Pasco, Richland now seemed like a good place to raise children. In light of the divergent purposes of the two places, only a few who lived in the construction camp went on to live in Richland. One who did, Nell MacGregor, noted the difference: "Those of us going into operations knew that Richland Village was a nice little place, unique of its kind, but it was going to seem pretty colorless after the roaring construction camp."[10]

The vast majority of the residents of Hanford Camp in 1944–45 either left the area altogether or went to live in off-reservation towns. Moving to Richland was not an option. From the start of construction at Hanford, Colonel Matthias had described the village as an exclusive place. Some supervisory "construction people" would be allowed to reside there until their jobs ended, Matthias wrote in 1943, but then they would have to leave to make room for operations personnel. He also instructed DuPont to set rental rates for dormitories high enough to discourage laborers from moving there from the Hanford Camp. The colonel figured that the working men and women who built the plant simply would not want the kinds of homes that operators required: "It appears that people who will not be required to live in the village, will be the people whose housing standards are none too high."[11] Matthias clarified the matter of class distinctions at the end of the war as he reconstructed the history of housing around Hanford for the Manhattan Project. He explained that to retain "key supervisors and essential office workers," the army and DuPont had had to build relatively good homes at Richland. The "houses, hotels, trailers, plain board shacks and tents" of Kennewick and Pasco, by contrast, "were, for the most part, occupied by mechanics and common laborers and their families."[12] By the end of the war, outlying towns had inherited the functions and population of Hanford Camp, while the army had created in Richland a kind of model community meant to satisfy the valued employees who resided there. Richland's segregation continued after the war. During the building

2.2 The town of Pasco was profoundly changed by the advent of Hanford during World War II. Until Hanford Camp and Richland were erected, it was the central town of the Hanford project. It had the area's main train depot and Hanford's initial offices. Over time, the administrative functions and most workers moved to either Hanford Camp or Richland. Yet Pasco's population more than doubled during the war. Many of the newcomers were African Americans and Mexican Americans, who were not allowed to reside in Kennewick or Richland. This photo shows black soldiers participating as drivers in a Pasco parade, likely at the end of World War II. Photograph courtesy of the Franklin County Historical Society and Museum.

boom from 1947 to 1955, up to 13,000 construction workers were housed in North Richland, another temporary community of trailers, prefabricated housing, and barracks.[13]

Taking whatever spilled over from the Hanford reservation during World War II, nearby towns grew like Topsy—and without the benefit of the planning that characterized Richland. Pasco's population ballooned from 3,913 in 1940 to about 8,500 in March 1945, while Kennewick's expanded from 1,918 to 7,500.[14] Much of Kennewick's wartime population lived in barracks and trailers, living arrangements associated with an unstable population of temporary workers. Meanwhile, Pasco's status had changed: it became the chief off-site playground for Hanford workers, with more than its share of bars and brothels. One wartime resident described the town this way: "Death. Crime. Squalor. Pasco Main Street: sewer of Hanford, greatest war project in the U.S."[15]

Pasco was also where minorities were assigned to live. When Matthias decided that Chicano laborers from Texas could reside in neither the black nor the white barracks at the segregated Hanford Camp (figuring that the resulting racial unrest would undermine productivity), they were shipped off to live on the outskirts of Pasco. African Americans who stayed past the construction phase found that Pasco's crowded black neighborhoods were virtually the only place where they could get housing. The army had no inclination to act as urban reformer during the war. It professed to be too busy fighting the enemy abroad to take time to deal with racism at home. But the army did recognize some of the burdens it had imposed on the towns surrounding Hanford. Matthias wrote of the "unbearable load [placed] on the facilities, both social and law-enforcing, of the Pasco area," and the federal government awarded some money to Franklin and Benton counties to help communities cope with the shortages of housing and school rooms.[16]

In light of Hanford's effects on Pasco and Kennewick, the army reconsidered its plans for Richland. When the Manhattan Project first came to south-central Washington, it did not intend to build much in the way of commercial, civic, or medical facilities at Richland, assuming that the surrounding area would provide them. It immediately became clear, however, that nearby towns had been overtaxed by the influx of people. By June of 1943 the army and DuPont had modified plans for Richland to allow for more shops and services.[17] Many more adjustments to designs for the village would follow.

The haste with which Richland was built worked against successful planning and control. The army started construction in early 1943. It condemned the townsite and surrounding farms and irrigation works, and saved more than 150 houses and some stores and utilities to include in the new town. Most vestiges of the farming community were bulldozed. One observer recalled passing through in October 1943: "There was a service station, a first-aid place, some contractors' shacks, and acres of land swirling with dust as trucks moved over it. Houses had been razed here and sagebrush scraped off, and building materials were being stacked ready for use."[18] Within a year, 15,000 people lived in the village.

Richland was thrown together in a hurry.[19] Consequently, its form evolved in much the same way as did the wartime reactors: builders debated and altered its design even in the midst of construction. The arguments and changes stemmed in large part from three considerations. First, the intended capacity of the village grew, from about 6,500 people in March of 1943 to over 15,000 by mid-1944, which meant that the army had to continually add new facilities and find more construction materials.[20] Second, the army's initial thinking about the village was ambivalent. On the one hand, fearful of having to justify expenses to Congress, the Manhattan Project aimed "to serve minimum needs and no more."[21] As at the construction camps of the region's New Deal projects, laborers lived in barracks and had a dining hall and basic recreation facilities. Rent in barracks or huts was $1.40 per week and included cleaning services; a twenty-one-meal punch card cost $12.98.[22] On the other hand, the "minimum needs" of operating employees and their families apparently amounted to considerably more than, say, those of construction workers living at Hanford Camp or in Pasco. Third, while the army sought substantial control over Richland, it was in no position to build and run the town by itself. It had to rely on contractors and subcontractors who, as it turned out, possessed a vision for Richland that differed from the army's. Because the contractors took responsibility for constructing and peopling Richland, they had a decided influence on the shape of the new town.

Orders to build as cheaply as possible came from the top. General Groves wanted inexpensive housing at all three Manhattan Project towns, and he continually reminded Matthias to keep homes in Richland "to the bare essentials." Groves's thinking was consistent with his views on the larger project, as one journalist noted in 1946: "If the atomic bomb was a dud, [the army] might have to account for every dollar to a cold-eyed congressional committee. So nothing went into Richland that wasn't 'necessary.'"[23] Yet the "bare essentials" required for operating employees resident in the town differed significantly from the minimums required for construction workers at Hanford Camp and for "general operating labor" living off site. The village was a relatively exclusive place. Roger Williams, a key DuPont manager responsible for the company's Manhattan Project work, sized up Richland's prospective population in April 1943 as "a distinctly higher type than that encountered in the usual war emergency project."[24]

Speaking for the army's contractor at Hanford, Williams gave voice to a more comfortable vision of Richland than the army possessed. As the organization responsible for building and running the village, DuPont had a great deal of say in decisions about what kind of housing constituted the minimum required for operators and their families. It took an even keener interest in the subject because those operators would be DuPont employees. The company was essentially charged with building a community for itself, using the army's money, and it saw no reason to skimp on the town. Its preferences were reinforced by the architect whom it subcontracted to design the village and its houses. The Spokane firm of G. Albin Pehrson brought its own ideas to Richland, which also called for more than the army initially intended.

Throughout 1943 DuPont and Pehrson negotiated with the army to upgrade the quality of housing and design in Richland. The army at first proposed erecting barracks instead of a town of houses, and several dormitories were built to help cope with a housing shortage, but the army eventually agreed with DuPont that it needed "a complete Village at Richland" containing "minimum facilities for comfortable living."[25] This meant a town of houses rather than apartments. Over the months, DuPont defined its minimums differently from the army's. In April the company proposed a housing scheme that included some three-bedroom homes, prompting a command from Groves to scale back planning "to the bare essentials." Groves complained again in June that Pehrson's designs remained too extravagant, and he instructed Matthias to reduce costs once more and eliminate such frills as a funeral parlor.[26] (At Los Alamos, Groves became notorious for trying to cut hot plates from the budget for housing.) As late as February of 1944, Groves continued to urge Matthias to keep housing to a minimum, insisting that "we make every effort to reduce family housing in Richland by whatever means practicable."[27]

When completed, wartime Richland represented a compromise between competing visions. As a result of Groves's efforts at economy and efficiency, the town wound up with inadequate numbers of sidewalks, garages, stores, and shopping areas, no civic center, roads too narrow for much auto traffic, few and small home fixtures, and a good measure of temporary housing. All of these savings would present problems when Richland was remodeled in the postwar period.[28] But DuPont's and Pehrson's influence were also unmistakable. Colonel Kenneth D. Nichols, one of Groves's top officers in the Manhattan Project, noted that the general got his way at Los Alamos, which had spartan housing, while Richland got nicer homes than Groves had wanted.[29] G. Albin Pehrson made sure that the homes had relatively big yards. Groves had advocated situating the houses close together, but Pehrson felt that a "too-cramped" plan would jeopardize the morale of

2.3 Richland Village was hastily planned and built during 1943–44. Meant to house the employees who would operate the Hanford Engineer Works, it excluded construction laborers, janitors, and many other blue-collar workers. General Groves of the Army Corps of Engineers argued with DuPont managers over just how comfortable the town should be. The resulting housing had more amenities than Groves had originally wanted, but was still rather plain and efficient. Garages, basements, and better heating facilities would not be added until the Cold War expansion began in 1947. Photograph courtesy of the U.S. Department of Energy.

employees. Hanford workers were being "transplanted to what will seem a strange country," the architect explained, and their housing needed to take the surroundings into account. "In the desert, where space is the key characteristic of the view, a cramped village of cramped houses would be out of character, a palpable and conscious discord."[30]

While the architect secured this relatively spacious layout, DuPont ensured that Richland housing would be placed and occupied according more to the company's criteria than the army's. Groves had objected to an early proposal to segregate the bigger and smaller housing units, and to some extent the different types of homes were mixed together. However, the largest and most desirable houses tended to cluster in the eastern part of town, in pleasant neighborhoods near the Columbia River, while the less appealing duplexes tended to be located in the western part of town.[31]

DuPont encouraged another kind of segregation by assigning employees to housing according to rank within the company rather than such factors as need. As a result, higher-level employees rather than those with more children generally got first chance at the larger homes. The company's "eligibility list," which ranked every employee's priority for housing, "was prepared from the organization chart by beginning at the top and continuing down the line through various job classifications until all eligible employees were listed." Thus the least desirable homes (type A and B duplexes and the "prefabs") went to people in the "Operator to Foremen Classifications"; small, single-family houses (types F and H) went to the "Shift Supervisor to Assistant Shift Supervisor Classifications"; and the better single-family houses (types D, E, and G) went to "Chief Supervisors" and their superiors. Once surplus housing became available, early in 1945, some people who had not before been eligible—including schoolteachers and truck drivers—were allowed to occupy houses and dormitory rooms in Richland. However, "laborers, janitors, and other manual workers" remained excluded from village housing.[32]

In this fashion DuPont reinforced the army's differentiation between Richland and the neighboring towns while superimposing its own corporate hierarchy on the village. The communities surrounding the Hanford Engineer Works (HEW) functioned as vital components in the system of production. When the Manhattan Project exerted influence over these communities, their capital and operating costs had to be calculated as carefully as those of any of the other inputs to the manufacturing process, such as labor, electricity, and raw materials. To be used effectively, these investments were subject to close management. Where managers judged that improvisational structures would suffice, as in Pasco, these were employed; where effective operation required greater investment in more permanent building programs, as in the case of Richland, they were undertaken—to the extent deemed necessary.

Perhaps DuPont's greatest success was keeping Richland from being fenced in. Groves and Matthias initially intended to build and patrol a fence around the town, just as the army did at Oak Ridge and Los Alamos. After all, one impetus for having a government town in the first place was to house "those for whom security requirements or emergency need dictated that they be constantly at hand and under control." Another aim was to prevent access to Richland from outside the project. But DuPont objected to such controls over town residents and apparently changed the army's mind.[33] The army still policed Richland and kept close watch over its population—wartime censors examined each departing letter; security agents listened in on phone calls; hotel porters acted as "counter-espionage agents"—but the village never became a city behind a fence. In fact, because it remained open for popular inspection, late in 1944 Matthias asked DuPont to treat the town as a window on to the rest of Hanford. The village, he wrote, was the only "point at which public attention will . . . be directed and the opinions reached by the general public as to the manner in which Richland will operate will extend in their minds to the rest of the project."[34] Richland was well on its way to becoming a model town. The fact that vices, minorities, and the lower classes had been relegated to other towns such as Pasco certainly enhanced the village's ability to play the role.

The layout of Richland mirrored two different sets of influences—the cost-efficiency of the army and the corporate paternalism of DuPont. When discussion turned from the design to the operation of the village, the contest between visions continued, but this time with roles reversed. Colonel Matthias was reluctant to let DuPont have unchecked authority over the completed town because he regarded the company as too hidebound. Under Matthias's guidance, a more proactive army took steps to help organize and socialize Richland residents. For its part, DuPont wished to run Richland "without any Army control or checking," and it complained that the military was too "liberal" in providing services to townspeople. W. O. Simon, the leading DuPont executive at Hanford, extolled the company's more "conservative" approach, which "would avoid any activity or endeavor not necessary to its immediate needs and would tend to follow rather than anticipate public opinion."[35] Consequently, the two sides clashed. DuPont, for example, came to Hanford expecting that the nature of the work there "preclude[d] the use of many women in operating areas" and assuming that mothers in Richland would stay home. But by 1944 the army faced an acute labor shortage, so it instructed DuPont to build the Richland nursery and extended day care school so that more women could accept jobs. When the army promoted the formation of Villagers, Inc., to organize community groups and sponsor social events, DuPont resisted the idea, perhaps seeing it as a threat to the company's authority in Richland. In the end Matthias professed to give DuPont a free hand in operating the town, but he retained final "veto power."[36]

Control over Richland was not monolithic; no single agency had the power to completely determine the community's shape. The army and

DuPont disagreed regularly over how the town should develop. Both entities claimed to act in the residents' interests and both clearly aimed to boost productivity and morale, but the two organizations felt different kinds of pressure in pursuing their goals. The army worried more about costs and about the political consequences of its decisions, while DuPont attempted to keep its employees content while preserving its corporate prerogatives. In the end, neither side could assert itself unilaterally. While the army and DuPont limited each other's ability to dictate the terms of life in wartime Richland, neither really felt compelled to consult with the residents of the town about important decisions. The army and DuPont controlled the village largely according to their own needs; apart from them, Richland "had no political powers."[37] Virtually no governmental machinery existed to express or implement the popular will. The town had no elected city council or school board, although residents could vote in state and federal elections.

This pattern continued after the war. With the dropping of atomic bombs on Japan, the need for secrecy around the Manhattan Project declined somewhat, and so did the need to control the town. For about a year Richland existed in a sort of limbo while the United States decided whether to continue making atomic weapons and, if so, under whose authority. The town population fell from a peak of 15,400 in March of 1945 to less than 13,000 for much of 1946.[38] Then in later 1946 and early 1947, with the hardening of Cold War tensions, the future of the village was secured and Richland once more began to grow. Its population rose to 14,000 in early 1947, on its way to almost 22,000 by 1950. General Electric (GE) replaced DuPont as the main contractor on September 1, 1946. In contrast to Oak Ridge, where one company operated the plant while another ran the town, GE would be in charge of both factory and community at Hanford. Then, on January 1, 1947, the army yielded to the new, civilian Atomic Energy Commission. The theme for the Richland Day celebrations in the summer of 1947 summarized the town's prospects: "We're here to stay."[39]

The AEC somewhat grudgingly kept Hanford in operation because of the beginnings of a sense of emergency that would soon define the Cold War, and because Hanford's reactors proved to be not just reliably productive but also unexpectedly resilient. At the same time that the American defense establishment determined that it needed to manufacture as much plutonium as it possibly could, the Atomic Energy Commission worried that the three reactors built at wartime Hanford might not survive much longer. In early April 1946, DuPont managers had informed the army of "the precarious position the plant is in at the moment, indicating a very indeterminate life, particularly for the piles."[40] The B, D, and F reactors all suffered from "graphite creep," a condition generated when the graphite blocks comprising the production piles began to expand in size as a result of their long-term exposure to heat and radiation. As the blocks expanded, they threatened to warp the aluminum process tubes that carried both fuel elements and cooling water through the pile. In response to this crisis, the AEC shut down the B reactor to hold it in reserve should the other piles fail. At that point, the B pile had expanded by roughly two inches on a side. Graphite creep continued to afflict the younger D and F reactors as long as they ran at wartime power levels.[41]

The specter of graphite creep threatened to undermine Hanford's ability to meet the nation's need for weapons-grade plutonium, but the problem's resolution reaffirmed Hanford's culture of production, extending and expanding the commitment to the graphite-moderated, water-cooled, wartime reactor design. For most of 1947, the AEC planned as if the B, D, and F piles would have to be shut down, and started working on replacement reactors. To minimize the time and cost of building these new piles, the commission's staff planned to build them where they could make use of the water-pumping facilities already in place. Needless to say, the replacement reactors (plus two additional reactors required to increase the site's output) would follow the wartime model. The proposed pile next to B was tentatively designated BR (for B's replacement), and the other two were designated DR and FR. The General Advisory Committee of the AEC, the commission's science advisors under the leadership of Robert Oppenheimer, recommended that construction of the replacement reactors proceed, and suggested methods of maximizing production. It also suggested that the research arms of the AEC and GE ought to investigate ways to address the long-term problems posed by graphite creep.[42]

The solution that saved the B, D, and F reactors, though, and that led directly to an increase in plutonium production, was developed on site. Engineers at Hanford discovered that more heat—not less—reduced the changes wrought in graphite by neutron bombardment. They then changed the mixture of gases circulating around each pile. Once most of the helium in the atmosphere surrounding each reactor was replaced with carbon dioxide, the graphite blocks retained more heat and as a result stopped expanding. When the engineers replaced helium with carbon dioxide in the D reactor, it went from being the first candidate for replacement to being a model of the long-lived plant. Hanford operators then used the F reactor as a laboratory for finding just the right combination of helium and carbon dioxide in the atmosphere surrounding the pile; they began with a mixture of 10 percent carbon dioxide and then increased the concentration while measuring the effect on the expansion of the pile's graphite blocks.[43] If it could be done safely and economically, running the reactors at higher power made them more productive while slowing the advance of graphite creep. Hanford's own cost studies showed that production peaked when the reactors operated at high power, "just on the edge of failure," with skilled operators at the reins.[44]

Solving the problem of keeping the wartime reactors and their successors in operation over the long term helped to establish Hanford's postwar presence. Not just a wartime operation, Hanford now seemed to have found a long-term footing; Richland did as well, and was rebuilt with a view to postwar settlement. The second burst of growth in Richland required renewed attention to matters of layout and design, and the resulting planning process revealed afresh just how little say residents had over local affairs. General Electric and the AEC hired engineers and architects to prepare a master plan for the village. In November 1948 the subcontractors issued a rather lavish report aimed at shepherding the town toward a larger, permanent, more attractive existence. The document stands out as exceptional in planning history because, in contrast to similar reports for other towns, it was adhered to rather closely.[45] The plan's success came in part precisely because the town had no politics as usual via which the ideas of engineers and architects could be opposed or diluted by local interest groups. Moreover, as a creature of GE and the AEC, the master plan in many ways simply echoed what Richland authorities had already decided.

By arrangement with GE and the AEC, planners conceived and implemented a master plan without significant input from residents. While the architects and engineers generally did not ask citizens about their preferences, they did consult with a "relatively small group" of AEC and GE officials. In delivering their final product, the subcontractors urged GE and the AEC to explain the plan to the community, thereby educating citizens so that they would appreciate its virtues and support its implementation. They also urged the AEC and GE to allow residents to serve on a proposed town-planning commission, albeit only in an advisory capacity. But the planners went no further than suggesting that the townspeople's approval be solicited. No provision was made for getting residents' input before the plan was drafted; no provision was made for modifying the plan after citizens had a chance to view it. Furthermore, despite advice to solicit residents' support, the AEC and GE were not eager to publicize the master plan. When journalists requested copies of the document in November 1949, they were told that it was not "advisable" to circulate the master plan beyond management circles.[46] Even more than in most towns of the time, planning was not a participatory process in Richland.

Creating a master plan using outside experts was to some extent a charade, because the government and its contractor had made key planning decisions long before the architects and engineers presented their report in late 1948. The firms hired by GE and the AEC made a point of criticizing what their predecessors had done during wartime, calling it unsatisfactory for the peacetime era. (Indeed, this reflected a broader pattern in the late 1940s in which the AEC and GE deflected criticism of their operations by blaming a perceived problem on what the army and DuPont had done in wartime, and explaining that they were committed to correcting their predecessors' mistakes.) The emphasis between 1943 and 1945, the master plan explained, had been on meeting a national emergency rather than addressing the long-term needs of a community.[47] It implied that postwar Richland would evolve in a more careful fashion, following the patient thinking of planners. But this view ignored the momentum of the ongoing boom at Hanford, which had been created by another national emergency.

The army and the AEC increased plutonium production and enlarged Richland beginning in late 1946, a full two years before the master plan appeared, and in the process they made a series of key planning decisions without consulting either residents or outside experts. At the end of 1946 they launched a planting program around the village to reduce dust storms. In the spring of 1947 GE and the AEC contracted again with the architect G. Albin Pehrson and approved his designs for upgraded houses. Richland's new homes would have safer, cleaner, larger basements; oil-burning furnaces rather than coal-fired heating; porches for the first time; individual driveways in place of communal parking compounds; and the town's first curbs and sidewalks. The village was being planned without the benefit of the master plan. In fact, in May 1947 the AEC had specifically decided that it could not afford to await completion of an overall plan before continuing the expansion effort. The AEC determined that the "original layout" of the town from before 1945, along with the ongoing building effort, had established "fairly definitely [the] direction of proposed expansion."[48] So the master plan finally issued in late 1948 merely tended to confirm decisions already made in the haste of 1943–45 and 1946–47.

Citizen participation in public affairs would eventually come to Richland. Indeed, by late 1948 circumstances in the nation's capital (although not in the community itself) had begun to increase pressure for the "normalization" of the town. But the AEC and its main contractor at Hanford still reigned, with a minimum of input from the townspeople. In a 1948 report devoted to radiation safety and industrial health, an AEC review committee paused to criticize the "dictatorship by service contractors" in atomic towns: "Many, perhaps most, of the residents are in general content with the existing order but many would like some degree of control over the homes they live in, and some degree of self-government."[49]

Official thinking about the town remained almost entirely instrumental at the onset of the Cold War. Richland existed to serve the Hanford plant, and the town would change only if it made production more efficient. General Electric and the AEC elaborated on this attitude when forced to defend themselves against charges of overspending. During the late 1940s, members of Congress questioned GE officials about cost overruns on a new plutonium fabrication facility for the plant and a new school building for the town. Construction on both projects had started before designers had actually completed the plans, thus driving up costs (just as had happened during World War II). In both cases, GE defended the extra expenditures as a direct result of the urgent drive to increase production under emergency conditions. One spokesman explained with reference to Richland, "This school building, while no plutonium is produced in it, nevertheless is a part of the whole facility which looks toward the maintaining of satisfactory output of the plant." A couple of years later the AEC defended a money-losing transit system in virtually the same terms: Richland's buses were "as much a tool of the manufacture of the plutonium as the atomic piles themselves."[50]

The army and DuPont, the AEC and GE treated Richland as part of an industrial process and calibrated its activity to maximize the efficiency of production. This instrumentalist view of community did not make Richland's managers altogether insensitive to the human needs of residents. In fact, officials paid a good deal of attention to townspeople's morale as they tried to wring more plutonium out of Hanford's plant. But for the army and the AEC, the town had no purpose or identity apart from the manufacture of plutonium. As AEC chairman Gordon Dean explained in 1950, "AEC communities exist solely and completely for the support of the atomic energy plants and laboratories in or adjacent to the communities."[51]

The Atomic City of the West: Richland as New Frontier

Early Richland's purpose and provenance prevented residents from forming attachments to it in many of the customary ways. They could not vote or run for local electoral office or contribute to city planning. They could not buy or remodel their homes. Federal agencies and their corporate contractors told residents where they could and could not live, sent security agents around to ask about their lives and those of their neighbors, and evicted them if they left their jobs at Hanford. Moreover, there was little in the way of local heritage or history to grab on to because most prewar landmarks had been bulldozed to make way for the new town. Additionally, Richland's relationship to the rest of the Northwest seemed precarious. Neither a state nor a federal highway linked it to other towns, and it lacked "direct representation in the state legislature." Early town resident Margaret Collins, recalling life in early Richland in 1993, claimed, "We weren't sure if we were part of Washington state."[52]

If residents could not relate to Richland and the rest of the Pacific Northwest in a conventional fashion, they nonetheless found ways to take mental possession of the town—to develop a sense of place and a sense of community, to create a shared past, and to imagine their hometown as something other than a cog in a federal machine. They drew upon a set of common experiences, and upon pervasive themes in American culture that helped explain their place to themselves and to others. The two things most prominently associated with Richland were the frontier and nuclear weapons, symbols that became closely linked. Richland became known as the Atomic City of the West.

The meanings of the atom and the West exerted a powerful and lasting influence on the minds of people in Richland and, later, the Tri-Cities, yoking the communities around Hanford to prominent national narratives and thus to historic American missions. They resonated with the actual, shared experiences of moving to and residing in Richland and working at Hanford. And they proved remarkably malleable, capable of being deployed in a variety of situations to mean different things. The inhabitants of Richland may not have been able to identify with their town in many of the customary ways, but they nonetheless came to form close attachments to the community. By the same token, they did not initially feel much a part of Washington State or the Pacific Northwest, but they had no difficulty seeing themselves as Westerners, pioneers on geographic and technological frontiers.

To develop a positive western identity, Richland had to overcome some of the negative connotations of "frontier." When architect G. Albin Pehrson first saw the rural settlement in 1943, he described an impoverished place where, because of the newness of settlement, "shelter had usually to wait for the productivity of the land": "Having no capital or barely enough to acquire property and the minimum tools for its development, most of the ranchers and their families were forced to live as best they could. In this respect they were not unlike the farmers of the Middle West who spent their money on barns or other productive improvements, but neglected their homes. Only the Richland residents were poorer; the country was young."[53] One of Pehrson's jobs was to eliminate these frontier conditions so that the army and DuPont could recruit and retain employees for the plutonium plant. Still, living conditions around Richland during the war remained rather primitive—although in ways different from those Pehrson had originally found—and therefore prompted people to think of the entire project as a frontier project. Hanford Camp in particular earned this reputation, but those living in Richland also felt that they were roughing it in an isolated, semispartan place. The town's initial shortage of amenities, state of perpetual construction, and demographic transience heightened the sense that it belonged to an older West.[54]

Typical of frontier conditions were the dust storms that plagued the town. During the 1820s fur traders left behind the first written accounts of winds blowing topsoil around the confluence of the Columbia, Yakima, and Snake rivers. Over the decades that nonnatives colonized the gusty, arid country, grazing, farming, and drought made the storms worse. The situation became critical during World War II, when construction around Hanford scraped away the topsoil, loosening more sediment for winds to carry away. But the army needed employees and residents to stay. Having planted Richland right in the path of the storms, it now had to find ways to prevent the "termination winds" and dust from driving people elsewhere—perhaps to jobs in shipyards or aircraft plants on the west side of the Cascades. It initially tried to solve the problem by handing out hoses and free grass seed to residents, asking them to plant lawns. It also

2.4 Residents of Richland found numerous ways to identify with the American frontier. The cover of the souvenir program for Richland Day in 1947 (organized by the Richland Junior Chamber of Commerce) featured a logo identifying the town as "The Atom Bustin' Village of the West." The image compared Hanford's work with the atom to rodeos (bronco-bustin') and other activities associated with the legacy of the American West. Highlighting the West was also a way of comparing Richland to an eastern counterpart, Oak Ridge, Tennessee. Authors' collection.

encouraged tree plantings around town. In the later 1940s efforts became more serious as town planners and managers developed a park system, advocated irrigated agriculture on adjacent lands, and built a shelter belt of trees around the west side of Richland in an attempt to limit the impact of dust storms on townspeople. Such programs may have reduced the severity of the dust storms in the town but could not prevent them altogether. Dust remains a pollution hazard in the region today as well as a link to the cruder conditions of the old West.[55]

If the dust storms struck Richland residents as reminiscent of the nineteenth-century American frontier, the experience of moving to the town and living there in the 1940s made them feel like pioneers. Many new residents came from DuPont's ordnance plants in Colorado, Tennessee, Utah, and Illinois, where production had slowed or stopped. Of the 238 arriving from Illinois, for instance, many had worked at the Kankakee Ordnance Plant and then spent eight to twelve weeks training at Oak Ridge.[56] Because newcomers during and after the war were literally westering overland migrants, it seemed natural to compare their movement to those who had crossed the overland trail a hundred years earlier. It also seemed natural to liken their settling amid the sagebrush to homesteading. It was easy to slip into a frontier identity. "We considered ourselves kind of like pioneers," wartime arrival Louise Cease recalled.[57]

Westward migration fostered a sense of community in Richland not only by evoking common American symbols but also by providing a broadly shared experience. The town grew not through the random movement of a cross-section of the eastern population to south-central Washington but rather by way of a highly selective process of migration—directed in large part by federal and corporate managers—that produced a homogeneous population. Besides sharing a migration from east to west and the occupation of a brand new town, adult inhabitants of Richland overwhelmingly tended to be white, between thirty and forty years old, and to possess good educations and skills that earned them relatively high wages. They had other things in common as well. "We were isolated [in Richland]," Margaret Collins later recalled. "People couldn't live there unless they worked at the plant or were part of its support services. There was a common bond. We all came from someplace else in the United States." She also called Richland "classless"—by which she meant that everybody belonged to the middle class.[58]

The townspeople expressed considerable satisfaction at helping to shape a community to their mutual tastes. As one reporter described Richland in 1946: "The majority of its men are scientists and engineers, with common interests and a common point of view. The town is their baby and they love it."[59] The phrasing is telling: although the army and DuPont monopolized political power at Richland, residents nonetheless somehow felt empowered by their experiences there. They felt the town belonged to them, too, and saw it as a kind of chosen place. As the destination for a cohesive group of migrants, Richland possessed some of the character of what geographer Wilbur Zelinsky calls "voluntary regions." These are places whose population consists not so much of people required to be there "by circumstances of birth and social heredity" but rather of "self-selected groups of like-minded, mobile, atomistic individuals."[60] Of course, many residents of early Richland had been sent there by the U.S. government and its corporate contractors. But they nonetheless saw their presence there as a product of freely made decisions to move west. "We have voluntarily chosen this area as our home," one civic leader explained in 1958. The sentiment strengthened a feeling of ownership.[61]

Those who arrived during and immediately after the war came to see themselves as a founding generation. In September 1945 the *Richland Villager* credited residents with creating the town (neatly ignoring the fact that the army and DuPont were the true builders):

> Villagers not only watched Richland grow from a country crossroads, they designed it, nailed the boards, poured the concrete, slapped the last shingle on the final roof. What the town is, they made it. . . .
>
> In Richland, the wanderers for once had a town all their very own. Here was no community with established organizations. Here the villager dreamed 'em up, organized 'em and usually got elected to office. . . .
>
> In Richland, where sagebrush took the place of neon lighted fun spots, newcomers stirred up their own excitement.[62]

Early residents of Richland behaved like pioneers when they staked a claim to the town through their widespread participation in its civic and social life.

A simple geographic calculation cemented Richland's western identity. Once the atomic bomb became public knowledge in August 1945, Richland was often contrasted to Oak Ridge, Tennessee, another government town spawned by the Manhattan Project. Oak Ridge became known as the Atomic City of the East, and Richland as the Atomic City of the West. This frequently repeated slogan had numerous variations. One was "the Atom-Bustin' Village of the West," as Richland styled itself during a summer celebration in 1947. People in Richland participated in atom-bustin' just as rodeo cowboys competed in bronco-bustin'.[63]

Describing Richland as a characteristically western place tapped into a deep reservoir of myths and images in American culture. Those myths and images called to mind an old West of untamed nature, farms and ranches, and unbridled individualism, while Richland residents lived in a new West of mastered nature, cities as well as industry, and preponderant government and corporations. Yet the traditional meanings attached to the West proved so powerful that it required no great act of imagination for the townspeople to conceive of themselves as pioneers and cowboys. These notions were already such staples in the popular mind that incorporating them into understandings of the new community may well have been an unthinking act.[64]

That this refashioning of identity took place within the context of mobilization for World War II and the Cold War likely made the process even more forceful. To place Richland's western imagery in wartime perspective, it is useful to recall how Frederick Jackson Turner summarized the American definition of frontier as "the meeting point between savagery and civilization."[65] The idea of savagery implied Indians who needed to be defeated if American civilization was to prevail, as well as wilderness or nature that required taming. Pioneers in this perspective were people from "civilization" who, by going to the frontier, placed themselves as a shield between savagery and the rest of society, performing the dirty work necessary for "civilization" to survive, expand, and remain vital.

Turner's language proved compelling for twentieth-century Americans as they interpreted the world. By bifurcating the world into savagery and civilization, the damned and the saved, wrong and right, the frontier has resonated especially well when Americans have mobilized to face foreign forms of "savagery" — German Nazism, Japanese aggression, Soviet communism. It has also provided a fertile source of imagery for Americans to associate with technological advancement. When Charles Lindbergh crossed the Atlantic by air in 1927, the U.S. ambassador to France introduced him to the French people by saying, "[T]his young man from out of the West brings you better than anything else the spirit of America." Colonel Theodore Roosevelt, the Rough Riders' son, proclaimed Lindbergh the "lineal descendent" of "Daniel Boone, Davy Crockett, and men of that type ... who made America." Commentators connected Lindbergh's success to virtues derived from his upbringing on a Minnesota farm: he came from frontier stock and opened a new frontier in the air.[66] Hanford perpetuated and drew upon this conflation of the geographic and the intellectual frontiers.

When the people of Richland and Hanford identified themselves as pioneers, they not only confirmed that they, like the rest of the country, had mobilized for national defense but also suggested that they were playing a leading role in the nation's efforts. Before the public and the workforce learned the purpose of the Manhattan Project, there had been questions about whether the work at Hanford was really contributing to the wartime effort. Describing Hanford in terms of the frontier left no room for doubt about the trail-blazing role of Richland residents in a mobilized America. It also associated the population with the idea of *sacrifice*. Living near Hanford exposed townspeople to risks and hardships that the majority Americans would never face. Who could question the community's character?

The frontier imagery that grew around Hanford between 1943 and 1947

Atomic FRONTIER Days

A NEW LIGHT ON THE OLD FRONTIER

RICHLAND
SEPT. 4-5-6

WASHINGTON

1948

LDS ETERNAL PROGRESS

2.5 and 2.6 During the late 1940s and the 1950s the city of Richland held an annual celebration called Atomic Frontier Days. The occasion provided an opportunity not only to emphasize the town's connection to the historic West but also to speculate on how Richland was helping usher in an atomic future (fig. 2.5). While the local economy revolved primarily around production for military purposes, the visions cultivated during Atomic Frontier Days highlighted such peaceful uses of "the atom" as providing cleaner energy for smokestack industries. Such celebrations also encouraged the people of Richland to behave as they imagined pioneers had, so men grew beards and wore cowboy hats while women donned bonnets and long dresses with aprons. Each festival featured a parade consisting of marching bands, military personnel and vehicles, and floats, such as the one in fig. 2.6, which was sponsored by the Church of Jesus Christ of Latter Day Saints and recalled the Mormons' handcart migration to Utah during the nineteenth century. Like many other parts of the inland West, Richland and Hanford had sizeable populations of Mormons who married their own vision of eternal progress to the atomic frontier. Program image from authors' collection; float photograph courtesy of the U.S. Department of Energy.

served to connect a new community to the past by calling to mind previous missions and mobilizations in American history. These historical connotations persisted even while the town itself became more permanent and refined and as new dimensions were added to Richland's western self-image. After 1947 or so, the identity of the village was captured in the idea of the "Atomic Frontier," joining the past to the future. Between 1948 and 1959 the city of Richland celebrated Atomic Frontier Days. This yearly festival paid homage to the community's imagined similarities to "the mining and lumbering towns of the early West." For example, Richland sponsored an annual rodeo during Atomic Frontier Days, complete with colorful Indians, and asked residents to dress up like cowboys and cowgirls. Publicists spoke of how, during the war, "the desert fought a savage fight against the invasion of highways, water-power, industry and man." The "unsure" and "unbelieving" could not take the harsh conditions, but, like earlier pioneers, "the plucky, the faithful, those who had unbounded belief in the cause, stuck it out." Now they were engaged in conquering "the Atomic Wilderness on this, our last frontier."[67]

While the idea of an Atomic Frontier called to mind familiar images of the past, it also conjured up ideas about the present and the future. The pioneers working at Hanford and living in Richland not only identified with earlier generations of Westerners, but also saw themselves as innovators in the forward-looking present. They used frontier imagery to comment on the rapid pace of change in front of them, and commended themselves for their role in shaping that change and, with it, the future. Existing on the cutting edge of industry and technology meant that Richland was pioneering tomorrow. Local residents imagined the Tri-Cities as a "Crossroads of the Future": "History will record a new Western saga when the story of the Tri-Cities is told . . . of a region triply blessed with water, sun, and land, plus the indescribable, fantastic, development of atomic power; Conestogas replaced by snarling trucks, horses by hotrods, falsefronts by graceful steel and stone buildings, and the legendary tent-towns by thousands of trailers—most of whose occupants will join a useful citizenry to build this modern colossus in central Washington state."[68] The Atomic Frontier Days festival of 1950 typified the forward-looking sentiment of the time. Focusing on flying saucers and rocket ships, it tried to predict life in the year 2950 and described "Richland, the Atomic City, as the key to the future."[69]

Richland's pioneering role (not to mention its sense of self-importance) stemmed in part from its association with technology. As Americans discussed the potential benefits of nuclear energy, the community felt certain that, by harnessing the peaceful uses of the atom, it was helping to usher in a "golden age" and a "revolution into all phases of life, a physical turning point probably dwarfing anything in human history." In the world of technology, Hanford was said to stand on "the industrial frontier." As a "most unique manufacturing concern of our time," with special "engineering knowhow . . . , construction skill [and] unusual operational methods," the plant had reportedly surpassed America's already considerable achievements in "mass production" and the "assembly line method." Hanford fabricated its product in a nuclear reactor, a futuristic technology with a science-fiction pedigree going back to H. G. Wells. The chemical separation facilities broke new ground as a processing plant, relying on remote control and television monitoring, both major industrial innovations. Its product, plutonium, also seemed path-breaking—"usable a thousand years from now for war or peace. . . . It is owned by a free people; it bears a union label."[70] Vannevar Bush, the electrical engineer and government science advisor who successfully pushed for a crash program to build nuclear weapons in 1941, and who oversaw the efforts of academic scientists in the program, prepared a report urging federal support for science at the end of World War II. He called the work "Science—The Endless Frontier." Hanford's workers were living at the frontier—in the iconography of Atomic Frontier Days, they and their families had left the smoky industry of the East behind and moved West into a nuclear-powered new day.

Futuristic interpretations of the frontier encouraged not only a western but also a high-tech or scientific identity. It was in this context that Richland became so strongly connected to atomic symbols. Newspapers often called it Atomic Town, A-Town, or A-City, and a welter of names around the community enshrined this spirit. Town managers dubbed one street Proton Lane, and in 1952 one entrepreneur opened a restaurant called Fission Chips. Cheerleaders at Columbia High School wore sweaters with picture of atomic bombs stitched to the front and the word PEP lettered on each bomb.[71] Even though Hanford was never designated a national laboratory and remained at some distance from AEC centers of research, Richland imagined itself as a frontier of science and technology. (Battelle-operated Pacific Northwest National Laboratory, established as Pacific Northwest Laboratory in 1965 and designated a national laboratory by the Department of Energy [DOE] thirty years later, is a creation of the period after Hanford's operational life came to an end.)

During the early 1950s still another kind of emphasis was layered atop the themes of the old and the new West. In this period the federal government undertook the long process of turning Richland over to the people who lived and worked there. The townspeople needed the best deal they could get from the United States, and as they contemplated buying their own homes and stores, they lobbied for continued federal subsidies for municipal government and local schools. In asking the government to extend certain benefits to their community, the people of Richland once more identified themselves as pioneers—this time emphasizing the sacrifices they had made on behalf of their country. Fred Clagett, a civic leader who would go on to become Richland's first mayor, advised a 1954 congressional hearing that he regarded the AEC's proposed donation of local utilities systems to the new municipality of Richland as "simply part of the payment for the hazards of our situation here."[72]

During the 1950s Richland residents argued that the nation owed them something extra in return for their role as pioneers, just as a century earlier it had owed special consideration to migrants on the Oregon Trail. Many claims revolved around the contentious matter of housing. Some sought permission to remain in the village after they had retired from work at Hanford. One emphasized the disruptions that a forced departure would wreak "in the settled habits a person develops in pioneering a home such as has occurred here."[73] The pioneers had apparently put down deep roots. When residents campaigned for lower housing prices, they highlighted the sacrifices made by early settlers as a reason why they deserved better treatment from the government. One woman demonstrated the sentiment in a 1954 statement before the Joint Committee on Atomic Energy: "Since Richland changed from a sleepy country village, I watched it grow, pioneered it with the rest of the oldtimers. We've lived through heat and dust storms with no flowers, no grass, no trees and we put in grass and had to water it constantly so it could live in the hot dry desert air. We had no stores, no entertainment facilities and there were many other handicaps. Now Richland is a beautiful place."[74] In highlighting their sacrifices, townspeople continued to slight the huge role that federal agencies and contractors had played. While relying on the common assumption that American pioneers had succeeded through hard work and individualism, not government handouts, the main reason for lauding the efforts and experience of Richland's pioneers was to gain better treatment from Uncle Sam. At the same time, Hanford and Richland represented only one group of claimants in the Columbia Basin. Farmers and veterans interested in developing the area's postwar agricultural potential also pointed to past service and sacrifice in pursuit of federal support.

Making Richland "Normal": Disposal and Incorporation

Taking *psychological* possession of Richland was one thing; taking *legal* possession was another. The latter required more than planting trees and gardens, supporting neighbors and schools, and embracing the imagery of the frontier. For Richlanders to acquire private property and the right to full self-government of their town, two things had to happen. First, the Atomic Energy Commission had to let it go. Accustomed to running the community according to their own instrumentalist logic since 1943, federal managers had to yield the reins to local residents. Second, local residents had to decide to take control of their own affairs, to purchase their own homes, and to drop the habit of deferring to the AEC and its contractor. In short, the AEC and the townspeople both had to reverse their initial positions on the issue of an independent, locally owned Richland. Getting to this point was no simple task. Richland had acquired a frontier identity within just a few years of being occupied by the army; it took a decade and a half for the town to attain its independence.

During the early Cold War years, most residents seemed content with the town. Their attitudes were conveyed in the meetings of the Richland Community Council, a group the AEC had asked GE to establish in 1948 as a means for the contractor to consult with the local population. The regular meetings of the council, which consisted of eight or nine members elected by the community, provided a forum for residents to raise questions, air complaints, and propose changes. However, the council served only as an advisory group. It had no legislative authority and received no more than a minimal budget from GE. Atomic Energy Commission and General Electric officials customarily attended council meetings, but not always with enthusiasm. One GE executive characterized some council members as "self-appointed champions of the people who are generally irresponsible."[75]

The Richland Community Council did question the status quo, but never in too hostile a fashion. For example, one of the recurring issues before it was housing. For much of the 1950s the community had too few single-family homes to meet the demand, and AEC policies toward housing remained fairly rigid. Because it worried above all about recruiting and retaining a qualified workforce for the Hanford plant, the commission gave top priority for housing to active employees. Consequently, when Hanford employees retired, resigned, or were fired, they and their families lost the right to live in Richland. Additionally, employees whose spouses died or divorced them, and those whose dependents left home, had to give up their houses and move into dormitories. On behalf of retirees, widows, and divorced and widowed employees, the council frequently urged GE to make exceptions to its housing policies, usually to no avail.[76] Clearly, the townspeople did not always get what they wanted from GE and the AEC; sometimes GE barely responded to council complaints. But it is telling that many of those complaints came from people who wanted to keep their homes in Richland, ostensibly because they liked living there. They and the council protested because they wished to stay in the town, not because they generally disapproved of the way it was run.

There were sound reasons for residents not to complain too much. Living in Richland was a bargain. For political reasons, the AEC could never afford to appear profligate, but at the same time it needed to keep its middle-class workforce content and on the job. Maintaining an "above-average community" was one strategy used to make that happen.[77] Richland got decent homes, good schools, and more services than a community its size usually had. And because the town had no tax base, the AEC had to pay for all the amenities the town required. It refused to let its support of the town fall below a certain minimum because it believed that any significant cutbacks would jeopardize the efficiency of Hanford production. Thus it subsidized the money-losing bus system and kept rents relatively low until after the Korean War. Funds spent on Richland, the AEC explained, "represent expenditures made as a means of getting the atomic energy job done."[78]

If townspeople did not complain too much because they found Richland comfortable, they also doubtless hesitated to voice discontent for other reasons. Most heads of household in Richland worked for GE (so did most council members), and many of these employees probably were quite reluctant to challenge the company line in public. Thus when some residents wrote letters to the *Tri-City Herald* in 1952 to criticize their lack of input with local government, they frequently withheld their names from publication. Newspaper reporters recalled that townspeople who disagreed with the AEC and GE either did not talk about their views or, when speaking to reporters, insisted on anonymity. Other employees probably sincerely shared GE's view. Indeed, some council members actually came from management ranks at GE, and perhaps they had helped to establish the company line.[79]

Finally, most Richland residents appeared to accept official reasoning that the urgency of plutonium production took precedence over potentially disruptive reforms. In 1951 the council reported the citizens' desire "to place the community on daylight savings time," but in the same breath conceded that if the Hanford plant remained on standard time the town would have to conform to it. Nobody lost sight of Richland's place in the scheme of production. A council member (and critic of "the socialistic trends" at Richland), upon pondering the implications of broader participation in civic affairs, summarized the prevailing sentiment: "The atomic program is Number 1. If it takes leaving [the town] as it is, let's leave it; if we can have more and more self-government, OK."[80] Criticisms of GE were always subordinate to the overall production mission. What bound the community together was more important than what divided it.

Outside observers proved far readier to criticize AEC towns. Particularly during the late 1940s, Richland became the target of numerous complaints, including many that stemmed from a perception of socialism at Richland. In a period of virulent anticommunism, government ownership and operation of the town struck Americans as anathema.[81] The absence of private enterprise was assumed to be a main cause of the wasteful spending perceived in Richland. But socialism implied not only faulty economics but also faulty politics. Critics characterized local government as a "benevolent dictatorship" and Richland as "completely a 'company town' . . . under the absolute and arbitrary control" of GE and the AEC. One report likened the community to a "police state."[82]

If such criticisms had been confined to a few articles in magazines and newspapers, they may not have amounted to much. But elected officials who joined the refrain pressured the AEC to change. Members of Congress generally knew little about the technical details of atomic bombs and, overall, provided little effective oversight of America's nuclear weapons programs. But nearly everyone felt competent attacking overspending on schools and buses, and many in Congress did not hesitate to criticize how the AEC ran its towns.[83] The House and Senate conducted a series of inquiries into federal spending and policies at Richland, Los Alamos, and Oak Ridge, and accumulated complaints about how the AEC handled things there.[84]

In response to criticisms from politicians, the AEC changed its approach toward towns under its control. The commission spoke proudly about those communities when it needed to recruit and retain workers for its facilities, but otherwise its rhetoric became more defensive and its policies more conciliatory.[85] And the AEC's changed outlook filtered down to its contractors. At Hanford, GE took steps to counter the criticisms, particularly the charge of excessive interference in people's lives. Again reverting to criticism of how the army used to do things, it allowed that Richland "had been operated in a somewhat autocratic fashion" during World War II, but asserted that under its management the town had become more normal. "Richland is not a town within a stockade," the company claimed. In 1949 GE loosened its regulations on commercial tenants to give them a freer hand at running businesses. In 1951 Lewis F. Huck, GE's manager for community planning at Richland, captured the contractor's reformist outlook by professing to include "a measure of 'uncontrol'" in his planning efforts as a way of demonstrating that GE did not intrude too much into townspeople's lives.[86] Like the AEC, GE increasingly strove to give the impression of normalcy in Richland. That town managers had to work abnormally hard to achieve a semblance of normalcy suggests the bind they found themselves in.

While altering how towns operated, the AEC undertook a more important change by adopting the goal of handing the communities entirely over to the population. This policy became known as "disposal and incorporation." "Disposal" meant selling town real estate to residents and business owners; "incorporation" meant the creation of local self-government, during which process federal towns converted into independent municipalities. The AEC developed this policy in deliberate phases during the late 1940s and early 1950s. It began selling real estate in Oak Ridge and Richland during the mid-1950s, and by 1960 both towns had incorporated. (Disposal and incorporation took longer and cost more in Los Alamos, where residences and laboratory buildings were much more intermixed.[87]) Until about 1970, disposal and incorporation were accompanied by considerable AEC subsidies to local school districts and governments. It took the AEC more than two decades to turn towns completely over to their residents.

The slow pace of this process suggests that having a program of disposal and incorporation in place was at times more important than actually completing it. The policy helped the AEC deflect political criticism by demonstrating that the agency was addressing the problems in its towns that outside observers had identified. But the commission never pursued disposal and incorporation speedily; for political purposes, having the policy was perhaps more important than implementing it. For example, not all officials within the agency wholeheartedly supported the goals. While some at Richland spoke eagerly about getting the "town off our hands," the greater concern appeared to be the disruptions the policy posed to community life and, thus, to the production of plutonium. At a public meeting in Richland in 1953, J. E. Travis of the AEC explained the agency's reluctance: "The pressure for disposal is from the outside." Congress, not the AEC, was insisting on making changes to atomic towns.[88]

Even if the AEC had embraced disposal and incorporation, it would have found it impossible to implement the policy quickly. Resistance from townspeople formed one impediment. The commission found itself trying to sell property to reluctant buyers, and promote self-government to reluctant citizens. When the AEC first considered withdrawal from atomic towns in 1948, it took a "dismal view" of the prospects for incorporation at Richland "because of the demonstrated lack of citizen interest."[89] (No more consistent than any authority, the AEC apparently spent considerable time discouraging meaningful action by the town council, only to be discouraged by the lack of civic activism.) It would take some time and some adroit maneuvering to convince atomic pioneers to purchase homesteads and become civic-minded. Political hurdles comprised another obstacle. To implement its policy the AEC needed support from Congress and from the state legislatures of Washington, Tennessee, and New Mexico. Getting all its political and legal ducks in a row would prove time-consuming. Finally, the commission's job would not end once disposal and incorporation had occurred. The newly independent towns would require ongoing federal subsidy until they could develop their own sources of revenue.[90] And even then the AEC could not afford to walk away from the towns entirely, because their well-being was expected to remain an important factor in sustained production at nuclear facilities. The commission needed to plan for a very long stay.

Atomic Energy Commission efforts to divest itself of Richland came to a head in 1958 when the town incorporated as an independent city under the laws of the state of Washington. The village had incorporated for the first time in 1910, but the army had changed that status when it suspended the operation of local government in 1943. Federal officials did not actually get around to legally dissolving the incorporated town until 1948, the same year the AEC began to think seriously about getting the people of Richland to incorporate once again as a municipality.[91]

The AEC program for attaining self-government in its communities was defined in stages.[92] In February 1947—in its second month of existence—the commission began addressing the problem of how to foster "democratic local self-government" in Richland, Oak Ridge, and Los Alamos without "jeopardizing AEC's stake in effective operation." It adopted the goal of making the communities as "normal" as possible, which implied abandoning the model of a federally supervised company town. By December 1947 the AEC had determined, "It is the desire and intention of the commission that, consistent with security and other requirements, residents at field installations shall enjoy those facilities, services, and activities which are properly a part of American community life." This ideal was at least something to shoot for, and "democratic control of a visible self-government" became a "long-run objective."[93]

The AEC retained Lyman S. Moore, city manager for Portland, Maine, to study the situation and make general recommendations about how it should proceed. The commission then contracted with Public Administration Service, Inc., a nonprofit firm from Chicago, to examine Richland's specific situation. Public Administration Service (PAS) reported in October 1950 that to accomplish its goals the AEC needed to overcome considerable resistance to disposal and incorporation from local residents. The populace needed to be persuaded to see change as a positive thing, and one way to do this was to make residents pay more of their own way. Public Administration Service conceded that Richland's relatively low rents and relatively abundant services may once have been crucial for attracting newcomers, but argued that it was now time to make citizens bear more of the true expense of living in the town. Only then, the consultants argued, would the demand for private homes increase.[94] The AEC accepted PAS's recommendations, pleased that raising rents would help placate congressional critics, and it increased housing prices by about 25 percent. Naturally many Richland residents complained, but the AEC would not reverse course. It pointed to a federal law requiring that it set rents so that they were comparable to those in surrounding communities, but its more important motive was to encourage tenants to accept the idea of disposal.[95]

With the studies completed and the process of "educating" residents underway, the AEC convened a Panel on Community Operations to lay out the path toward disposal and incorporation. The panel's findings, released in August 1951 and known as the Scurry Report, specified a sequence of steps toward normalizing the towns of Richland and Oak Ridge: passage of enabling legislation by Congress and the respective state legislatures, development of a program to sell real estate to residents so townspeople would become owners before they became voters, incorporation of towns as independent municipalities, and ten to fifteen years of continuing subsidy until Oak Ridge and Richland could generate enough revenue on their own to support the level of services that residents had come to expect. The process would take twenty years, the Scurry Report predicted, and over that period, disposal and incorporation could not be permitted to interfere with production.[96]

In outlining the future of AEC towns, the Scurry Report assumed that "the AEC necessities are identical with the necessities of the people" in Richland and Oak Ridge—in other words, that the commission and its tenants generally wanted the same things. Both wanted "good communities," pleasant, amenity-filled towns that attracted and retained employees and sustained their morale. Both wanted the towns to be self-governing, and both wanted residents to become homeowners. And both wanted to eliminate the "paternalistic federal control" that perpetuated a "'company-town' situation." The Scurry Report anticipated widespread agreement, speculating that only "the less articulate residents" might diverge from the consensus. And even they would come around once the commission started making changes.[97]

In fact, consensus did not prevail anywhere. Some AEC officials, worried that it would affect production, felt uneasy about letting go of the towns. At Hanford, the primary contractor expressed similar ambivalence. In 1952 one GE spokesperson explained, "The problems involved are complex and difficult and the objective must be attained without adverse affect [sic] on the ability of the company as a principal contractor to attain the prime objective of producing plutonium in ever increasing quantities and at a lower and lower cost." Both General Electric and the Atomic Energy Commission still tended to see Richland as a tool for accomplishing their respective goals rather than as a community with its own needs and identity; having structured the Tri-Cities with a view to their smooth function as components of Hanford's operation, self-determination for Richland implied less effective management on one part of the production mission. General Electric, however, did not feel the same political pressure that the AEC did. Rather, it feared disruptions in plant operations and the demoralization of employees, which could undermine its ability to fulfill contractual obligations.[98] Consequently, on several occasions GE sided with Richland's population and against the AEC in negotiations over the town's future.

Finally, the townspeople of Richland agreed neither among themselves nor with the AEC about the prospect of disposal and incorporation. Many citizens strongly favored the city's independence. One man spoke for many when he wrote to Senator Henry M. Jackson: "I like my job here—I like the climate and the people. I like the idea that what I am doing represents a contribution to the welfare of our country." He wanted to stay put, and he hoped to be able to purchase his own home.[99] Still, considerable opposition to disposal and incorporation persisted during the early and mid-1950s. For some, resistance stemmed from the sense that Richland could never become a conventional hometown, as one woman explained to Jackson: "This is definitely not a normal community and never will be. . . . The truth is that people are here for one reason only, the job, and not because this is a place for 'gracious living.' After living here eight years, I have never felt that this is anything but temporary and look forward to the time we can move to a normal community."[100]

For most of the early 1950s, opinion in Richland was divided over the issue of disposal and incorporation. As late as 1955 a poll of Richland residents produced a rejection of AEC plans for disposal and incorporation by a three-to-one vote.[101] Nonetheless, within two years the townspeople had begun to buy their homes and businesses from the AEC, and within another year they had voted to incorporate as an independent municipality. They may have objected to the AEC's plans for their town, but their minds were not closed on the subject.

Most opposition to disposal and incorporation hinged not on principle but on the very practical issue of home ownership. People resisted disposal and incorporation because of the risks they perceived in buying real estate in a town with a questionable economic future. To begin with, Richland's housing stock was rather plain, with most units built in haste out of inexpensive materials. Moreover, many residents felt reluctant to invest in local homes because they feared that property values could easily plummet. Richland relied on a one-industry economy, and the future of that industry remained too uncertain. Hanford could become technologically obsolescent, it was suggested, or peace could break out around the world, thereby curtailing production and employment. Or capricious federal politics could undermine the local economy, as one critic worried: "Home ownership in a one-industry town subject to the vagaries of congressional action is a luxury only the incompetent can afford."[102]

Aware of Hanford's boom-and-bust past, the townspeople imagined many ways that their economy could collapse. Among the things they did *not* consider when appraising home values and the town's long-term prospects were issues of public safety and environmental health. Asked whether the "manufacture of highly dangerous materials in the area" might reduce the value of housing in Richland, H. E. Bolin from the Spokane office of the Federal Housing Administration replied, "Value is in the mind. Danger would reduce value only if the people feared it. Here they are used to it."[103] The hazards of working at and living near Hanford—so much a part of residents' pioneer identity—were not among the reasons they resisted disposal and incorporation.

Despite objections to AEC plans for their town, residents developed a sense of inevitability regarding disposal. After having been pampered by the army and the AEC, Richland residents seemed in no hurry to embrace the reality of self-sufficiency, which was certain to raise their taxes and housing costs. Yet few could oppose disposal and incorporation on moral grounds, because a government-owned and -operated town contradicted the American way. Those who questioned independence for Richland were criticized as people who wanted "something for nothing" and who supported "a socialistic idea"; such comments, at the height of McCarthyism, truly stung. In the end few believed that they could avoid disposal and incorporation. A 1952 poll showed that, while townspeople regarded home ownership as a risky proposition, most expected to buy their homes, if forced to make a choice.[104] The trick was to get the best possible terms.

Through 1955 or so, disposal and incorporation remained general proposals, not specific plans. Eventually the AEC would make concrete proposals and respond to residents' concerns about housing prices. In the meantime, a growing number of townspeople recognized that until it attained its independence, Richland remained in limbo, and they wanted its awkward status resolved. Apart from the price of homes, the restrictions on who could live in them remained prickly issues that prodded some to accept the idea of disposal. In June 1955 an electrician at Hanford wrote to Senator Warren Magnuson asking for help. This man had purchased property just outside the old town of Richland in 1922, built a house, and operated a farm until 1943, when the army confiscated his holdings. He went to work at Hanford and consequently was permitted to rent his house back from the government, and he earned income on the side by raising chickens and selling eggs. But by 1955 he was approaching retirement age. He wanted to stay in the house that he had built and where he had lived for thirty-three years, partly because he needed his earnings from poultry to supplement his retirement income. But AEC policy required that he vacate the house once he left his job at Hanford.[105] Most Richland residents were not close to retirement, but throughout the early 1950s an increasing number reached retirement age, wanted to stay where they were, and found themselves running up against AEC policies. They became supporters of disposal.

In 1956 a Benton County sheriff's deputy wrote to Senator Jackson with another housing problem. The man had recently married and moved into his wife's home in Richland. She had worked at Hanford for a year and rented a house in town. The deputy wanted to remain in Richland, largely because the sheriff hoped to have one of his deputies on hand in that part of the county. Shortly after the couple began to keep house together, however, GE informed them that they would have to leave their home. Because the woman had gotten married, she was no longer considered the head of the household. Her husband was the head of the household, according to official policy, and because he was not "project-connected," the couple lost the right to live in a Richland home. Jackson complained to the AEC about the situation, but the commission refused to yield and carried out the eviction.[106] Once again, people who wanted to put down roots in Richland were forced to leave.

Atomic Energy Commission policies, the slow pace and uncertainties of implementing disposal and incorporation, and the changes occurring in the local population steadily eroded the community's opposition to independence. Individuals in leadership positions increasingly urged Congress to act. In 1955 the rector of Richland's Episcopal church, William Greenfield, wrote Jackson to describe how AEC and GE policies were hurting the town: "In 1954 some three hundred families left Richland and either bought or built [nearby], the majority of them in Kennewick. Needless to say, this state of affairs continues and will continue until such time as land is made available in Richland for the purpose of home construction to the owner's design. This is removing from Richland the solid dependable type of citizen, and the man who is about to retire."[107] Indeed, it was during the 1950s

2.7 Expansion at Hanford during the Cold War not only led to additional growth in Pasco and Kennewick but also hastened their economic reorientation toward the nuclear industry. This photograph looks east toward Kennewick in the center, the Columbia River, and Pasco in the upper right. The fields and orchards in the foreground make it clear that, while the Tri-Cities were increasingly dependent on Hanford manufacturing, many in the area still raised crops as well as livestock. Photograph courtesy U.S. Department of Energy.

that Kennewick began to leave trailer parks and barracks behind and take shape as a permanent community. In 1959 a geographer commented that many Kennewick residents were "ex-Richlanders who got tired of waiting for the government to sell the Richland property." Occupying housing that had been built atop lands once used as orchards and farms, these settlers preferred Kennewick over Pasco because it lay closer to Hanford and had better river views, and because its demographic makeup more resembled Richland's.[108] At the same time, even as residence became more fluid, Hanford's culture remained deeply inscribed on the surrounding communities. Richland was most closely associated with management and administration and served as the site for the Federal Office Building. Kennewick had more of the character of the working engineers and builders who lived there, people who made things and solved material problems rather than policy ones.

Perhaps the strongest local advocate of disposal and incorporation was the Richland Community Council. More than other residents, council members dealt regularly with the practical limitations imposed by the town's relationship to Hanford and the AEC. In their work they discovered over and over just how Richland's dependent status prevented the council from enacting its and the residents' wishes.[109] They consequently became early advocates of disposal and incorporation—not because they regarded the policies as ideal but rather because they saw them as inevitable and necessary. The council adopted its position in defiance of the popular will; the townspeople had asked it to stop supporting AEC efforts. But the council remained adamant about pressing forward, and perhaps about ending its own powerlessness as a merely advisory body. In May of 1955 it petitioned Congress to deal promptly with disposal and incorporation so that the community could learn its fate and so that the flow of "desirable families moving out of Richland" would stop.[110]

That summer Congress finally passed the Atomic Energy Community Act of 1955, providing for the transfer of property and government in Oak Ridge and Richland to town residents. The AEC then asked the Federal Housing Administration to conduct appraisals of homes in Richland in order to establish sales prices. When the FHA released its figures on May 10, 1956, the townspeople loudly protested that the proposed prices were too high, and inundated the state's congressional delegation with complaints. In response the Joint Committee on Atomic Energy sent George Norris, Jr., its executive counsel, to investigate. In a critical report back to Washington, D.C., Norris found that most residents were "anxious to buy and own their homes but did not want to do so at these prices."[111] The Joint Committee scheduled a hearing the following month and invited a delegation of Richland citizens to participate.

In testimony before Congress the residents of Richland reiterated their apprehensions about owning property in Hanford's bedroom community. They argued that the local economy—and hence the value of homes—relied too much on one industry, which in turn depended too much on political rather than market considerations. They contrasted their situation to that of Oak Ridge, which, having been designated a national laboratory, seemed to have a more secure future. They reminded Congress that Richland's housing stock was generally mediocre and would require additional expenditures on remodeling to become satisfactory. Finally, they played their trump card by making lower housing prices a matter of national security. Some valued employees had already left the project, and more would follow if they did not get fair treatment in housing matters. Morale would decline, they warned, Hanford would lose some ability to attract and keep workers, and rates of production would suffer. Officials from General Electric echoed the townspeople's claims, warning of "recurring unrest in the work force" if something was not done about the price of homes.[112]

In the course of the hearings, Washington's congressmen and senators came to the aid of their constituents. A town with "no political powers" suddenly acquired clout. Senator Warren Magnuson and congressmen Hal Holmes and Don Magnuson demanded that Richland residents be placated, and Senator Jackson sharply condemned the FHA officials responsible for the first set of appraisals. (On the same day as the hearings, not coincidentally, Jackson also introduced legislation to authorize construction of a dual-purpose reactor at Hanford, one step toward diversifying the local economy. More on the politics of Richland follows in Chapter 3.)[113] With the support of these elected officials, Richland residents won the day. The AEC contracted with another appraiser, who in early 1957 recommended reducing the sales price on about half the homes in Richland. Townspeople and GE accepted the new appraisals, and in June 1957 the first home changed hands. By July 1958, the government had sold 4,200 Richland homes.[114]

Once the AEC had disposed of Richland's private property, it turned to the matter of self-government. This time it settled financial issues quickly by assuring the town of ample subsidies for municipal operations and public schools. With these promises, town residents approved incorporation by a five-to-one margin in July 1958. The town drew up and approved a charter, and achieved its independence on December 12, 1958. Citizens celebrated with an evening of fireworks. Richland set off aerial bombs at 8:00 and 8:15, and at 8:20 began a countdown for a "simulated H-bomb explosion" at 8:30. The exploded device, consisting mainly of dynamite with napalm added to create a fireball-and-mushroom-cloud effect, broke two windows in the uptown shopping district. Then at 8:40 Miss Richland used a "uranium-tipped wand" to ignite a bonfire, the evening's final pyrotechnic event.[115]

Having started off with a bang, the new city of Richland thrived for the next few years. In 1961 the town was named an All America City in an annual competition sponsored by *Look* magazine and the National Municipal League. The recognition confirmed Richland's status as a model community and rewarded the town for its initiative in rejecting "the easy paternalism of government operation." *Look* praised Richland as "a city that faced a problem of growing up and standing on its own two feet,"[116] and local residents and boosters took enormous pride in the accomplishment. Of course, nobody wanted to mention that Richland stood up on its own two feet only after it had been pulled upright by the hand of the AEC before 1958, and only because the hand of the AEC continued to steady the town after 1958. In this respect, though, Richland was distinctive more as a matter of degree than of kind. After World War II the federal government made payments in lieu of taxes to many communities, especially those in the West—timber towns reliant on national forests, for example, and communities where schools educated military dependents.

The townspeople toiled right away to improve the town. By 1963 residents had spent roughly $17 million remodeling and enlarging houses that had sold for $28 million between 1957 and 1958. By 1965 Richland had topped 25,000 in population, added 816 housing units, annexed seven square miles of land, and built twenty-eight miles of new roads.[117] Richland had advanced further along the path toward being a model community. On the occasion of winning the All America City award, the *Tri-City Herald* editorialized that "Richland has no industry or annoying residents." (As in the navy city, San Diego, defense spending obviated the need for both.) Added town mayor Joyce Kelly, "Richland is debt free. It has intelligent citizens, a low crime rate, good fire protection and the area has abundant electric power." Richland's population was youthful and family-oriented, its schools and churches had good reputations, and it had no parking meters.[118]

However, dark clouds remained on the horizon of the All America City. Over the course of the 1960s, its municipal government received annual "assistance payments" from the AEC averaging $332,200, and the local schools received additional sums.[119] Richland needed to prove that it could generate the revenue needed to support the high level of services to which residents were accustomed. That task depended in turn upon the fate of Hanford, the town's narrow economic base. As it turned out, those residents who had expressed reluctance during the 1950s about buying real property in a one-industry town had good reason to be concerned. Between 1964 and 1971 the AEC cut Hanford operations severely, closing eight of the reservation's nine reactors. The shutdowns threatened not only the economy around Hanford but also the distinctiveness of Richland.

An Above-Average Community: Richland and Its Neighbors

A student of Richland described the town in 1952 as "an above-average community."[120] When building the village during World War II, the army had understood the need to make the town relatively comfortable and appealing, even though it could not always muster the resources or commitment to make the place special. When the AEC and GE took over the town after the war, they worked to improve the physical setting. The planners they hired emphasized that Richland should become a city of homes for a relatively prosperous and stable population of Hanford employees. To that end, they increased the portion of land devoted to single-family housing, commercial buildings, community services, and parks and playgrounds, while decreasing the proportion of industrial and vacant land. The planners also supported the AEC's policy of shunting postwar construction workers off to North Richland, a cluster of trailers, barracks, and mess halls that could expand or shrink as needed. When North Richland was abandoned during the mid-1950s, some of its occupants moved to trailer parks in West Richland, two miles away from the AEC town.[121] Richland remained the antithesis of the construction camp. For this reason, "above-average" implied not only an attractive, planned setting but also a certain type of population. "There are no unemployed, no slums, no marginal businesses," one booster pamphlet explained in 1949. "Crime is tightly controlled."[122]

Cognizant of the need to recruit and retain a well-trained workforce, managers ensured that Richland possessed an "above-average" level of services for a town its size. The municipal quality of life in Richland stood out especially clearly when contrasted to what its neighbors had to offer. In terms of medical services, Richland had a pediatrician in 1948 while Pasco and Kennewick did not. It also had a hospital that treated Richland residents and Hanford employees exclusively. During the cold winter of 1950, when pipes froze throughout the Tri-Cities, the plumbers who worked in Richland for GE completed repairs much more quickly than did plumbers in Kennewick and Pasco.[123] Because they were integral to the urgent task of plutonium production, the people of Richland were looked after faithfully.

Observers of Richland's ample comforts described the town as "ultra-modern," even futuristic. The perception stemmed partly from the fact that little remained to remind people of earlier, less affluent times. "We have none of the ills of cities," one AEC official explained. "We haven't inherited any horse and buggy ordinances."[124] Functioning as a kind of blank slate for planning values of the time, Richland took on a decidedly suburban appearance in the years after 1945. Strictly speaking, Richland could not be called a suburb. Most of its inhabitants did commute to work at an industrial plant, but there was no large nearby city with a dominant central business district. The village itself was the dominant city, performing residential, administrative, commercial, cultural, and recreational functions. Yet its layout increasingly borrowed from suburban models, in common with other postwar high-tech industries competing for technical manpower. Historian Carl Abbott summarizes the suburban goals underlying Richland's postwar design as "neighborhoods focused and organized around schools; densities and designs that assumed and forced use of automobiles; large-grain separation of activities; and use of space to protect class distinctions." Typical of the new order was the uptown shopping district, a distinct species of the evolving suburban mall.[125] An above-average community required an appropriate setting for the consumption of goods and services, though many residents took weekend shopping trips to Seattle, Portland, and Spokane. (Even some of the negative markers of suburbanization began to manifest themselves; as early as 1950, observers noted increased traffic congestion, greater reliance on private automobiles, and declining use of public transit.[126])

During the early Cold War years, Richland impressed many observers because it was new, planned, affluent, and suburban in appearance—a "model" community.[127] Of all the factors that marked Richland as "above-average" at the time, its social, economic, and racial composition probably contributed the most. The population was relatively youthful, healthy, well-educated, well-paid, economically secure, and overwhelmingly white. These traits stood out especially boldly before disposal and incorporation, but they persisted well into the 1980s and 1990s. And they stood out clearly when Richland was contrasted to its neighbors.

Richland's demographic distinction stemmed primarily from selective federal employment practices at Hanford. Managers at the plutonium plant required operations personnel with certain skills and educational backgrounds, screened applicants carefully for security and safety reasons before hiring them, paid workers relatively well, and operated Richland with an eye to keeping employees and their families content. A 1948 survey of the AEC "rank and file" produced a telling summary of male Richland residents: "The average employee is superior to the average industrial

2.8 As the nation's demand for plutonium grew and as Hanford added more reactors and processing plants during the early Cold War, the AEC and GE expanded Richland's housing supply. In many ways the new subdivisions resembled suburban tracts being built throughout the arid West. This 1952 photograph shows both the mass-produced new homes and the adjoining desert land. In between is a thin green line of trees planted as a shelter belt that was meant to reduce the impact of the steady winds and occasional dust storms on the new neighborhood. Photograph courtesy U.S. Department of Energy.

employee due to careful screening for security and for job requirements. He is 10 or 15 years younger than in industry. He has many advantages such as good housing at low rentals, good medical and hospital service and excellent recreational facilities. He enjoys good wages without the controversies often associated with unions in industry. He is safety conscious and not only wants an active safety program but demands it."[128]

Operations employees at Hanford—but not the janitorial or construction workers who were ineligible to reside in Richland—had to be relatively well educated. For example, by 1957 all those who operated reactors had to have at least a high-school diploma (along with four years of experience). Many in the town had even more schooling, however. Both in 1950 and 1960, the adult population of Richland had attended school for an average of more than 12.5 years, with males likelier to have gone further. In 1960 more than 40 percent of men over 24 years of age had attended college; the figure for Washington State was 22 percent. Richland parents expected more of the same from their children. About two-thirds of the 1960 and 1961 graduating classes from Columbia High School enrolled in college—most of them in four-year institutions.[129]

The relatively high level of education correlated with high-status jobs and good pay. Richland had a sizeable proportion of "administrative, technical, and engineering personnel" living within its borders. The majority of these professional employees worked, of course, for GE and the AEC, either on the Hanford Site or in town (Richland had very few self-employed residents), and they earned above-average salaries. In 1959 the median family income for Richland was $8,368; for Washington it was $6,225. The town's affluence did not result from the presence of a highly paid elite, however, and the population was not deeply stratified by income. Richland had neither a substantial lower class nor much of an upper class, but rather consisted mostly of what one observer termed "the middle of the middle class." In 1950, roughly 26 percent of all American families had annual incomes under $2,000, 53 percent had incomes between $2,000 and $5,000, and 21 percent had incomes over $5,000. In Richland, very few families had incomes under $2,000, most had incomes between $2,000 and $5,000, and perhaps 10 percent had incomes over $5,000.[130] And prior to 1956, of course, Richland's unemployment rate was minuscule, because heads of household who lost their jobs at Hanford had to move out.

By the standards of the time, Richland seemed above-average in two additional ways—its relative youthfulness and healthiness. A town of young families, postwar Richland underwent a baby boom that also exceeded the national norm. In 1948 the town's birth rate stood at 34 per 1000, compared to the national figure of 20 per 1000; in 1950 15.3 percent of the population was under five years old, compared to the national figure of 10.8 percent. Although Richland's birth rate gradually leveled off during the later 1950s, the town retained a disproportionately large number of schoolchildren.[131] By contrast, only 2.2 percent of the town's population was over 65 years old in 1960, compared to 10 percent of all people in Washington, again reflecting the fact that no retirees remained in town. One husband and wife remembered that there were "no funerals" and "no old people" in Richland during the late 1940s, and that when the wife's elderly mother visited she felt out of place in such a youthful population. Ten years later, when Richland joined the rest of the country in worrying about juvenile delinquency, one reason for the perceived shortage of supervision for children may have been the absence of grandparents. Most adult residents had moved there from some other part of the country and had left their own parents behind.[132] They also left behind much illness. A young, prosperous, well-educated populace, served by its own doctors and hospital, made for a healthy community. Richland far surpassed national averages by attaining lower rates of worker absenteeism (1.0 percent, compared to 3.7 percent nationally), deaths in population over five years old (2.7 percent, to 10.0 percent), and infant mortality (18.9 per 1000, to 32.6 per 1000). Master planners saw no need for a large cemetery in town; few people were expected to die there, and many of those that did would be buried in the towns they came from.[133]

Residents naturally liked to hear and to believe that they were part of a model community. Observers who wanted to construe Richland as an icon of Cold War American life, and AEC and GE officials who needed proof of their organization's success, also championed the idea that Richland's population was exemplary. But if the town could be called above-average, another word for it was homogeneous. In part because the FBI had to clear all Hanford employees, there would be few ideological or political "nonconformists" in the population. There were also few religious "nonconformists." Planners in 1948 counted "23 active religious groups" and 6,561 Christian church members, but no Jews.[134] The town's families resembled one another—socially, economically, and culturally. Strangers upon their arrival in Richland, residents quickly discovered that they had a lot in common with each other. One former resident recalled in 1990, "[W]e were all young when we first hired in there. There weren't any old people to speak of in the town, and everybody was in the same boat. We all came in there with . . . the job that they wanted us to do and we all had our families there and everybody got along."[135] At the end of an overland journey, it doubtless proved reassuring to find others with like interests and backgrounds.

Richland's homogeneity and supposed superiority, however, were not universally admired. Hanford had introduced to the Tri-Cities changes that were not uniformly welcome, and nearby towns did not always appreciate their new neighbor.[136] These other towns were being overshadowed by the newer, bigger, wealthier Richland, and the inhabitants of Pasco and Kennewick sometimes regarded the upstart community as aloof or arrogant.[137]

There is evidence that Richland attained its "above-average" status at the expense of other towns. Very few of the construction workers employed in Hanford's postwar expansion, for example, lived in Richland. In 1950 fewer than 10 percent of Richland's wage-earners worked in construction, while 22 percent of Pasco's, 38 percent of Kennewick's, and virtually all of North Richland's wage-earners worked in construction.[138] Moreover, because working-class families amounted to a small percentage of Richland's population, labor's voice was not prominent during Richland's negotiations with the federal government over the terms of disposal and incorporation in the 1950s.[139] During the 1960s, when residents were polled about the direction of the city, 60 percent of the respondents opposed the idea of Richland becoming involved in providing housing assistance to the impoverished or programs for the elderly.[140] The implication was clear: the needy in the Tri-Cities, and the public-sector programs that served them, belonged in Pasco and Kennewick, not in Richland. The Atomic City of the West wanted government support, of course, but it preferred that the money go to propping up industry at Hanford and maintaining schools and services in Richland—not to assisting the poor or the aged. It aspired to perpetuate its homogeneity, if it could.

During World War II, the army and DuPont had created Richland as a town apart from its neighbors. Pasco and Kennewick got more of the social problems associated with an economic boom and with a more diverse, transitory, disproportionately male population, while Richland got middle-class homes, morals, services, and inhabitants, and less demographic diversity. This pattern continued into the 1950s, 1960s, and 1970s. If Richland seemed to be an above-average community, Pasco and Kennewick distinctly did not.

To be sure, Richland could not stray too far from Kennewick and Pasco because all three came under the influence of Hanford. Prior to World War II, Pasco's economy had revolved around the railroad and supplying farmers, and Kennewick's around the processing and shipping of produce. While these orientations were not completely erased by World War II and the Cold War,[141] both towns were influenced dramatically by the advent of plutonium production. Between 1940 and 1960 Pasco and Kennewick grew rapidly in population and became more affluent, partly because each of them—especially Kennewick—housed an increasing number of Hanford employees and their families. In 1963 one GE report estimated that 80 percent of Tri-City families depended upon Hanford for their livelihoods.[142] The plutonium plant had introduced a manufacturing economy that increased the prosperity of the whole region, so in some ways the three towns developed along similar paths.

Having said that, the differences between Pasco, Kennewick, and Richland—first understood and reinforced by the army during World War II—remained prominent through the decades. Richland was the biggest, wealthiest, and best educated. Its median family income for 1950 ($4,864) and 1960 ($8,368) substantially exceeded Kennewick's ($3,750 and $6,598) and Pasco's ($3,972 and $6,590). Whereas one-quarter of the households in Pasco and Kennewick earned less than $2,000 annually, according to the 1950 census, less than one-twentieth of Richland's did. In 1960 about 60 percent of Richland's labor force worked in manufacturing, compared to Kennewick's 19.8 percent and Pasco's 9.6 percent. Three-quarters of Richland's adults had completed high school, compared to slightly more than half the adults in the other two towns.[143] Finally, the distribution of racial minorities differentiated the cities, although in this regard Kennewick had more in common with Richland than with Pasco. The proportion of nonwhites in Richland and Kennewick was tiny; the population of Pasco, by contrast, consisted of about 10 percent nonwhites in 1950 and 1960. In regard to people of color, the sharp, lasting distinctions between the Tri-Cities reinforced economic differences between the towns (see Appendix 3).

Pasco had been identified as the town to which nonwhites were assigned during World War II. The handful of African Americans who lived there before 1941 had marked the town as the place for minorities. The black population swelled during wartime, and the army also directed to Pasco those Mexican Americans recruited to work at Hanford. People of color were not really welcome in the town; one white resident complained about the African Americans and Mexicans that wartime Hanford had attracted to the area (as well as the "men out of the South swamp . . . illiterates . . . hillbillies . . . poor white trash").[144] An educator concerned with the uplift of African Americans in Delaware wrote directly to DuPont management, reporting in 1944 that information circulating among social workers indicated that conditions in the "Pasco-Hanford-Kennewick Production area" were "leading to disentegration [sic] of family life, increased delinquency, illegitimacy and large labor turnover certainly limiting plant operation." Sympathetic to the problems faced by DuPont, the educator continued, "I can well understand how difficult . . . standards are to attain in a far flung area like the State of Washington. . . . But, frequently such problems do not

2.9 For most of the 1950s, Richland had very few African American residents and Kennewick had none. The vast majority of blacks lived in Pasco—but not Pasco proper. Rather, they dwelled almost entirely in the segregated, impoverished neighborhood across the railway tracks known as East Pasco. This photograph, taken at some point during the 1950s, documents the kind of substandard housing available to residents of East Pasco. Courtesy of *Tri-City Herald.*

wait especially when they involve racial tensions which can become combustible as I understand exists in Benton and Franklin Counties."[145] With thousands of construction workers filtering through Hanford, though, neither the army nor DuPont worried much about long-term structural remedies.

During the war, the army's policy was to subordinate social concerns to the mission; do what was necessary to drive the construction labor force to complete the Hanford Works and then create, directly and indirectly, patterns of community life that echoed the division of labor in the operating facility, with occupational groups housed in distinct communities. The division of labor overlaid divisions based on race. Pasco became the locality where people of color lived, but it did not offer them much comfort. African Americans were segregated on the "wrong" side of the railroad tracks, in a blighted neighborhood called East Pasco, where they lived in cramped, unsanitary housing. In 1951 one observer counted 483 homes for nonwhites in Pasco, 95 percent of them "substandard." Among the living areas was a trailer camp that had no water or sewer lines. Blacks could neither obtain home loans in the town nor patronize many businesses. They attended segregated classrooms in a school system that employed no black teachers, administrators, or staff. Such conditions persisted basically unabated until the 1960s, when the local chapter of the National Association for the Advancement of Colored People, the Urban League, and the federal government began to apply pressure to overturn them.[146]

As bad as Pasco was, it was the only choice for most African Americans in the Tri-Cities. During the early Cold War period, Kennewick relied on a combination of formal and informal means to keep out virtually all blacks. Whites reinforced an ordinance requiring African Americans to leave town by sundown with pressure imposed to keep real estate out of black hands. The sheriff of Kennewick explained the racial homogeneity of his community by reporting that "if anybody in this town ever sells property to a nigger, he's liable to be run out of town."[147] Attitudes in Richland were apparently different. Katie Barton, an African American who moved from Texas to Pasco in 1947, recalled that "Richland was the least prejudiced of all" the Tri-Cities.[148] Yet Richland had almost as few African Americans as did Kennewick in 1950, and even in 1960 the percentage of blacks remained minute.

In this regard, too, the makeup of Richland's population derived from hiring practices at Hanford. During World War II, military service and wartime industry recruited many more African Americans to the Pacific Northwest than had lived there before.[149] Although some left the region after the war, many remained behind, even though the number of jobs for them in industry diminished. Beginning in 1947, Urban League officials in Portland and Seattle encouraged African Americans to apply for work in the booming economy around Hanford, but the League had little success getting blacks hired at anything other than unskilled positions and construction work. E. Shelton Hill of the Portland chapter of the National Urban League in 1948 found no blacks on the AEC's payroll, while GE employed only a few African Americans as janitors for Richland dormitories. Union policies also prevented the employment of African Americans in some building capacities, yet construction firms hired hundreds of blacks during the postwar boom, expecting them to reside and eat in largely segregated conditions in North Richland. Those African Americans who were hired to do Hanford-related work thus encountered such abysmal living conditions—especially in East Pasco and North Richland—that few wanted to stay. Other employers were hardly better. Many private businesses in the Tri-Cities did not want African American workers (or patrons), and municipal, state, and other federal agencies also hired few blacks. "I have the feeling that they don't want to give any colored the jobs," one African American observer reported to the Seattle Urban League, "they want them to leave." Responding to these conditions, the Seattle Urban League asked the AEC to help conduct a survey on race relations in the area around Hanford, but the AEC refused to support the effort unless "the survey's findings and report become the confidential property of the Atomic Energy Commission." The AEC also wanted no community group going over the survey's findings. Unwilling to accept such terms, the Urban League found its own funding and commissioned a researcher to produce a report in 1951 that they called "Review of the Economic and Cultural Problems of the Tri-City Area, As They Relate To Minority People."[150]

At Hanford the Urban League found institutions that proved quite reluctant to address the matter of race relations. When faced with the same issues during World War II, the army had refused to play the role of reformer because it did not want to divert attention from more important tasks at hand. "The responsibility of the Office of the District Engineer," one Oak Ridge officer explained in 1944, "is not to promote social changes . . . but to see that the community is efficiently run and that everybody has a chance to live decently in it."[151] When an Urban League investigator appeared at Hanford in 1949 to survey race relations, he received basically the same response from the AEC's deputy manager: "We have enough trouble here without having to cope with a Negro problem. We've got to think of our white majority, many of whom are southerners and would not stand for Negroes here."[152] Atomic Energy Commission and General Electric officials felt that they could placate either the white Southerner or the African American, but not both. Fair conditions for blacks were sacrificed in the name of efficiency, as an Urban League investigator explained in 1949:

> Race relations are in a very poor state at Richland. . . . There is evidence
> that . . . persons in top positions have had little or no exposure to persons
> of minority groups and that they have permitted their thinking to be influenced
> by stereotypes and information that they might have received from other
> sections. Hence, this has permitted persons in lower positions, who might for
> any number of reasons have had racial feelings, to more or less develop policy
> as they saw fit. This policy has been one of expediency. There has been a great

amount of segregation in areas such as housing and recreation. This has been done from the point of view that persons of different races get along much better together when they are separated; that is, there is less occasion for friction with this type of arrangement, and that the efficiency of the operation is assured in this respect.[153]

With production entrenched as its undisputed priority, the AEC only hesitantly addressed matters of inequality at Hanford. In October 1949 it quietly approached the Seattle Urban League for help. It hoped to prod GE to employ more blacks, but first had to deal with the embarrassing fact that there were no African Americans on its own payroll. The commission began to change conditions at Richland, but only very slowly. Only one black person lived in Richland as of March 1950; by the end of the year, two more African American families had moved there. In 1951 two blacks worked for the AEC at Hanford—one a typist, and one a clerk—and "less than a dozen Negro clerks and custodians" were employed by General Electric. (Construction contractors employed about 250 African Americans.) A 1951 study by the AEC revealed that there had been "occasional instances where the use of eating and recreational facilities by non-whites has been discouraged" in Richland, but housing and schools were not segregated.[154] Of course, it would have been hard to establish separate facilities for two or three African American families. This fact became something for which the AEC congratulated itself in March of 1953 after President Dwight D. Eisenhower ordered the desegregation of schools on all U.S. military bases. Anticipating that it, too, would receive instructions similar to the military's, the commission surveyed conditions at Oak Ridge, Los Alamos, and Richland. It decided it needed to worry only about Oak Ridge because, with no more than a handful of nonwhite pupils, "there is no school segregation at Hanford and Los Alamos."[155]

The AEC initially proved complacent about its hiring policies at Hanford and the low number of African Americans in Richland, but the agency changed over time. During the 1950s the number of blacks in town grew slowly, no doubt assisted by the process of disposal and incorporation, which dissolved the linkage between Richland's population and Hanford's hiring practices. The 1960 census recorded 189 African Americans in Richland. Yet the number of black employees at Hanford remained small. In 1966 blacks amounted to about 1 percent of the total—or 82 out of around 8,000—so the AEC instructed its contractors to hire more African Ameri-

2.10 From the early 1940s through the 1960s, a number of factors produced social and racial segregation within and among the Tri-Cities. Pasco became home to most African Americans in the region—although within the town, blacks mostly lived in the worst neighborhoods. Kennewick became known as the community most hostile to integration by African Americans. In each of the Tri-Cities, different forms of discrimination began to break down during the 1960s. This undated photo, likely from the late 1960s, shows a protest led by the NAACP against Kennewick's exclusionary policies. Photograph courtesy of the Franklin County Historical Society and Museum.

cans.[156] Federal pressure altered race relations around Hanford during the 1960s, and so did initiatives in each town. The NAACP and other civil-rights activists led protests against discrimination in Pasco and Kennewick throughout the 1960s; Pasco began desegregating its schools in 1965 and elected its first African American to the city council in 1967; and Richland hired its first black policeman in 1969.[157] In all three towns, conditions changed hesitantly.

After the 1960s the racial composition of towns continued to correlate with their economic fortunes. Pasco remained the most diverse and the least prosperous. Its minority population dwarfed those of its neighbors. The African American population of Pasco went from 1,334 (9.6 percent) in 1970 to 1,414 (7.9 percent) in 1980 and 1,126 (5.6 percent) in 1990, and the percentage of Hispanics increased from 3.8 percent in 1970 to 20.8 percent in 1980 and 41.2 percent in 1990. Over the same period, Pasco retained the smallest percentage of owner-occupied housing, the lowest per capita income, and the highest level of unemployment. By 1990 Pasco contained less than half the population of Kennewick, earned less than half the per capita income of Richland, had about one-third of its households living below the poverty line, and was the only one of the Tri-Cities with double-digit unemployment.[158] Although some of its residents worked at Hanford, Pasco was the least touched by its prosperity and the most willing to challenge further nuclear development. In July of 1986 the Pasco City Council passed a resolution against importing additional radioactive wastes to Hanford, saying that the idea "has placed an economic cloud on this area of irreparable and disastrous proportions" and threatened the agricultural and transportation-oriented economy upon which Pasco still depended more than Richland and Kennewick.[159]

Such a resolution never would have passed in 1986 in Kennewick or Richland, both of which remained more tightly in orbit around Hanford. Richland gained a reputation as the "Gold Coast" of the Tri-Cities while Kennewick became "a sprawling shopping area and bedroom for Hanford's blue collar workforce."[160] By 1980 Kennewick had become the largest of the three towns, although it remained the least diverse, with African Americans and Hispanics comprising but 1.1 percent of its population. Kennewick grew rapidly during the 1970s because, in contrast to Richland, it had more available land for development and a higher housing density. Kennewick also grew by becoming the leading retailer of the Tri-Cities with the completion of the Columbia Center shopping mall in 1969. According to the 1980 census, Kennewick housed the lion's share of both retail and construction workers in its urban area (with many of the latter recruited to work on Washington Public Power Supply System projects).[161] During the next decade, while the populations of the other two towns mostly leveled off, Kennewick's grew by almost 8,000, from 34,400 to 42,200. Many of the newcomers, at long last, were minorities; in 1990 Hispanics amounted to 8.7 percent of the population (see Appendix 4).

Richland changed more slowly than its neighbors. Much of its future had been determined during the 1940s and 1950s when, for example, planners had limited the amount of space available for housing and established a relatively low density. The town's population increased from 26,290 in 1970 to 33,578 in 1980, then leveled off to 32,315 in 1990. The turnover of the population proved relatively slow, which perhaps helps to explain why

2.11 For all their differences, Richland, Pasco, and Kennewick forged a common identity not only as the Tri-Cities but also as communities affiliated with nuclear technologies. One manifestation of this shared orientation was their sponsorship of the Atomic Cup, a summer race on the Columbia River for hydroplanes. Since 1966, the Tri-Cities have been an annual stop on the national circuit of races for unlimited hydroplanes. For the first seven years, the event was known as the Atomic Cup race. In 1973 the name changed, and in most subsequent years the event was known as the Columbia Cup. (In 2006, on the fortieth anniversary of the original event, the race was again called the Atomic Cup.) This photo captured the 1968 race along with viewers on the Kennewick shoreline. Courtesy of *Tri-City Herald*.

a town that had once been so youthful had by 1980 attained the highest median age within the Tri-Cities. Similarly, the proportion of nonwhites remained small. By 1980 the number of African Americans had increased to 471, and the number of Hispanics to 714, yet together the two races amounted to no more than 3.5 percent of Richland. By 1990 the figures had hardly changed: blacks and Hispanics comprised 4.4 percent of the population.[162]

In many ways Richland remained what it had been since 1943 — a prosperous and homogeneous city that, more than Kennewick or Pasco, relied upon the U.S. government for an economic base. This federal cushion insulated the town somewhat from the blows taken during busts at Hanford after 1964. Richland suffered its share of hard times, to be sure, but as headquarters for the federal effort at Hanford it could always fall back on the steady employment offered by the offices of the AEC and DOE and their contractors. Kennewick and Pasco, by contrast, with their populations of construction workers, retail employees, farm laborers, and generally less skilled workers, had lower per capita incomes and fell into deeper valleys during slowdowns.

The towns remained distinctive, then, despite efforts to minimize the differences among them. The local newspaper had promoted a Tri-City identity since 1950, and after 1963 boosters framed economic-development campaigns in terms of the Tri-Cities rather than individual cities. During 1968, the same boosters campaigned to have the U.S. Bureau of the Census recognize the three communities as one "standard metropolitan statistical area," the thinking being that this status would increase the towns' visibility and attract more investment and immigrants.[163] The effort perhaps helped define a single urban entity for outside observers, and even local residents used the phrase "Tri-Cities" without a second thought. But within the metropolis the distinctions between Richland, Kennewick, and Pasco remained forceful. Indeed, they were so strong that attempts to merge the three towns into a single municipality, promoted by Richland reformers in the 1960s and 1980s with the goal of distributing the virtues of their ideal community more widely, failed handily. The towns continued to go their own ways—even though in some ways they remained yoked together by the economic and political power of, and increasingly by the ecological problems at, Hanford.

Yet even the connection to Hanford changed. Before 1980, association with things nuclear had been primarily a positive thing. Afterward, Tri-City communities and even boosters increasingly began to distance themselves from the region's atomic identity, although not without a fight. Since the early 1960s the leading regional boosters organization, the Tri-City Nuclear Industrial Council (or TCNIC), had sought to attract more nuclear-oriented industry to the region. In 1979, in the wake of the incident at the Three Mile Island power reactor, Sam Volpentest of the TCNIC recoiled at the suggestion that the council eliminate "Nuclear" from its title: "Hell no. . . . We're not going to drop it. We're stubborn. We've got nothing to fear from nuclear energy." By early 1985, however, the group had renamed itself the Tri-Cities Industrial Development Council, dropping the association with atomic energy in order not to scare prospective businesses away. Volpentest did concede that "the emphasis has to change" in the local economy.[164]

2.12a, 2.12b Richland's attachment to a nuclear identity changed over the years. Upon incorporating in 1958, Richland adopted a city seal that featured the atom prominently, along with representations of the sun and the Columbia River. By 1990 the community required different imagery. The new city seal continued to feature the river and sun but dropped all association with the atom. Nonetheless, the lettering of the new logo suggested that Richland remained as forward-looking and technologically advanced as it had always claimed to be. Images courtesy of the City of Richland.

Richland proved hesitant to shed its atomic identity. In the face of considerable criticism from outsiders after 1980, citizens refused to eliminate the mushroom-cloud logo from the uniforms of athletic teams of the Richland High School Bombers. One reason may have been an attachment to local history. Residents doubtless recognized the polarizing effect of the mushroom-cloud symbol in the late twentieth century, but the high school logo nonetheless captured forcefully crucial aspects of the town's past. Richland no longer had to borrow a heritage from a mythical western past lived by someone else; it had its own story now. As late as 1985 one journalist noted that the town wore "its atomic heart on its sleeve. . . . Both the city and the Chamber of Commerce logos include depictions of the atom." In some ways, however, the town had to yield to changes in American attitudes toward nuclear power. Holding on proudly to history was one thing; employing controversial symbols for promotional purposes was another. In 1990 the Richland City Council sponsored a contest to develop a new city logo to replace the old one because "city officials felt a new image is needed in light of the thrust to diversify the economy and reduce the emphasis on Richland's ties to Hanford." The winning design dropped the atom yet still conveyed a futuristic image. A stylized portrayal of the sun, river, and mountains, the image was perhaps meant to remind people not only of local natural amenities but also of aspirations to clean up the surrounding environment.[165] Richland saw no reason to jettison its mushroom-cloud past, but it also saw no reason to advertise the atom as part of its future; the Tri-Cities hydroplane race, run from facilities built by Hanford employees on the banks of the placid Columbia River, underwent a name change. The Atomic Cup became the Columbia Cup.

If the town's identification with the nuclear industry waned, its attachment to the West remained. By drawing attention to climate, landforms, and the Hanford mission of cleanup, the new logo gave heightened attention to the classically western natural setting. It conveyed boosters' perennial message—that tourists, investors, and new residents alike would find Richland an attractive place. Moreover, as the cleanup efforts proceeded at Hanford, entailing the development of new technologies and substantial consultation with a wide variety of "stakeholders," the rhetoric of the old West endured. Richland and the Tri-Cities no longer comprised an *atomic* frontier, but they were back in the saddle again as pioneers for America. Thus one local scholar extolled the cleanup effort: "In the new frontiers of waste remediation, environmental restoration and the preservation of democratic principles through open public involvement, the Hanford Site leads the nation."[166]

Richland's western identity had emerged before the town became associated with the atom in 1945, and it persisted after the nuclear identity had become obsolete. Similarly, the town's economic dependence on the federal government never really wavered after World War II. Although the AEC had disposed of real estate and made the community self-governing during the 1950s, Richland remained in many ways a ward of Uncle Sam. The cleanup program gave new vigor to a very troubled economy during the late 1980s and the 1990s by pumping enormous federal sums into Richland and Hanford, but the program required vigilance, particularly because progress was so slow. The DOE had promised to devote at least thirty years of effort and money to cleanup, but boosters from Richland and the Tri-Cities had to lobby federal agencies and Congress to maintain a high level of spending and employment.

The dependence of Richland on the federal government had hardly diminished. While Uncle Sam no longer owned or ran the town, its residents' prosperity depended heavily on decisions made in the other Washington. So did their identity. As the U.S. Department of Energy adopted new missions with the end of the Cold War, Richland diverged from the path of Perivale St. Andrews. It no longer helped make bombs; it instead devoted itself to cleaning up the mess that bomb makers had made. Outsiders might lump the town and Hanford together as problems for the nation to solve, but the residents of the town preferred to see themselves as solving problems for the rest of the country. In their minds Richland remained a model of sorts.

For all the differences among communities and segments of the population in the Tri-Cities and their environs, one significant source of cohesion came from the feeling that all were being dismissed by outsiders. Richland High's mushroom cloud logo, defended in response to the criticisms of the bombing of Nagasaki with a plutonium-fueled nuclear weapon, sustained the community's explicit association with the bomb itself. Hanford High in Kennewick adopted a school symbol reminiscent of that of the U.S. Air Force Strategic Air Command, which had as its motto "Peace Is Our Profession." Outsiders did not seem always to understand Hanford's mission: to win World War II and to prevent World War III by arming the United States. Environmentalist critics of Hanford in the 1970s and 1980s could be dismissed as ignorant, as lacking the technical expertise possessed by those who actually did the work and understood the technology. But it is important to recapture more than a sense of the conflict between Hanford and outsiders. As a futuristic industrial template and as a technologically utopian community, Hanford and Richland also, at times, attracted a great deal of support from the citizens of Washington, which had once styled itself "the nuclear progress state."

Three

THE POLITICS
OF HANFORD

Warfare and Welfare

IN 1951, ON THE FLOOR OF THE HOUSE OF REPRESENTATIVES, Henry M. "Scoop" Jackson, a young congressman from western Washington State, urged his colleagues to support a proposal for a six- to tenfold increase in spending on atomic weapons. His reasoning was simple: The nation faced a grave challenge from the Soviet Union, which not only possessed formidable conventional forces but was also steadily increasing its own nuclear capabilities, and to meet this threat, America had to mass-produce nuclear bombs:

> I confess to being struck by the irony of having to advance complicated and
> detailed arguments in support of an all-out atomic program. This is the best
> weapon we have—it is our one real hope of deterring Stalin. It is the natural
> weapon of a country weak in brute manpower but superlatively strong in sci-
> ence and technology. How can we conceivably afford not to go all out? How can
> we conceivably not want to make every possible atomic weapon we can?
>
> I believe that reasonable men can differ only on the degree of expansion
> that is now physically possible. In my own mind I am positive that we can
> immediately undertake to quintuple our expenditures on the atom—to spend

six billions annually. But it may well turn out that we should now increase our spending to 10 billions a year.

I cannot, however, imagine any Member of this House going before his constituents and saying that he is not in favor of making every single atomic weapon it is within our power to produce.[1]

Just as Americans committed themselves to the enormous program of mobilization against communism, as spelled out in the National Security Council's policy paper number 68 (NSC 68), Jackson's statement summed up the thinking behind the development of Hanford and America's nuclear weapons complex: mass-production of nuclear weapons represented essential protection of American democracy. Moreover, in the context of politics, making an adequate number of bombs was both a requirement and an achievement of American government. At the height of the Cold War it was simply assumed that the nation, expressing its will through the democratic system, stood virtually unanimous in its support for the rapid, large-scale manufacture of nuclear weapons. Certainly the Truman administration had held to this policy, which in different ways also resonated with political conditions in Washington State.

Jackson's view of the necessity of urgent, high-volume production of bombs stands in sharp contrast to more recent sentiments. Since 1980 or so we have been much more likely to wonder about the extent to which widespread production of nuclear weapons affected the public health and surrounding environment, and to wonder whether the nation will ever manage to clean up the wastes generated in the manufacture of nuclear weapons. In these more recent contexts, Cold War production of nuclear weapons represents not only a crisis for American democracy but also the failure of it.

Regarding Hanford, this line of thinking emerged particularly during the late 1980s after the public release of previously classified documents detailing the legacy of radioactive emissions from nuclear weapon facilities. People increasingly construed the risks presented by those emissions as something the government had imposed upon its own people without consultation or consent. "It sounds like something done in Russia," commented Tom Bailie, a farmer born and raised just downwind from Hanford who attributes lifelong health problems to his childhood exposure to radioactive emissions.[2]

What a difference forty years make. Jackson called for as many new nuclear weapons as the nation could produce, and for him the nation's response represented a triumph for democracy. For Tom Bailie, the nation's response to international threats represented a failure of the political system. A truly democratic government would never have poisoned its own people—at least not without consulting or warning them about the danger they faced. That it appeared to have done so suggested to Bailie that the United States had come to imitate the very enemy that Jackson had meant to keep at bay.

The contrast between Scoop Jackson's and Tom Bailie's sentiments exemplifies the polarization characterizing much of the discussion surrounding America's nuclear weapons program.[3] If the history of Hanford has become polarized, it is partly due to the way that the questions being asked tend to oversimplify the story. How could a democratic government poison its own people? The query reduces the history of Hanford to a futile debate over ends and means. Did the goals of the national security state—winning World War II and the Cold War—justify the methods used to attain them? The question also grossly divides the cast into those who acted (ostensibly monolithic federal agencies imposing their will in secret) and those who were victimized (the unsuspecting people who either did nothing about or simply reacted to the government's actions). The question of means and ends is posed as if Hanford's neighbors were of one mind, had no connection to the government, took no initiatives regarding Hanford, had little power to influence events, would have left the vicinity had they known what federal agencies knew about health risks, and could not really have either contributed to or been in any measure responsible for what transpired at Hanford.

If Bailie's query reduces the debate over Hanford to a simplistic dualism, Jackson's question—"How can we conceivably not want to make every possible atomic weapon we can?"—allowed for no debate at all. After all, he said, "reasonable men" could differ only on "the degree of expansion that is now physically possible," not on whether the mass-production of atomic bombs was necessary. Those in charge of American nuclear weapons programs employed a monopoly on classified information and spoke from a position of what they assumed was unassailable authority. In doing so they prevented or discouraged outside observers from asking whether other options had been considered or safety issues had been explored adequately. Moreover, such justification no doubt encouraged people to overlook the fact that advocates of such weapons sometimes had interests

other than the defense of the nation. Nuclear weapons sites created interest groups that lobbied for increased defense spending to buoy regional
economies.[4] In urging that the Atomic Energy Commission (AEC) spend
more money on building bombs, Henry Jackson argued for more federal
dollars to be spent in his home state. He did so for complex reasons: to
oppose Stalinist expansionism, to contribute to developing the economic
potential of the Columbia River system, to serve his labor and industrial
constituencies, and to insulate himself against accusations that he was soft
on communism—a routine charge made in Washington's bare-knuckles
political environment. Production of nuclear weapons was never solely a
matter of military preparedness.

While the significance of Hanford is intimately tied to the workings
of the American political system, the questions that have framed investigation of that significance have generally oversimplified the politics of
Hanford by focusing on national ends and means. Instead of positing that
the country either successfully stood up to the challenges presented by its
enemies or that it failed by not protecting its citizens from the hazardous
emissions associated with plutonium production, we can achieve a more
satisfactory picture of Hanford by looking at political relationships among
government agencies and among federal, state, and local interest groups.
By posing different questions, we can move away from oversimplified,
polarized answers.

Rather than using national concerns as the primary lens through which
to view Hanford, we begin with "the other Washington" and consider
regional and local perspectives. We argue that the people of the Pacific
Northwest were never entirely passive about Hanford. It is true that at times
they knew little about the plant and that for years they remained generally
uninformed about the health risks that Hanford presented. Nonetheless,
they came to see Hanford as an economic asset worth exploiting, and they
acted through their national and local political leaders to do so.

As citizens of the American republic, Hanford's neighbors supported
nuclear weapons production for national security. As individuals seeking
economic security, they embodied the longstanding drive in the American West to overcome the region's economic underdevelopment. They
embraced Hanford as a key to generating economic and demographic
growth in the Columbia Basin. Moreover, to maximize the region's potential, they relied on political channels to ensure that Hanford served their
interest in development. In encouraging local economic development

through federal government spending, the American political system worked remarkably well for selected local constituencies that secured a striking array of economic resources as a result of their association with Hanford. (There were numerous groups, including Native Americans, who generally were not part of these discussions.)

For the people of Washington, Hanford did not exist in a vacuum. Rather, many viewed it in the context of other ongoing initiatives to develop the state. When Scoop Jackson first considered the impact of Hanford on Washington, for instance, he saw it through the lens of a Columbia Valley Authority—a New Deal–era proposal, modeled after the Tennessee Valley Authority, to plan development of agriculture, industry, electrification, recreation, and other activities along the Columbia River and its tributaries. Later proposals for Hanford envisioned the site as a component of Washington's system of higher education and as part of the Northwest's response to the "energy crisis" of the 1970s. In other words, while Hanford is customarily viewed as a federal imposition on a western place, some Washingtonians were eager to make it part of their own agendas. And while Hanford is customarily viewed as an extension of the national security state, it also needs to be viewed within the context of the emergence of the American welfare state in the years after the New Deal.

The politicization of Hanford as a local and regional asset occurred in distinct phases. Between 1942 and 1945 the federal government did impose its atomic weapons facilities on local populations, and the army's secrecy and urgency prevented any serious negotiation between nation and region or locality. This relationship continued into the late 1940s. Early Cold War dangers resembled those that had prevailed during the war, so AEC policies regarding weapons production went largely unchallenged. There was no serious questioning of AEC plans to add five new reactors at Hanford. As Jackson explained, the nation needed to manufacture as many weapons as it could, and the congressman's constituents were not liable to complain if military preparedness meant a new regional industry pumping hundreds of millions of dollars into the Columbia Basin. To be sure, their tacit consent to the buildup of Hanford was never well informed, and the secrecy surrounding the site prevented Pacific Northwesterners from appreciating the risks to their health and environment that accompanied those dollars. As a result, the army and the AEC were able to act without much restraint in south-central Washington between 1943 and 1950, and they built and operated Hanford accordingly.

Then, during the 1950s and 1960s, the federal government's power to act unilaterally diminished, especially in regard to the plant's effect on the local economy. Hanford's future became a subject of negotiation among federal, state, and local interests, and state officials regarded Hanford as an economic asset to be protected and nurtured through political channels. Changes in the politics of Hanford stemmed from several factors. First, with the end of complete atomic secrecy after Hiroshima, Northwesterners tended to embrace Hanford and were prepared to see it as an opportunity for economic development. Second, the management of nuclear weapons programs was transferred in early 1947 from the army to the AEC, and so became a more visible focus of congressional politics. With the Joint Committee on Atomic Energy (JCAE) leading the way, Congress increasingly regarded some AEC activities as matters of conventional debate. It was Congress, for example, that pressured the AEC to "normalize" life in towns such as Richland. In order to thrive, the AEC had to placate the senators and representatives who controlled its budget and regulated its activity. This need made the commission more receptive to the wishes of regional and local interests. Third, the people who lived in areas affected by AEC activities developed into constituencies that used the political process for their own ends. Elected officials in Washington State, especially its delegation in Congress, soon discovered their Tri-Cities constituents and toiled increasingly to serve their needs, as residents of Richland, Pasco, and Kennewick developed political voices and grew steadily more confident about using them. Henry Jackson ranked as the most accomplished legislator when it came to funneling federal revenues to Hanford; mass-production of nuclear weapons had a very positive effect on the economy of his home state, and doubtless reinforced his popularity among voters.

Finally, the evolution of Hanford itself between 1950 and 1975 presented a series of "crises" for the Tri-Cities that tightened relations between local constituents and their elected officials and produced new lobbying groups that brought greater focus and energy to political efforts. In the 1950s, the transfer of Richland from the U.S. government to its inhabitants, along with the campaign for a dual-purpose reactor, allowed the actors to rehearse their roles. Lobbying on Hanford's behalf peaked in 1963–71, when the AEC shut down eight of Hanford's nine production reactors as federal spending in the Tri-Cities shifted. Until 1964 nearly 100 percent of federal spending at Hanford supported weapons work; by 1975 expenditures for nonmilitary purposes reached 75 percent.[5] Over time Hanford passed from the almost apolitical sphere of secret military necessity to the highly public realm of government-sponsored investment in domestic and local economies. In other words, it became more a creature of the welfare than of the warfare state.

To understand the politics of Hanford, then, we need to pay close attention to not only the efforts of federal agents but also the activities of state and local politicians, including governors Daniel J. Evans (1965–77) and Dixy Lee Ray (1977–81), and Senator Warren G. Magnuson (1945–81). But nobody became more closely associated with Hanford's fortunes than Henry M. Jackson, first as a congressman (1941–53) from Everett, and then as a U.S. senator (1953–83). In 1951 Congressman Jackson supported increased AEC spending at Hanford to ensure national security; in 1971 Senator Jackson lobbied for continued AEC spending at Hanford to prevent local economic collapse; in 1991 Jackson's successors advocated increased federal spending on Hanford to clean up the environment and stimulate the local economy. In every instance, the state's elected officials secured considerable federal spending for the state and the Tri-Cities, to the point that the communities near Hanford became dependent on government subsidy and regarded such spending as a right. To identify the consequences of one or the other of these federal investments or to characterize any other policy or consequence regarding Hanford as simply a "failure" or a "success" of American democracy is to neglect a host of questions about how the political system worked and which interests it served. It also overlooks the extent to which Washington's politicians and politically active citizens—especially in the late 1950s and the 1960s—spoke out on behalf of an active production program at Hanford. In this period, arguably, representative democracy worked by delivering to Washingtonians the kind of federal support for Hanford for which they had asked.

Hanford during Wartime, 1942–50

When the United States government appeared in south-central Washington in 1942–43 to build a plutonium factory, it was perceived mainly as a disruption. The army suddenly acquired 670 square miles of land,[6] condemned more than 2,000 parcels of land, and told approximately 1,500 inhabitants to leave. Many concluded that the government had behaved arrogantly. M. Grace Merrick of Dearborn, Michigan, who lost fifty-nine acres to the army, asked in a letter to the army, "Could a totalitarian gov't be more high handed?" (The army answered her question in part by launching an investigation of her character and background.)[7] Some of the dispossessed took their anger to federal court in Yakima, where they requested greater compensation for their property than the army had offered. Local juries often sympathized with the plaintiffs and awarded them more money.[8]

Racial prejudice heightened the bitterness of eviction. On May 2, 1943, the *Seattle Times* reported incorrectly that the army was building a new city at Hanford to be occupied exclusively by "Negroes . . . from the South." Some landowners doubtless objected to being kicked out to make space for African Americans. The same article alleged that there were "hundreds" of displaced people "protesting that they were getting 'worse treatment than the Japs,'" referring to those Japanese and Japanese Americans who had been confined in internment camps. "Evacuated Japs, they pointed out, received train fare, traveling accommodations, living quarters and food, but the farmers 'are getting tossed out without any consideration.'"[9] Fifty years later, some among the dispossessed families continued to contrast their experiences unfavorably to those of the incarcerated Issei and Nisei. Bernard Warby of Yakima, when arguing, "The government stole our property," claimed that "many of the displaced residents feel they are entitled to the same kind of payments that interned Americans of Japanese descent received from the government."[10]

Hanford's neighbors hardly welcomed the project with open arms. Local politicians hastened both to speak up on behalf of voters antagonized by the disruption and to deflect the blame from themselves. U.S. Congressman Hal Holmes, whose district covered southeastern Washington, made a special effort to distance himself from the Manhattan Project. Within weeks of the army condemning the land for Hanford, Holmes visited Colonel Franklin T. Matthias to explain that, because "he had not been involved in the selection of the site, he did not want any credit for it and asked that [Matthias] convey that attitude to the local people whenever the occasion demanded."[11]

As months passed the disruptions caused by Hanford seemed only to grow. The plant threatened to absorb too many workers and to drive wages too high. Holmes again visited Colonel Matthias, this time pleading the case of local growers, who feared for their ability to find and pay farm laborers. Other politicians viewed Hanford as an uninvited intrusion into the state. In 1944 U.S. Senator Mon Wallgren, a Democrat, protested that

3.1 During World War II, Senators Harry S. Truman and Warren G. Magnuson inquired about the Hanford Engineer Works in Washington State, but because of the army's secrecy were not able to learn all that they wanted to know. A few years later, in 1950, President Truman (at the bottom of the steps, looking down) and Senator Magnuson (at the top of the steps, in a lighter-colored suit) visited the Tri-Cities together. Politicians were just beginning to identify local voters as important constituents. Photograph courtesy of the U.S. Department of Energy.

project workers got too little time off work to vote in the November election in which he was running for governor. Earlier, Wallgren's opponent, Republican Governor Arthur Langlie, had worried that the depression might return at war's end. Afraid that the state might have to support any workers left behind by the federal government—particularly nonwhite workers—Langlie urged Matthias, once the work was done, to arrange "to return most of the construction workmen back to their original centers of activity, particularly the negroes."[12] No politician wanted his career damaged by whatever was being built at Hanford.

Suspicion of Hanford stemmed partly from the fact that local communities and Washington State had virtually no recourse for coping with its impact on them. It also grew from the army's secrecy. Nobody knew what went on there, therefore nobody was able to express local pride in Hanford's contribution to ending the war. Secrecy encouraged rumor and speculation, to which the army then had to reply. In January 1945 new U.S. Senator

Warren G. Magnuson told a Spokane audience that after the war DuPont would produce "nitrates, plastics, and nylon hosiery" at Hanford. This wishful thinking represented perhaps the first effort by a state politician to identify Hanford as a lasting economic asset. Boosters from Wenatchee and Pasco, thinking along similar lines, approached Colonel Matthias during the first half of 1945 to learn what the government intended to do with Hanford after the war. The army deflected such inquiries and tried to suppress Magnuson's kind of talk about Hanford's current uses.[13]

Magnuson was not the only U.S. senator to wonder about Hanford Engineer Works. An old friend of Senator Harry S Truman, U.S. district judge (and former senator) Lewis B. Schwellenbach of Spokane, was assigned to preside over the Yakima hearings regarding compensation for the lands condemned to make way for Hanford. In July 1943 he wrote Truman about "a DuPont project" at Hanford that was "the largest condemnation, so far as the area taken and the number of persons involved, that has ever been started any place." As a pretext for getting the senator to visit the Northwest, Schwellenbach invited "a personal investigation" of Hanford "by the chairman of the Truman Committee." He asked, "Why not come out and conduct it next month. It could be handled from Spokane and I still have a nice cool comfortable house where you can get a good sleep every night." Within a week or so, Truman replied, "I know something about that tremendous real estate deal, and I have been informed that it is for the construction of a plant to make a terrific explosion for a secret weapon that will be a wonder. I hope it works." In the spirit of Schwellenbach's invitation, Truman added, "I sure hope . . . that I will have an opportunity to make an investigation of this real estate deal some time in the near future."[14]

The senator was not entirely joking. Later that year an inspector from Truman's Special Committee to Investigate the National Defense Program showed up unannounced at Hanford's gate and demanded to survey the project. Matthias detained the inspector until, after checking with Groves, he refused to allow an investigation.[15] Only upon FDR's death did Truman learn the full story of the Manhattan Project. When on August 6, 1945, he announced to the world the atomic bombing of Hiroshima, he had himself known the whole secret for less than four months.

Once Hanford came partially out of the closet, public attitudes shifted dramatically. The Manhattan Project soon became a source of pride to Americans (as well as a source of anxiety).[16] Pacific Northwesterners joined in the sentiment. A *Seattle Times* story of August 10, 1945, with the headline "This State Produced War's 3 Most Destructive Weapons," touted the pride of "industry and labor" in Boeing's B-17 and B-29 bombers and "the atomic bomb." It was in this context that Mrs. J. W. Nichols had informed Colonel Matthias, "Your bombs are certainly wonderful."[17] It seemed natural for Hanford's neighbors to adopt the plant as a symbol of their own contributions or sacrifices toward winning the war. At the same time, they began to learn about its economic potential. Journalist William L. Laurence assessed Hanford's prospects in 1946: "The atomic pile is actually a three-in-one plant: It creates large quantities of plutonium. It produces a host of valuable new radioactive elements. It liberates a vast amount of atomic energy, which today goes to heat the Columbia, but promises more utilitarian applications for tomorrow."[18]

Pacific Northwesterners showed interest in such tantalizing predictions, but for most of the 1940s they exerted little or no influence over atomic activities in the region. Indeed, during World War II the army had gone out of its way to specify to Governor Langlie that it "would not accept concurrent jurisdiction with the State or County over the Hanford Area."[19] The federal government assumed complete hegemony over the site. Decisions about atomic energy, made at the top echelons of American government, revolved almost entirely around perceptions of the need for the United States to mobilize for war. If the army or the AEC considered the local contexts of its facilities at all in this era, it was to maximize the efficiency of their own production and to minimize costs. The decisions thus made led to considerable expansion at Hanford. As the AEC added five new reactors between 1947 and 1955, the population and economy of the Tri-Cities expanded to keep pace. Consequently, there was as yet little need for local interests to lobby the federal government to direct greater support to their communities. Moreover, local interests were not yet well defined or organized. Richland's population was in many ways "captive" to the AEC, while Pasco and Kennewick were still struggling to catch up to the changes wrought by the war. In contrast to Seattle, which launched a successful campaign to boost the production of bombers for the Air Force at its Boeing plants, the Tri-Cities lacked the requisite resources, incentive, and perhaps even self-awareness to lobby effectively in Washington, D.C., for "their" Hanford interests.[20]

During the early years of the Cold War, the exercise of power at Hanford remained similar to the top-down methods used by the army during World War II. The AEC and its prime contractor, General Electric (GE), assuming that the urgency and importance of their mission justified the approach, made decisions about Richland and Hanford with little or no consultation with local and state interests. The absence of full and open consultation with local and state interests prevailed in many realms of activity, including policies regarding publicizing knowledge of radioactive emissions from plutonium production. Planning for the growth of Richland during the late 1940s exemplified the imperious management style of the era. Planners basically ignored local residents and business owners, and consulted mainly with the AEC and its contractor. George W. Wickstead, the "chief planner" for GE, reflected that planning efforts for the community had been enhanced by the presence of "a central controlling, non-elective government" and by the absence of "private ownership problems."[21] In certain respects Richland was a planner's dream, for democratic politics and the market economy could not interfere with the design and layout of the town.

Through the late 1940s and early 1950s, residents of Richland remained virtually voiceless in state politics and generally isolated from other parts of Washington. Security considerations, the absence of private land ownership, and the AEC's insistence that no other entity interfere with Hanford all served to divorce the community from its political surroundings. Fred Clagett, a chemist at the plant who was active in town affairs, observed in 1952, "Richland is not a city; as far as Benton county and Washington state are concerned it is a rural area. It is not served by any national or state highway. It does not have direct representation in the state legislature. Although roughly half the population of Benton County is located in what we know as Richland, the elected office holders of Benton County do not include a single resident of Richland. For six Project years Richland has had no local self government except for the School District."[22] So long as Richland remained tightly harnessed to Hanford's production mission, its population could have little effective say in local or state politics. Increasingly, however, Tri-City residents were becoming aware both of their potential significance in state politics and of the state's politicians as advocates.

A Local Voice in National Affairs:
Hanford, Washington State, and Washington, D.C., 1947–63

Once the postwar boom was underway, the political pattern slowly began to change to allow for steadily more input from state and local interests into decisions about Hanford and Richland. Nobody outside the national-security elite had much sway concerning Hanford's basic mission, the production of plutonium for weapons, but in more peripheral areas of activity, state and local interests started exerting some influence. Furthermore, the managers of the Hanford Works paid greater attention to the sensibilities of state and local constituents. For example, Lewis F. Huck, in charge of community planning, explained in 1951 that GE would now work to "give a measure of 'uncontrol'" over town activities to convey the idea that GE and the AEC did not dominate Richland too much.[23]

The basis for the new arrangement was the 1946 decision to place atomic energy under civilian control and to allow Congress some oversight and budgeting power over the AEC. In stark contrast to conditions during the Second World War, when almost nobody knew what was happening at Hanford, by the late 1940s the daily life of Richland and the annual appropriation for Hanford became matters of public discussion. Members of Congress seldom had enough expertise or information about the AEC's technical activities to comment on them, but many considered themselves well qualified to discuss town management and to spot wasteful spending. In October 1951, for instance, a Texas congressman wrote the chairman of the AEC to complain that Richland's new master plan made public too much information about the town's layout—and also that the document itself seemed too luxurious and expensive.[24]

Such comments encouraged the AEC to rethink its relationship to towns. The commission found it awkward to be in the business of operating towns, in part because government-run communities invited criticism. Observers characterized the AEC as dictatorial, socialistic, and profligate and denounced the communities themselves as generally "abnormal." As such criticism mounted before Congress, so did potentially disruptive proposals to "solve" the problem. For example, Senator Joseph C. O'Mahoney of Wyoming suggested in 1950 that the responsibility for governing the atomic towns should be turned over to residents by the end of fiscal year 1951.[25]

In response, AEC spokespeople increasingly made it appear as if the commission was not eager to run communities and was seeking a means to relieve itself of "the burden of managing" them. Carroll Wilson, general manager of the AEC, pointed out that his agency "was not in the business of operating towns because we like to." It had not chosen or built Oak Ridge, Los Alamos, or Richland, but had "inherited" them from the army "along with the rest of the operation." Commissioner Gordon Dean claimed in 1950, with some hyperbole, that the AEC towns had generated "the largest number of headaches for us." Operating and governing Oak Ridge, Los Alamos, and Richland was costly in terms not only of money and time but also political goodwill. It is no wonder that the AEC, when it built new facilities in Idaho and South Carolina during the late 1940s and early 1950s, explicitly located its operations where existing communities, lying outside federal reservations, could absorb the influx of workers.[26] It is also no wonder that, as early as 1947, the AEC began considering how it could best divest itself of its towns.

Yet the AEC dragged its feet on actually abandoning communities because it regarded the smooth functioning of existing towns as crucial to fulfilling its mission. The commission pursued two strategies simultaneously. Over the long term and to deflect criticism, it acted publicly to free itself from the problems that town management entailed by disposing of Richland, Oak Ridge, and Los Alamos and by avoiding entanglement in any new atomic communities. It also became more responsive to townspeople's and Congress' concerns about how the communities diverged from the American norm. Meanwhile, over the short term, the AEC continued to operate towns as vital appendages of atomic-weapons work.

Once certain AEC activities became a matter of public debate in Congress, it was inevitable that the government's operation of Hanford and Richland would change. And as change took place, local and state interests attempted to influence events. The opportunity to gain a voice in municipal government began late in 1947 when the AEC, responding to pressure from Congress, announced the goal of making its communities as "normal" as possible. More specifically, the commission instructed contractors to ensure that, "consistent with security and other requirements, residents at field installations shall enjoy those facilities, services, and activities which are properly a part of American community life." The new policy implied, among other things, an eventual end to federally owned company towns and the growth of self-rule and property ownership by residents. The AEC made "democratic control of a visible local government" its "long-run objective" for Los Alamos, Oak Ridge, and Hanford.[27]

To prepare residents for self-government and solicit their opinions on issues, the AEC asked GE to create the Richland Community Council in 1948. This body of eight or nine elected members represented the townspeople's interests in consultations with GE and the AEC about community operations. One key limit on the council was that it possessed only advisory powers. The council could and did express opinions about GE and AEC policies and actions, but those opinions had no bearing on events unless GE and the AEC wanted them to.[28] Another limitation was that many council members were employed by General Electric, and it seems that they already accepted the company line or else did not wish to risk the consequences of challenging it. One early councilwoman, E. T. "Pat" Merrill, recalled that service on the advisory body "was kind of like playing house."[29]

The community council possessed little direct power, but indirectly it contributed to the successful assertion of influence by local interests. For one thing, its formation and operation suggested that there was such a thing as a local interest. Richland residents seldom got what they asked for from the AEC and GE, but because "their" council now had at least some disagreements with town managers, they could see more clearly than before that the town had needs of its own apart from what GE and the AEC wanted. For another, the council offered prospective leaders a chance to step forward and meet the public. Additionally, the monthly meetings of the council provided a forum for airing issues. In sum, the council gave the community and its concerns a focus and visibility that had not been present before.

The significance of the Richland Community Council became apparent between 1951 and 1956 as the AEC unveiled plans to "normalize" Richland. Among other things, the commission needed to establish prices for real estate in the town, devise a method for transferring local government to citizens, and commit to a program of subsidy for municipal services and schools until the new city began collecting an adequate amount of its own revenue. All of these ends, furthermore, had to be attained without unduly provoking or alienating townspeople, because the AEC continued to rely upon their good morale to ensure a high level of plutonium production at Hanford.[30] At each step in the process of disposal and incorporation, members of the community council—either individually or collectively—found themselves mediating and directing the sometimes spirited discussions over the terms under which Richland would become "normalized." In some instances the council proved more accepting of AEC and GE proposals than the rest of the town, partly because its members felt especially strongly about the need for greater autonomy in making local-government decisions.[31]

Working through the council, the inhabitants of Richland ensured that their voice meant something in the deliberations over the town's future. But by themselves the residents could not exert enough pressure to alter conditions significantly. They needed allies. They were able to cultivate these allies in part through the efforts of council members. The General Electric Company in several instances concurred with the grievances of townspeople because it wanted to keep its Hanford employees content and on the job. Just as DuPont and the army had not always seen eye-to-eye during the building and operating of Richland during World War II, so GE and the AEC disagreed occasionally about how to manage the town during the Cold War. Surely the most important allies recruited by Richland residents, however, were Washington State's elected officials in Congress. These senators and representatives frequently assumed the lead in dealing with the AEC over issues at Hanford and Richland, and they enjoyed considerable success in bending the preferences and policies of the commission to local interests and needs.

The cultivation of the AEC by local and state leaders constituted an important part of the broader transformation of the economy and politics of Washington State after 1940. Federal spending, particularly on defense and high-technology industries, has had a primary role in leading the Evergreen State away from the orientation toward extraction of natural resources that largely continued to characterize the economies of neighboring Oregon, Idaho, and British Columbia. This fundamental change—evident as much in Boeing's Seattle as in Hanford's Tri-Cities—was partly a matter of timing and luck and geography. But it was also a result of politics and, more specifically, the ability of Washington's congressional delegation to direct defense-related spending to the state and, no less important, to develop other sources of funding when military spending diminished.[32] It is true that the national security state imposed Hanford on Washingtonians without really consulting them, and that by doing so, in some sense, the political system may have "failed" regional and local interests. At the same time, however, the people of Washington and the Tri-Cities made the political system work very well for them, by deriving substantial local and state benefits from federal spending on Hanford.

The stage for their success was set during the 1930s. As in the rest of the West, particularly in the Pacific Northwest, the Republican Party had controlled most elections in Washington since statehood in 1889. The calamity of the Great Depression, however, had realigned partisan strengths. Using

the issue of public power in particular, Washington's Democrats took lasting control of the state's congressional delegation. While Oregon and many
other western states reverted to Republican control by the 1940s and 1950s,
Washington's senators and representatives remained largely Democratic
through the 1970s. This put the state in a favorable position with the party
that controlled Congress for all but a very few years between 1932 and 1994.[33]
Furthermore, the seniority of a select few politicians placed them in powerful positions in Congress where they could exert their influence effectively
on behalf of what they perceived as Washington's interests. With these connections, the Tri-Cities and the state managed to ensure that federal money
spent at Hanford went not only for warfare but also for regional and local
welfare. Because Red-baiting was a normal feature of the state's postwar
politics, it was a matter of political necessity as well as foreign policy conviction for Henry Jackson to be a strong anticommunist. At the same time, the
New Deal template of regional development, in which the Columbia and its
tributaries might play the role that the Tennessee River and its tributaries did
in the hydro-nuclear regional economy of the Southeast, was important as
well. Washingtonians' strong tendency was to mix the two.

Many Washington politicians participated in lobbying Congress and
the AEC on behalf of state and local interests. The state's three governors
between 1957 and 1981—Albert D. Rosellini (1957–65), Daniel J. Evans
(1965–77), and Dixy Lee Ray (1977–81)—eagerly supported atomic energy.
Evans had been trained as an engineer, and Ray had been a university scientist and—under President Nixon—chair of the Atomic Energy Commission. As enthusiastic as these governors were, however, they never exerted
much influence over development in the Tri-Cities. Until 1989 Washington
State failed to provide the one major public institution requested of it by
the Tri-Cities—a university.[34] The vast majority of money available to Hanford came from the U.S. government. Consequently, the Tri-Cities long
remained a federal enclave, much more attuned to affairs in the nation's
capital of Washington, D.C., than to those in the state capital of Olympia
or in western Washington generally.

[handwritten margin note: federal enclave]

At the federal level, Hanford had many strong friends. A stream of
representatives from Washington's fourth congressional district, which
included Hanford and Richland, labored on behalf of Hanford and the
Tri-Cities.[35] The most devoted may have been Mike McCormack, the Richland Democrat who earned the nickname "Atomic Mike" while serving
in Congress from 1971 to 1981. And once Senator Magnuson figured out

that Hanford would not be producing nylon hosiery after World War II, he too made the site a focus of his enormously successful campaign of pork-barrel politics on the Evergreen State's behalf. One key to his effectiveness was longevity. Elected to serve in the Senate for six straight terms, between 1944 and 1981, Magnuson rose to become chair of the Committee on Appropriations. Walter Mondale (D-MN), who had served in the Senate alongside "Maggie," once described Magnuson's formula "for evenly dividing the federal budget among the states: 'Fifty percent for Washington State, and fifty percent for the rest of the country.'"[36]

Nobody, however, surpassed Henry M. "Scoop" Jackson in advocating for Hanford and the Tri-Cities in the nation's capital. Jackson served as congressman from western Washington from 1941 to 1953, and then moved to the Senate, where he remained until he died in 1983. During the war, Congressman Jackson served on the influential House rivers and harbors committee, and there he began to articulate a view of the Columbia as an economic engine for the state. Because of his interest in both defense issues and Washington's economic development, Jackson sought and won a seat on the Joint Committee on Atomic Energy in 1949. The next year he began appearing in eastern Washington at Tri-Cities press conferences to announce new appropriations, construction, and jobs for Hanford.[37] Ambitious to gain election to statewide office in the 1952 senatorial election, the representative was already milking the Tri-Cities for votes. His 1951 call for a six- to tenfold increase in the AEC's budget doubtless played well in Richland, Kennewick, and Pasco.

Jackson lost his seat on the Joint Committee in 1953 as he jumped to the Senate and as the Republicans took brief control of Congress, but he regained it in 1955. From that position, and from his post on the Senate Armed Services Committee, he maintained a keen interest in the nation's nuclear weapons facilities. This interest, of course, reflected Jackson's deep-seated concern about a strong defense for the United States, but it also served him well in his efforts both to use federal spending to develop Washington and to represent effectively the interests of state voters who lived in the Tri-Cities area. Known as "the senator from Boeing," Jackson could equally well have been called the senator from Hanford. Local boosters marveled at his and Magnuson's abilities to advance or protect legislation favorable to Hanford: "Somehow, Senator Henry M. Jackson . . . and Senator Warren G. Magnuson always pulled a rabbit from their hats."[38]

Jackson's labors regarding Hanford and the Tri-Cities took many forms.

3.2 During the Cold War, as elected officials from Washington state came to recognize Hanford as an important economic asset, no politician became more closely identified with Hanford than Henry M. Jackson. This undated photograph shows Jackson at a banquet in Richland's Desert Inn; next to him is Senator Brien McMahon (D-CT). Both belonged to the Joint Committee on Atomic Energy in the U.S. Congress, which Jackson had joined in 1949. The photograph most likely dates from the period 1949–1952, when Jackson was still a congressman representing a district centered on Everett on Puget Sound—hundreds of miles away from the Tri-Cities. Only in 1953 did Jackson become a U.S. senator and start to represent the entire Evergreen State. Photograph courtesy of the U.S. Department of Energy.

He responded to letters from state voters aggrieved by one AEC action or another. When farmers just outside the Hanford Site wanted access to the lands that had been set aside as a buffer around the reservation, Jackson led their 1957 campaign to have the AEC release the acreage.[39] Staff members of the commission became accustomed to his inquiries and learned over time the need to respond promptly, and in duplicate, to the senator's requests.[40] On occasion Jackson criticized AEC policies or procedures roundly, and he also proved willing to push the commission into areas in which it was reluctant to go, including domestic uses of atomic energy.

Perhaps Jackson's most striking single day of lobbying on behalf of Tri-Cities interests occurred on June 21, 1956. First, he helped Richland residents buy their houses from the AEC. The townspeople had been in an uproar since May, when prices for the homes had been announced.

The prices were too high, and residents had protested in dozens of letters fired off to Magnuson and Jackson. The senators had arranged for hearings before a JCAE subcommittee, where Jackson launched a blistering attack on the FHA procedures that had been used to develop the appraisals of Richland houses. Another round of appraisals was ordered that lowered the prices for about half of the homes in town.[41]

On the same day, Jackson introduced legislation to authorize construction of the dual-purpose N reactor at Hanford.[42] Tri-Cities residents were eager to diversify their economy; one of the reasons they wanted lower housing prices was because they feared that the Tri-Cities depended too much on Hanford. So long as Hanford produced simply weapons-grade plutonium using comparatively obsolete reactors, its future seemed limited. The N reactor promised new options, with a newer technology for producing plutonium that would make it less vulnerable to shutdown. Moreover, the N reactor was designed to generate electricity. While producing fuel for nuclear bombs, it could generate kilowatts for distribution via the Bonneville Power Administration (BPA) network. Over time, it was thought, Hanford could also become a center for research and development in the atomic power field.

Awareness of Hanford's ability to contribute to regional prosperity dawned on the Pacific Northwest during the 1950s. During the spring of 1956, Fremont E. Kast, a doctoral candidate at the University of Washington's College of Business Administration, wrote, "One of the most spectacular developments in Washington's manufacturing economy, has been the growth of the Hanford Operation. . . . It has transformed an isolated region into one of the major centers of economic activity within the state. Benton County has advanced from a position of little importance as an industrial region . . . to the fourth largest manufacturing county in 1953. The project has created entire new communities and new markets."[43] Part of the significance of the site, Kast continued, lay in the magnitude of the nation's investment in nuclear technology, calculated by the AEC to be $13.158 billion between 1943 and 1955. At Hanford, while construction employment waxed and waned, nuclear technology supported stable "manufacturing employment" for over 9,000 workers. "Even with a decline in international tensions," Kast concluded, "the industrial potentials for atomic energy are sufficient to justify the continuation of this operation. . . . Hanford and the community of Richland have assumed a permanent place in the state's economy."[44]

Washington State's congressional delegation came to view Hanford in part through the lens of the development of the Columbia River. Scoop Jackson found plans for a Columbia Valley Authority, modeled on the Tennessee Valley Authority, in train when he arrived in Washington, D.C., in 1941.[45] By 1945 Jackson had joined Washington Senator Hugh B. Mitchell in proposing legislation to establish the CVA. Even though Jackson represented Washington's second congressional district, west of the Cascades, he committed himself to the bill. As a member of the House Committee on Rivers and Harbors, he felt that the proposal fell within his purview. He also received letters of support for the proposal from all over the state, many from his constituents in northwest Washington who believed that a CVA would guarantee the continued development of public power. Jackson took credit for tying Snohomish County to the Grand Coulee generators via an electricity transmission system, guaranteeing industries in his district a low-cost power supply. Mitchell, the Senate sponsor, and Jackson, the measure's sponsor in the House, argued that the CVA would promote employment among veterans and war workers, who were leaving shipyards and airplane factories at the end of 1945. A press release from Mitchell's office promised the state's citizens that "important decisions will be made at home rather than in Washington, D.C., three thousand miles away. With a CVA great regional progress can be made, encouraged . . . through the cooperative development of the resources of that great river basin."[46]

Though the proposal died early in 1946, Jackson would revive it in different forms throughout his career. Rooted in the New Deal effort to build the great dams at Grand Coulee and Bonneville, in the establishment of the Bonneville Power Authority as a distributor of low-cost power to Northwest cities and industry, and in a vision of a more densely populated and prosperous region after World War II, the CVA was a legacy of the 1930s carried over to the postwar world. Planning for use of all of the Columbia's resources, and particularly the development of its hydropower potential, was vital, CVA's backers argued. As a congressman, Hugh B. Mitchell (having lost his Senate seat and having joined Jackson in the House in 1949) made the point to a committee considering the CVA proposal: "Our region rests on a too-narrow economic base. Possibly 7 out of every 10 gainfully employed in the Pacific Northwest are directly or indirectly dependent upon forest industries for their livelihood. . . . Our people are keenly conscious of the need for economic diversification, or broadening the base for industry, for agriculture and for commerce."[47] The Columbia drainage represented a resource for the region's industrial development.

CVA proposals attracted an assortment of proponents and opponents. Labor unions and Grange chapters joined public power advocates in endorsing the plan. Jackson reported in 1950 that chambers of commerce were against such plans, while sportsmen's groups were divided, "although a majority of them seem to feel that CVA would interfere with fish and wildlife protection." The earliest and most enthusiastic constituency for the plan were aspiring farmers—particularly those veterans who by law were given priority in selecting irrigated acreage.[48] As a consultant for the Bonneville Power Administration during the Depression, social critic and regional planner Lewis Mumford had sketched out a future of family farms, well-dispersed industry, and democracy that the region's hydropower would make possible. This vision persisted; after articles on the CVA in *Liberty* magazine and the national journal of a World War II veterans' organization, *AMVET*, Jackson's Washington, D.C., office was deluged with letters from veterans seeking information on homesteading in the Columbia Basin. While the bulk of newly irrigated lands would not be available until 1951, Jackson promised in the fall of 1946 that 6,000 acres of land irrigated directly from the Columbia near Pasco would be available in 1947. By 1952, demand for newly irrigated land remained so high that only veterans had been able to secure land allotments, leaving out others who hoped for farms of their own. Ultimately, Jackson and the Bureau of Reclamation announced, there would be 5,000 new eighty-acre farms in the Columbia Basin.[49]

"The conversion of waste lands into fruitful homesteads," in the words of one of Jackson's correspondents, would require juggling the interests of all claimants to the resources of the Columbia Basin, including Hanford— for which an important resource had always been a large, depopulated area in which to conduct its operations. Looking away from Hanford and Richland, Pasco quickly declared itself "The Largest City in the Columbia Basin Project." The Quincy Chamber of Commerce wrote to Jackson protesting the idea of federally planned cities in the Columbia Basin—in this case, planned by reclamation authorities, rather than the army—which threatened to divert business away from the area's established communities. "As a member of the Joint Atomic Energy Committee I am keenly aware of the removal of farm land from production in your area because of the Hanford Plant," Jackson wrote to a concerned farmer in the Kennewick Irrigation

District in 1951. "The Kennewick Project is an excellent way to compensate for this loss," while also providing a "12,000 [kilowatt] increase in firm power." Hanford was a crucial part of the picture, but only one piece in the complicated jigsaw puzzle surrounding the Columbia's big bend.

Finding a place for Hanford in his plans for the Columbia Valley, Jackson began to develop his stance as a Cold War liberal. Opposed to communism abroad, he would embrace a measure of "socialism" at home, in the form of public power. "I have not had any contact with [the electric utilities] lobby for the simple reason that my views and interests do not coincide with theirs," he wrote a Reed College student who sought his help with a term paper on the subject. "Their primary interest has been to further their hold on the private power field and, to my knowledge, have never made any attempt to help . . . the great public power development in the Columbia Basin." Finding a postwar political identity during President Truman's run for reelection in 1948, the young congressman defined postwar liberalism as FDR's successor did, as "vigilantly internationalist and anticommunist abroad but statist at home, committed to realizing the New Deal–Fair Deal vision of a strong, active federal government presiding over the economy." Running against Washington senator Harry P. Cain in the 1952 elections, Jackson argued that Cain was weak on national defense and an opponent of public power.[50] In contrast, Tri-Cities boosters such as *Tri-City Herald* editor Glenn C. Lee saw their community as the potential "powerhouse of the Pacific Northwest" because of public power and the defense industry at Hanford; the paper depicted itself as connected to both the nuclear plant upstream on the Columbia and the McNary Dam downstream.[51] Hanford, in the context of a rationally developed and managed Columbia River system, was part of a coherent larger vision, which conditioned Jackson's support for particular programs on the site, as well as his concern for Hanford's agriculturist neighbors. CVA legislation gave pride of place to Hanford; in the event of conflicts over water resources, they said, "preference shall be given to atomic energy requirements for national defense and to domestic, irrigation, mining and industrial purposes."[52]

Hanford could even have been useful in inoculating Jackson against criticism from the right. In July 1950, Jackson rose on the floor of the House to attack the editors of *Reader's Digest*, who had called the regional BPA administrator "another socialist" in an article critical of the CVA plan. The author of the piece, a former Wyoming governor, insisted that he had not made the charge and that the words must have been inserted after the text left his hands. Attacking the magazine's editors for a lack of journalistic integrity, Jackson reported that he had seen polling data collected by the National Association of Electric Light and Power Companies, an anti-TVA private power organization. The data showed, Jackson said, that Americans supported public power and the federal development of hydro resources, but they were against "socialism." *Reader's Digest* now made private power's strategy plain, Jackson asserted: attacking valley authorities as "socialist" would serve selfish interests, no matter who was defamed in the process, and would serve as a coded assault on public power.[53] But surely neither private power concerns nor Republican insurance agents, such as Jackson's electoral opponent, could oppose federal industrial development on the Columbia for the purpose of producing fissionable material.

Mixing Hanford into the CVA could help keep the latter project afloat. But ultimately the CVA had to do with a populous, ever-more-wealthy Washington. Some Washingtonians saw the struggle as one between a capitalist East and a democratic West. "History is running true to form in *Reader's Digest*," wrote the Pasco postmaster, canceling his subscription after the magazine's anti-CVA article. "It is the Industrial East battling the fast growing West. The fight was on at the time the Constitution was adopted and seems to never die . . . we out here cannot support that which is trying to curb our growth and lower our standard of living."[54] Across the Cascade range, the head of the Seattle longshoremen's union was just as adamant. "We want C.V.A.," he declared. "We want to cease being a colony for Eastern interests to utilize as a source of supply for their industry. We want to see our part of the nation industrialized in its own right."[55] These sentiments would be appealed to again at the end of the decade, when the public power ideal was called upon in support of Hanford's N reactor, as a regional investment facing defeat by the forces of Eastern capital. "Jealousy, Selfishness Only Threats to Hanford Power Plan," editorialized the Wenatchee *Daily World* in June 1962, echoing local sentiment in support of the CVA from twelve years before.[56] Washington politics, until the early 1960s, was defined by Red-baiting; for Jackson, supporting Hanford in the midst of the Columbia Valley was a prime example of mixing the values of the New Deal liberal and the Cold War liberal. The N reactor in particular would combine both of these by producing the raw material for nuclear weapons and cheap, abundant electricity for the Bonneville Power Administration.

Proposals for a reactor to produce both plutonium for weapons and steam for electricity generation, though, depended not just on regional demand but also on industrial interest in the mid-1950s. Here Hanford commended itself as a suitable test site for plans that would use pluto-nium production to subsidize the cost of nuclear power generation while the civilian technology was under development. As early as February of 1945, DuPont executives had seen the nuclear industry as too risky to venture into with the firm's own capital, but they hoped to be retained to do research and development with government funding, remaining well positioned to enter the field when its commercial potential was better realized.[57] In 1954, General Electric, planning to develop civilian nuclear plants to produce electricity at costs comparable to coal-burning facilities, lobbied the AEC for a contract to build a dual-purpose pile at Hanford, arguing that "the Company's plans would benefit the Northwest by making Hanford a source of, rather than a drain on, the power supply . . . further reducing the cost of plutonium for defense purposes. . . . Furthermore, we will learn to operate on an economical basis a truly large atomic fueled power station supplying a large block of dependable power."[58]

While boosters of nuclear energy believed that experienced engineers could easily render atomic piles into heat engines for electrical generating stations, those with more realistic experience were not so sure. The uncer-tainties associated with the technology encouraged some in the nascent nuclear industry to look at Hanford to provide a model for the economi-cal production of electricity; the site's reactors might be technologically backwards, but they had well-established construction and operating his-tories, and their reliability, in turn, suggested that the Hanford staff could be relied upon to get the best out of a new design.[59]

While Hanford's reactors operated at ever-higher powers after 1949, they seldom operated at temperatures sufficiently high for steam power generation. The water in the Hanford piles' process tubes served as a sec-ondary moderator, and turning it directly into steam would lower its neu-tron absorption, leading to instabilities in the chain reaction. By putting the water in the pipes under a pressure of 3,206 pounds per square inch, it could be prevented from turning to steam, and superheated water could be run through a heat exchanger to provide steam for a generating plant. But such pressures would require beefing up the tubes themselves, again interfering with the dynamics of the reaction. In turn, the higher tempera-tures would increase the degradation of the aluminum itself, again threat-ening the integrity of the process tubes as well as the mechanical strength of the whole pile—a difficulty that could be met with additional structural support, if it were known how the necessary braces would behave under neutron bombardment and how the pile would respond to the presence of the new material. These interlocking problems were so complex that Hanford had done well to buy its electricity (either in the form of coal or hydropower from the Bonneville Power Administration) and stick to the production of plutonium alone.[60]

Still, business advocates saw dual-purpose reactors, designed to produce both plutonium and electrical power, as the point of entry into the private nuclear arena. They argued that the legally mandated monopoly held by the federal government over fissionable materials should not extend over all of nuclear power. Given sufficient attention to the technical problems, and given the fact that production piles naturally produced heat along with plutonium, cogeneration schemes seemed like an obvious way to bring the atom into the electricity marketplace. Under such a plan, a private company would build a reactor and power generating station, and the federal government would provide uranium fuel and retrieve it, process it to extract plutonium, and then pay the operating company for the plutonium while outside customers paid for electricity. Such plans were commonplace in the industry literature at the beginning of the 1950s.[61]

But while Hanford produced plutonium according to a government plan, how could plutonium producers have confidence in the existence of a market for their wares? Following the lead of the head of the Joint Committee on Atomic Energy, Senator Brien McMahon (D-CT), in 1952 Jackson agreed that "there is no such thing as 'enough' atomic bombs." This truism, which had underlain Hanford's continuing production mission, in Jackson's thinking now provided a justification for its extension to a new product. "This means," he continued, "that industrial firms which build dual-purpose nuclear reactors have an unlimited market for the plutonium they produce." Four years later, using the same argument, Jackson brought plans to the Senate floor to force a reluctant AEC to build a dual-purpose reactor at Hanford, in line with GE's plans.[62]

Getting authorization from Congress and the administration to start building the new reactor at Hanford proved to be no simple task. For its part, the Eisenhower administration wanted to see the development of civilian nuclear power as the centerpiece of its Atoms For Peace initiative, a campaign to redeem the military atom in competition with the Soviet Union. Eisenhower insisted, however, that the new technology be developed under the aegis of private industry, not public power. Moreover, the idea of a dual-purpose reactor built with federal funds clashed with the interests of private power companies and the coal industry. While the AEC opposed a "kilowatt race" that might mortgage the long-term rational development of reactor technology to compete in the short term with empty achievements overseas, in 1956 Senators Albert Gore (D-TN, representing Oak Ridge and the TVA) and Henry Jackson (of Hanford and the Columbia River valley) insisted that government should take the lead in building a range of prototype plants, including a model built on the Hanford graphite-moderated model. The reactor development leader, Walter Zinn, who was retiring from Argonne Laboratory, noted that the Hanford reactor type had been overlooked by industry but might be an option for the economical production of electricity, since it certainly represented a well-understood technology. An ongoing series of studies predicted that a dual-purpose reactor at Hanford would produce plutonium at a lower cost than a single-purpose production pile, although also at a slower rate.[63] The "new production reactor" (NPR), later known as the N reactor, with its cooling water pipes to run at greater pressure to allow electrical generation as well as plutonium production, was funded for planning purposes in 1957 and for construction in 1958.[64]

Even this limited success required modifying the proposal for the dual-purpose reactor to make it more politically acceptable. Jackson and others built a stronger case that the government needed more plutonium, and that the dual-purpose reactor would be an economical method of adding to the supply because the sale of its steam power would help offset the costs of producing the fissionable material. Advocates of a new Hanford pile also deleted from their proposals the power-generating portion of the design, designating the reactor as "convertible" without securing commitment of funds for building a power plant at this stage. With these changes, with more encouraging engineering reports on the feasibility of the project, and with a greater desire to appease the JCAE and encourage nuclear power plants to be built on any terms, the AEC changed its position and supported Jackson's bill.

Once construction of the new production reactor had been authorized, Jackson and legislators from the Pacific Northwest continued to press for the addition of power-generating facilities, here appealing explicitly to the ideology of public power and regional industrial development. In March 1959, Washington's state senate passed a resolution asking for an electrical generator to be added to the new production reactor, noting that such a

3.3 This is a westward-looking view of the N reactor under construction, perhaps in 1960. The subject of intense debate in Congress, N was conceived as a dual-purpose reactor, capable not only of making weapons-grade plutonium but also of generating electricity for regional markets. Once political opposition to the kilowatt-producing side of the project was overcome after 1960, the Tri-Cities and Washington's elected officials felt confident that the local economy was on its way to becoming successfully diversified. The photograph also gives a sense of how Hanford's successive, massive construction projects re-engineered the land within the Hanford reservation. Photograph courtesy of the U.S. Department of Energy.

facility would take Hanford from the position of draining 300,000 kilo-watts annually from the BPA to contributing 700,000 kilowatts to the system. But the AEC pulled away from the idea of a convertible reactor in early 1959, insisting that such a facility would not produce electricity at a competitive rate, and so designing and building for such an eventuality was a waste of resources. "Our studies have convinced me that only the most optimistic assumptions . . . will make this project attractive as a power producing unit in the Northwest," the AEC chairman informed Jackson. The senator's staff, for its part, did not agree. "The AEC hired an eastern engineering consulting company . . . which produced an adverse feasibility report," reported Russ Holt, one of Jackson's aides, to the editor and publisher of the Longview *Daily News*. "Their findings were based on several questionable assumptions, including one that the power needs of the Pacific Northwest would not increase appreciably in the years ahead and that there would be little need for the electricity from this reactor during a major portion of its working life." One answer to the AEC's recalcitrance was to mobilize Northwest voices in support of the project.[65]

Fighting to recast the NPR (new production reactor) into a DPR (dual-purpose reactor), Jackson and his staff could draw upon the same rhetorical template employed in the unsuccessful battle for the Columbia Valley Authority bills. In the minds of Northwesterners, resistance to funding for a dual-purpose reactor at Hanford had deep roots. Now that Washington's atomic industrial plant might go into the power business, the old opponents — veterans of New Deal struggles over hydroelectricity-generating dams — appeared as predicted: private utility interests and legislators representing eastern coal-producing states. In the first month of the AEC's existence, Hugh B. Mitchell, a former U.S. senator and Seattle financier, had written to David Lilienthal urging him to be true to the tradition of the TVA and to keep eastern corporations out of nuclear power; the new technology, he believed, could only live up to its potential of social transformation if not shackled to the old economic structures of corporate capitalism. Public-power utilities in the West had been early proponents of government-sponsored nuclear power projects, in the tradition of the hydro projects of the 1930s. Following the cues of Holt's letter to the *Daily News*, newspaper editorials in Washington and Oregon were quick to identify this as a struggle over the Northwest's rightful destiny, a destiny which should have rested on abundant electrical power without the attacks of jealous Easterners, who too quickly assumed a static economy for the region.[66]

Public-power aspirations took a concrete institutional form in the Washington Public Power Supply System (WPPSS), which had been founded in 1957 to develop hydropower sites on behalf of the state's public utility districts. With headquarters in Kennewick, the WPPSS lobbied for the development of the Columbia's hydro potential, urging development of the Ben Franklin Dam near Richland, for example. In the absence of a formally established CVA, the Columbia Valley continued its piecemeal development by federal agencies, requiring lobbying and legislation for each project to go forward. Faced with the AEC's declaration that a DPR was not feasible, within a month Jackson turned to WPPSS managing director Owen Hurd, whose staff prepared a proposal of its own: that the NPR would produce weapons fuel for two years; plutonium for weapons and electricity for eight years; and power plus plutonium reactor fuel for the rest of the facility's life, about twenty-five years. Hurd acknowledged that both public and private power companies would have to have access to the resource, anticipating objections from private-power interests in Oregon and the Spokane area. Disappointed that the AEC would not allow the utility to operate the reactor while it was producing weapons fuel, Hurd still argued, against the AEC's view, that using a reactor to produce electricity subsidized by plutonium production would pencil out.[67]

The debate was not settled until after the 1960 election. The Kennedy administration put through funding for a generating facility for the N reactor, but only after significant further revisions to the plan and continuing pressure from figures such as Mark Hatfield, Oregon's Republican governor, who urged that "Hanford's electric energy potential should be harnessed for the benefit of the people of the Northwest." First, in the final architecture of the project, local public utilities, through the WPPSS, would build and pay for the generating facilities on land leased from the AEC. Second, the utilities would buy the steam produced by the reactor from the AEC. And third, as Hurd had proposed, the WPPSS would offer to sell half of the DPR's electrical output to private power companies, which assumed none of the risks of building and operating the plant. With these added conditions, Congress finally approved the dual-purpose role for the N reactor in 1962.[68] By 1964 the new reactor had begun to produce plutonium, and by 1966 a steam generator operated by the WPPSS had begun operations. Through the exercise of local and regional political clout, Hanford had added both new technology and a new mission of generating electricity.[69]

Tri-Cities support for the N reactor had stemmed largely from the expectation that the new project represented a form of economic diversification. More than 30,000 people found reason to celebrate the new departure on September 26, 1963, when President Kennedy visited Hanford to dedicate the N reactor and to help launch construction of the steam-generating facility. In a ceremony reminiscent of Richland's independence day nearly five years earlier, Kennedy waved another "uranium-tipped wand over a Geiger counter to activate a huge clamshell bucket which performed the actual groundbreaking." The president used the occasion to link Hanford's atomic frontier with new frontiers for the nation. First, he said, it was fit-

3.4 On September 26, 1963, President John F. Kennedy visited Hanford to dedicate the N reactor, which had recently begun to manufacture plutonium, and to help launch the construction of the steam-generating facility that would produce electricity for the Pacific Northwest. Kennedy stands at the center, with his head bowed. Senator Henry M. Jackson is directly left of Kennedy, holding papers in his hand. Washington State's Governor Albert Rosellini stands to the right of Kennedy, and Senator Warren G. Magnuson's head appears second from the end on the right. As members of the Democratic Party, all of these politicians were sympathetic to the idea of a dual-purpose reactor. Photograph courtesy of the U.S. Department of Energy.

ting for the N reactor to "strike a blow for peace" in a place "where so much has been done to build the military strength of the United States," signaling a turn toward the peaceful atom. Second, Kennedy declared that the dual-purpose reactor represented the capacity of science and technology to help the nation conserve its resources, by which he meant "use our . . . resources to the fullest." Low-cost nuclear power would help Americans meet their ever-growing demand for electricity, which Kennedy predicted would create the need to double the electricity supply every ten years. Third, as a new source for electricity, Hanford's N reactor would help Americans maintain and increase their affluence. "This country," the president promised, "will be richer and our children will enjoy a higher standard of living."[70]

By the time the N reactor began operations in the mid-1960s, however, the prospects for economic diversification in the Tri-Cities were seriously threatened. In January 1964 President Lyndon B. Johnson announced the closing of several of the AEC's plutonium-producing facilities, and Hanford's reactors stood atop the list of those marked for shutdown.[71] One count against Hanford was the durability of its product; the half-life for weapons-grade plutonium is about 24,000 years. By the early 1960s— barely one decade after Jackson had insisted that the United States needed to manufacture as many nuclear bombs as it possibly could—the United States calculated that it had enough plutonium to last awhile. Another count against Hanford was its dated technology. The product it had manufactured would not become obsolete anytime soon, but the piles that made the plutonium were aging severely and, unlike the newer AEC facilities at Savannah River, South Carolina, they remained relatively unsuited to producing valuable isotopes besides plutonium.[72] Thus it was not a difficult decision to plan to close Hanford's reactors. Between 1964 and 1971 the eight piles built between 1943 and 1955 were shut down. Between 1971 and 1986 only the N reactor remained in operation, producing mainly electricity rather than weapons-grade plutonium.

The N reactor may have helped diversify the local economy, but it had not helped diversify it enough. The anxieties of Richland residents spoke loudly to the economic risks faced by towns near Hanford. In the weeks after President Johnson's announcement of reactor shutdowns, one federal observer warned that without additional help from the AEC, the cutbacks proposed for Hanford would create in Richland a downward "spiral of delinquency, slums, crime, broken families, poorer health, and greater unemployment."[73]

Such threats seemed to justify many of the concerns townspeople had expressed a decade earlier. When the AEC had set about "normalizing" Richland during the 1950s, it had met considerable resistance from residents who feared for the future of the Tri-Cities' one-industry economy. Richland residents were not eager to buy homes in a place where local prosperity was subject to the unpredictable nature of politics, diplomacy, and technological obsolescence. In the 1960s, even with the diversification that accompanied the dual-purpose N reactor, it became apparent that these concerns had been justified. The closure of eight of the site's nine reactors exemplified precisely the kind of instability residents had worried about when they considered purchasing their houses.

In downplaying local fears in the 1950s and selling off Richland's homes and businesses, the AEC had created a community of homeowners and businesspeople with a much more profound economic stake in the area—individuals who cared not only about a rising unemployment rate but also about falling property values. Consequently, in the 1960s, when the AEC began to shut down reactors, its actions represented a more sizeable economic threat than would have arisen without disposal. Had Richland's residents remained tenants of the government, they might have found it easier to adapt to the eventual AEC cutbacks at Hanford. But as owners of homes and businesses, the people of Richland had a much more urgent reason for demanding that the AEC work to mitigate the impact of its reactor closings. By the 1960s the AEC had created a class of property owners who pressured the commission for the subsidies necessary to protect the value of Richland's privatized real estate.

Citizens from Pasco and Kennewick readily joined Richland residents in their lobbying effort. A defense department report in late 1963 had estimated that 80 percent of the jobs in the Tri-Cities area depended directly or indirectly on Hanford. An official with the Department of Defense, experienced at assessing the local impacts of military spending, found the Tri-Cities region "more completely dependent on Government payrolls than any other with which the Office of Economic Adjustment has had contact with." He also remarked upon "another factor of great importance": "the community's total *psychological* dependence upon Federal activities."[74] Now it seemed that the basis for economic sustenance and psychological identity was in serious jeopardy. To solve the crisis, local and state interests turned once more to politics and to Senators Jackson and Magnuson. Lobbying the AEC and Congress on behalf of Hanford as an economic asset to the Pacific Northwest reached a new level.

Local Boosters, Federal Programs, and "Nuclear Pork": Segmentation and Diversification at Hanford, 1963–71

Local efforts at lobbying were spearheaded by Tri-Cities boosters who, having emerged gradually in the late 1940s and the 1950s, stood fully prepared during the 1960s for the kind of political campaign necessitated by the closure of reactors at Hanford. These boosters possessed the resources required to make an impression at the federal level, including a chamber-of-commerce organization, a newspaper, and funds with which to hire professional lobbyists. Once they had cultivated Senators Jackson and Magnuson, too, they were assured of some success in mitigating the effects of the reactor closures. Their efforts resulted in bringing to the Tri-Cities what one historian calls "nuclear pork," that is, federal subsidies negotiated through the pork-barrel politics of the U.S. Congress.[75]

The boosters leading the diversification crusade got their start locally in 1947 with the founding of the *Tri-City Herald*. The *Herald* had two early competitors—the *Richland Villager*, a weekly subsidized and dominated by GE and serving the company town, and the *Columbia Basin News*, a Pasco paper published between 1949 and 1963 that, even after merging with the *Villager* in 1950, remained more oriented toward the agricultural district than the urban and manufacturing complex created by Hanford.[76] The *Herald*, after some acrimonious competition, outlasted the other papers in part because of its devotion to an urban readership and its focus on the industrial and metropolitan aspects of life along the Columbia River.

But the *Herald* did far more than cater to readers in an urban center. It helped create that urban center by devoting itself to boosterism, and it encouraged readers to accept for themselves the idea of a single, metropolitan identity. From its beginning, the *Herald* was never satisfied merely to "write about momentous events" affecting the communities, one early employee recalled. The paper did not think it was doing all it should unless it was making news, too. Its mission was not only "to promote the development" of Pasco, Richland, and Kennewick but also "to provide a balance and diversity in the economic background" of the area.[77]

The very naming of the newspaper exemplified its efforts to boost the metropolis and shape its future. Before 1943 Pasco and Kennewick had occasionally been called the "Twin Cities," and the surrounding area had sometimes been dubbed "Three Rivers" in acknowledgment of the confluence of the Columbia, Yakima, and Snake. But few thought of Richland, Pasco, and Kennewick as having much in common until publishers named their paper the *Tri-City Herald*. The publishers insisted on this local neologism partly because they aimed to advertise and circulate in all three towns but also because they aspired to treat Pasco, Richland, and Kennewick equally and to encourage "unified action" among them.[78] The newspaper even lobbied the U.S. Bureau of the Census to recognize the Tri-Cities as a "standard metropolitan statistical area" in order to advertise the region, and supported unsuccessful initiatives that would have joined the three towns (located in two separate counties) into a single municipality.[79] More than perhaps any other factor, the *Herald* was responsible for inscribing the idea of a "Tri-Cities" in local residents' minds.

3.5 The *Tri-City Herald* saw one of its tasks as promoting the metropolitan area of the Tri-Cities to prospective immigrants and investors. Without any apparent irony intended, this 1958 ad identified the urban region as Washington State's "hottest" market, based on the completion of additional dams on the Columbia and Snake rivers, the growth of irrigation from the Columbia Basin Project, the privatization of real estate in Richland, and continued expansion at Hanford, including authorization to build the N reactor. University of Washington Libraries, Special Collections; courtesy of *Tri-City Herald*.

The publishers of the *Herald* initially adopted a somewhat antagonistic stance toward GE and the AEC, partly out of rivalry with the *Villager* and partly because they possessed their own independent vision for local economic development. In this way, the paper, like the Richland Community Council, helped residents of the Tri-Cities identify community interests distinct from those of the AEC and its contractors. To fulfill its boosters' agenda, the *Herald* also lobbied Congress on behalf of the dual-purpose reactor in the late 1950s and early 1960s. The experience gave publishers and editors more exposure to the political process and solidified commitment to diversification. A January 1962 editorial summarized their concerns: "The atomic field is expanding but Hanford is standing still,

anchored to aims set nearly two decades ago. . . . Without diversification . . .
we are doomed to see . . . the atomic age pass Hanford by."⁸⁰

The Atomic Energy Commission, cognizant of the problems facing
the communities it had created, had also begun considering the matter of
diversification and in June 1962 held hearings on the subject in Richland.
The hearings produced ample testimony about the community's economic
dependence on the AEC. The survival of the municipality of Richland was
one focus of concern. Without expanded AEC subsidies, speakers testified,
the town would have a hard time raising enough revenue to provide an
adequate level of services if Hanford reactors closed. There was too little
private land—especially industrial and commercial real estate—on the
tax rolls, and too few businesses for a town of its size as well. Richland
appeared to lack a strong commercial core, both financially and spatially,
and the prevailing wage rates, set by the plant, struck many as too high to
assist in recruiting new types of industry. Both businesses and homeowners had reason to fear for their investments in the one-industry town.⁸¹

As a result of these hearings and other investigations, in November
1962 the AEC announced a new policy on "cooperation in community
industrial development efforts." In a report named after William H. Slaton,
head of the AEC committee studying the issue, the AEC began by noting
that "industrial development of communities is not a statutory objective
of the AEC," but "it is AEC policy to extend reasonable cooperation to
the economic development efforts of communities in which AEC activities constitute the major economic force." The AEC wanted to provide this
assistance in part because it admitted to a "moral obligation to prevent
[Richland] from becoming a depressed area." More practically, the AEC
needed to sustain "a reasonably stable and healthy economic environment"
around its plants to ensure their efficient operation and their continued
viability in the event that the commission needed to reverse course and
expand production.⁸²

Although the AEC understood the risks of reactor closures for local
communities, the Slaton Report did not offer the Tri-Cities much in the
way of assistance. Hoping rather optimistically that tourism would become
a profitable industry in Richland, it announced the AEC's consent to "individually prearranged bus tours through its Hanford area under controlled
conditions." It also commented that Hanford featured "extensive laboratories" staffed by "skilled scientists and engineers" who could help attract
additional high-tech industry. The AEC itself took some limited initiatives

to assist Richland and offered to make some excess land available for other uses, such as a 400-acre parcel requested by the Port of Benton. To help publicize the town to prospective investors, the AEC changed the name of the Hanford Operations Office to Richland Operations Office.[83]

The Slaton Report disappointed local boosters in more substantive areas. The AEC would not take the lead in bringing new business to its former communities. Committed to offering only "reasonable cooperation," the commission expected the communities themselves to assume the primary initiative and virtually all expenses. This disappointed those who had hoped for more federal support. John T. Conway, executive director of the JCAE, complained to Senator Jackson, "The proposed action by the commission . . . involves no major area of assistance." Glenn C. Lee, publisher of the *Tri-City Herald*, deemed the Slaton Report "not helpful, not encouraging, and not workable." Lee told Jackson that community leaders had virtually none of the technical expertise needed to understand opportunities available in the nuclear field, "no access to AEC files, and no permit to visit and inspect the plant." Without more prodding, Lee believed, the AEC and GE would not keep community leaders informed about opportunities for economic diversification. Fatefully, Lee appealed especially to the senator to "champion the cause" of the Tri-Cities. Jackson responded by taking up the challenge.[84]

Through the 1960s, the effort to diversify the local economy progressed through national political channels, wherein Jackson, Magnuson, and the Tri-Cities' representatives in the House wrung considerably more support for Hanford's diversification from the AEC than the Slaton Report had envisioned. To do so, members of Congress had to develop a close alliance with local businessmen.[85] Tri-Cities boosters, responding to the AEC requirement that communities take the initiative in diversification efforts, organized themselves early in 1963 into the Tri-City Nuclear Industrial Council (TCNIC) to promote diversification of the local economy. The group vowed to identify and attract new private businesses and government programs to the area, as well as to publicize the resources and amenities of the region. It aspired especially to move the Tri-Cities into "the civilian atomic field" and "space and missile work."[86]

Other communities affected by AEC cutbacks also organized in response, but the TCNIC differed from them in at least two respects. First, historians claim that the groups that emerged to cope with the economic crisis were "grassroots alliances."[87] That may have been true in other towns, but it was not altogether the case in the Tri-Cities. For one thing, it is not clear that the Tri-Cities possessed much in the way of "grassroots" yet. For another, a business and political elite dominated the TCNIC, while the interests of labor went largely unrepresented. Second, in contrast to the experience of like-minded groups from other towns, the TCNIC grew to have considerable influence.[88] One reason for the council's success was its connections to the local media. The Tri-City Nuclear Industrial Council's president was Robert F. Philip, also president of the *Tri-City Herald; Herald* publisher Glenn C. Lee served as the TCNIC's secretary-treasurer. Needless to say, the council's viewpoint was well represented in the local press, and the newspaper kept "the issue and results of diversification . . . on page one." Furthermore, as one admirer of the TCNIC explained, "a clear policy of attitude formation was embarked upon by the *Tri-City Herald*" to encourage local residents to support the TCNIC's efforts.[89]

The Tri-City Nuclear Industrial Council's success also depended heavily on contacts in Washington, D.C. The most important connection, of course, was Jackson. The senator met with Philip, Lee, and banker Sam Volpentest, the TCNIC's vice president, in Ritzville, Washington, during January 1963, and there they proposed to diversify the economy of the Tri-Cities through "segmentation," that is, replacing Hanford's single major contractor with several companies, each of which would be required to invest in broadening the local economy. "The result of the meeting," recalled one employee of the *Tri-City Herald*, was a "formula" whereby each new firm contracting to do work for the AEC would "make a tangible contribution" to diversifying the Tri-Cities economy "as a price for taking over part of Hanford." The following month Jackson helped arrange a meeting at Richland attended by TCNIC officers, chairman Glenn T. Seaborg of the AEC, top officials from the AEC Richland Operations Office and GE, and Jackson himself. The senator returned once more in July to monitor developments, and over the ensuing months, he was especially involved in developing a program of diversification that was tailored as much as possible to the community's needs. Suddenly, TCNIC members noticed, they enjoyed direct communication with AEC headquarters. "For the first time in the history of Hanford," one *Herald* editor recalled, "a group of local citizens had direct connections with Washington, D.C."[90]

Members of the TCNIC soon learned that, in addition to having members of their congressional delegation visit the Tri-Cities, they themselves needed a regular presence in the nation's capital. One of their first actions was to hire an "atomic-consulting firm" to lobby in Washington, D.C., on behalf of the Tri-Cities. They also eventually saw that they themselves needed to travel eastward on occasion. In May 1964 Sam Volpentest, the Richland banker who served as TCNIC vice president, wrote a telling thank-you note to Senator Magnuson: "Just a line to express my appreciation for the time you gave to Glenn Lee and myself. . . . We now realize the importance of coming to Washington [D.C.] regularly and will be more constantly in touch with you. . . . Warren, we appreciate all that you and 'Scoop' [Jackson] are doing for us and will attempt to help you all that we can from this end of the country."[91] Magnuson's help took perhaps its most tangible form in the new Richland federal building, dedicated that year and intended to give the community an "image of permanence." Magnuson accomplished in two weeks the process of authorizing the building's construction, which usually took about fifteen years. He blithely informed Volpentest, "Missoula doesn't want one."

The Tri-City Nuclear Industrial Council's "help" of the two senators took a number of forms. The council sent Jackson and Magnuson a steady stream of clippings from the *Herald* to update them on local attitudes and developments. The clippings were also intended to illustrate how the local press had handled a particular story. Members of the TCNIC were especially eager to point out that *Herald* articles portrayed the senators in the most favorable light. "I am enclosing another nice editorial pointed in your direction," Volpentest wrote Magnuson in 1964. "More will be forthcoming I promise you." The Tri-City Nuclear Industrial Council also enlisted in the senators' reelection campaigns; Volpentest had begun fundraising for Magnuson in 1956, having realized that "we had to have government connections to survive and Maggie was chairman of the Commerce Committee." By 1970 Volpentest had reportedly "raised thousands of dollars for Magnuson's and Jackson's political campaigns, and this, he feels, has helped assure him of cordial entreé to their offices."[92]

The involvement of Senators Jackson and Magnuson in the diversification effort proved crucial. The Slaton Report had tried to steer the AEC away from committing expenditures to its former communities. Moreover, when announcing the AEC cutbacks in 1964, President Johnson had vowed that the affected reactor sites would not become a "WPA nuclear project, just to provide employment when our needs have been met." Yet throughout the 1960s the commission either spent appropriations on Richland's diversification itself or required that Hanford contractors invest their own money in the local economy. Efforts to prop up local economies "did indeed smack of the New Deal rationale that Johnson formally eschewed."[93]

There is no doubt that Jackson and Magnuson figured heavily in pushing the AEC to take such proactive steps. The two senators responded faithfully to calls from their Richland constituents for further involvement. They also visited Hanford and Richland regularly, and their comments indicated that they were both informed and concerned about Richland. That the *Tri-City Herald* eventually dubbed Senator Jackson the "father" of Richland diversification suggests as well his importance in getting the AEC to pay close attention to the wishes of local leaders. No other site received such intense support from the commission specifically for its diversification efforts.[94] But Jackson's commitment to the region had always encompassed a variety of interests: national defense, industrial development, agriculture, and the needs of returning veterans. Indeed, in light of his support for the CVA, Jackson was committed to economic diversification long before the AEC, a fact that helps to explain the federal agency's inability to erase the area's history and build from a clean slate, despite the use of bulldozers. Farmers, fishers, and established communities remained, and became mixed in with the new nuclear political economy.

Local boosters and Washington State's elected officials planned a program of economic diversification in deliberations throughout 1963, long before the public learned of cutbacks at Hanford. Thus they were quite prepared when on January 8, 1964, President Lyndon B. Johnson announced a 25 percent reduction in plutonium production. Glenn T. Seaborg, chairman of the AEC, followed up the president's message with a statement that the commission would shut down four of the nation's fourteen production reactors to comply with LBJ's order. Because Hanford's plant included some of the nation's "oldest and smallest reactors," three of its reactors were among the four targeted for shutdown. Seaborg estimated that 2,000 positions, or 24 percent of the Hanford workforce, would be lost as the three reactors were phased out in 1964 and 1965.[95]

President Johnson's and Chairman Seaborg's statements of January 8, 1964, were followed on January 21 by the equally momentous announcement that the General Electric Company would withdraw from Hanford. This decision stemmed in large part, of course, from pressure on the AEC by local leaders and the state's elected officials, especially Jackson. General Electric presented at least two significant obstacles to diversification, according to local business interests. First, although GE was interested in commercial nuclear power, it did not perform research or development in that area at Hanford. Instead, it concentrated much of its effort in nuclear power at its growing facilities in and around San Jose, as it had earlier at the Knolls Atomic Power Laboratory in upstate New York. Since 1954 there had been a "somewhat steady stream of people transferred from Hanford to other GE installations, particularly in San Jose," according to one Hanford employee, and the flow accelerated anew in 1964, depleting the Tri-Cities' talented workforce. (This was due, at least in part, to policy set by the AEC in Washington, D.C., which contracted separately with GE for research and development and for production.) Lee complained about the situation to Jackson in November 1962, "It is a sort of a 'captive situation' where if any new ideas are generated here at the plant by the General Electric Company they send their men with these ideas in their brains down to California and develop the ideas there. Our region suffers accordingly."[96] The Tri-City Nuclear Industrial Council wanted contractors who would be more willing to help diversify Hanford's economy, not only by developing commercial nuclear power but also by adding more research and development to the production mission.

General Electric's image constituted the second problem. The presence of a single powerful contractor discouraged other companies from moving to the Tri-Cities. Frederick H. Warren, an engineer with the Washington, D.C., lobbyists hired by the TCNIC, urged the Atomic Energy Commission in September 1963 to dispel the "prevailing opinion, even among major industries, that the General Electric Company essentially has all prime opportunities 'sewed up'" at Hanford. Warren and others urged that the commission offer work to other contractors: "Continued insistence by AEC that no 'segmentation' of the operating functions is possible will be construed as thinly disguised preservation of the GE monopoly."[97]

Responding to critics' requests and especially to pressure from Jackson, the commission changed its policies. Instead of dealing with a single prime contractor, the AEC now divided GE's work into six separate realms and set about hiring a different firm to preside over each. Having several smaller contractors, rather than a single giant one, would in theory encourage businesses to relocate in Richland. More importantly, according to new AEC contracts the companies arriving to take over Hanford operations were required to commit specified sums of money to the local economy for the purpose of diversification. Following the plan hatched by Jackson and the TCNIC in early 1963, the price of doing business at Hanford now included a sizeable investment in new industry for the Tri-Cities.[98]

Hoping to support federal efforts and make its own response to the cutbacks at Hanford, the state of Washington stepped forward in October 1964 with a "master plan" for economic participation in the nuclear age. It regarded nuclear energy (like hydroelectric power before it) as safe, cheap, potentially "limitless," and therefore an ideal way to attract new industries. It promised to create a more favorable business climate for nuclear industry, strengthen the state's higher education programs in nuclear engineering, and develop a "site for storage of nuclear by-products and waste materials." The latter effort would be housed on 1,000 acres of Hanford land leased in September 1964 by the state from the AEC for "a peacetime nuclear industrial park." Washington intended to make a name for itself as the "Nuclear Progress State," as one booster dubbed it in 1971, but its ability to cushion the cutbacks at Hanford was minimal.[99] By itself it had few resources with which to affect the AEC policy of segmentation and diversification. The nuclear game was played mainly within the federal realm.

Still, the resonance between Hanford and the state's political culture emerging from the struggles over the N reactor, which had mobilized a state-wide political constituency in support of Hanford, continued, reinforced by concerns over the site's economic future. This was perhaps best symbolized by the election in 1964 of Daniel J. Evans as the state's governor. While the country voted for a Democrat, Lyndon Johnson, to continue in the White House, Washingtonians voted for a Republican to lead in Olympia. The antithesis to Johnson's persona as a volatile Southern pol, Evans was a Boeing engineer. His election ushered in a sixteen-year technocracy, a period during which the state would be led by governors with professional technical training.

Evans's signal political achievement was the purging of the John Birch Society from the ranks of the state's Republican party. Just as Jackson had forged a political identity merging the values of the New Deal and the Cold War, Evans appealed to an ideal of managerial efficiency that excluded the crude jingoism and prejudices of the Birchers, attempting to turn the page on an era of witch hunts and loyalty-oath controversies. Governing Washington in the aftermath of Seattle's futuristic World's Fair of 1962, Evans appealed instead to a technocratic political culture, congenial to Boeing, Hanford, and the denizens of Seattle's growing suburbs as well as the households overlooking the Columbia in Richland, all dominated by technical managers and engineers. The politics of Forward Thrust in Seattle and King County—the application of technical expertise to solve such problems as the pollution of Lake Washington and the threat of urban unrest—expressed the same political values as did the politics of nuclear progress in the Columbia Basin.

Local interests shaped diversification efforts much more profoundly than did the state. As the largely self-appointed leaders of the local communities, the Tri-City Nuclear Industrial Council tried to influence both federal policies and the public's reception of those policies. The Tri-City Nuclear Industrial Council professed to be a conduit, merely passing along the unadorned facts to "our citizens." But in fact the council had better access to information than other organizations, and it used its special relationship with the *Tri-City Herald* to manage the news so as to heighten confidence in diversification and segmentation. In September of 1963 the lobbyists hired by the TCNIC had advised the council to find ways to minimize "the psychological effects" of pending cutbacks.[100] So, the day after President Johnson and Chairman Seaborg announced the closing of three Hanford reactors over the next two years, the *Herald* ran stories intended to calm people's fears. One, headlined "Hanford Has History of Cutbacks," hinted that the plant had absorbed similar reductions in previous years. Another quoted the president of the Kennewick Chamber of Commerce predicting that the loss of 24 percent of Hanford's jobs would, "in the long run, . . . have a salutary effect in accomplishing sound diversification at Hanford."[101]

This style of coverage continued later in the year, on the occasion of the first actual shutdown of a Hanford reactor. Writing as publisher of the *Herald*, Lee explained to Clarence C. Ohlke, director of the AEC Office of Economic Impact and Conversion, exactly how he planned to handle the news: "We will treat the layoff from the shutting down of the first Hanford reactor as a 'single story' and go no further than that. . . . Then we will drop it." In order to manage the news more effectively, Lee asked the AEC to send its news releases to the *Herald* a little earlier because the newspaper, which planned a carefully timed column praising AEC management at Hanford, needed more time to "get our editorial ducks in a row."[102]

At times the TCNIC liked to pose as the AEC's guide, partner, and publicist in the Tri-Cities.[103] One goal behind such posturing was to manufacture an image of cooperation and consensus so that everybody—the AEC, its contractors, members of Congress, Washington State, local citizens, and Hanford employees—appeared to be working together on diversification. With so many potential malefactors outside of this circle to worry about—including "peace-mongers," environmentalists,[104] and other states and towns competing for the very economic resources that the Tri-Cities wanted—the TCNIC hoped to keep the various players in the diversification game in line. The Tri-City Nuclear Industrial Council never managed the AEC nearly as much as it had hoped to, however. In addition, it seldom succeeded in securing the cooperation of organized labor.

Working people did not always cooperate with diversification, no doubt in large part because they were expected to bear the brunt of any changes.[105] The *Herald*, for example, in May of 1964 called for "cooperation" between management, labor, and community groups in reducing the relatively high wages of the Tri-Cities, which supposedly discouraged outside investment in the local economy. Since workers were the only group being asked to give up something, it is no wonder that the unions seemed uninterested in cooperating. The Hanford Atomic Metals Trade Council (HAMTC) quickly announced its opposition to substantial portions of the AEC's and the TCNIC's plans. Its business agent, Dave Williams, criticized the TCNIC proposals because they shortchanged union members when it came to wages and jobs. Organized labor in general, responded Glenn Lee, lacked the proper "attitude" for attracting new industry to the area. Lee blamed unions for the community's "bad reputation from a standpoint of labor disputes and high labor costs," and wrote Senator Jackson about the need to "straighten out the thinking of Dave Williams."[106]

As the complaints about organized labor suggest, the TCNIC and the *Tri-City Herald* by no means spoke for, or exerted significant influence over, all those party to or affected by diversification programs. However, as a contest over Hanford's future developed between local interests and the federal government, the TCNIC and the *Herald* represented the most insistent and effective community voices. The Atomic Energy Commission did listen to the nuclear chamber of commerce as well as to local citizens, and it often acted favorably on their requests. The AEC's respect for Senators Magnuson and Jackson helped account for its responsiveness to Tri-Cities inquiries. Atomic Energy Commission officials in Washington, D.C., warned one another of the need to deal promptly with correspondence from the towns around Hanford. They also had learned to check with Jackson before moving forward with certain initiatives.[107]

Replying to local requests was one thing, but agreeing to them was another. In implementing the policy of segmentation and diversification, the AEC hewed to its own agenda as much as possible, and what resulted was a tug-of-war between the commission and Tri-Cities boosters, with Washington State's elected officials in the middle. On some occasions the AEC agenda coincided with the aims of the TCNIC and Senators Jackson and Magnuson for the Tri-Cities region, but on other occasions it did not. For example, the commission did not want to establish precedents for other AEC sites by going too far beyond the bounds of "reasonable" cooperation with local efforts, so it played its hand carefully. Like the *Herald*, it tried to manage news about the project. In January 1964 GE prepared a press release that explained its withdrawal from the Hanford project largely in terms of the prospective plight of Richland's economy and of the people to be laid off during cutbacks. The AEC thought that "overemphasis on this point can be misinterpreted," likely because it wished to project a more limited sense of responsibility for the local community. So the commission buried GE's press release and prepared its own explanation of segmentation and diversification, which emphasized, first, the effort to expand research activity at the Hanford Laboratories; second, the desire to encourage "commercial diversification of industry in the Tri-City communities"; and third, the pursuit of the best interests of both GE and the federal government.[108] No mention was made of the problem of unemployment.

Following the example of the TCNIC and the *Herald*, the AEC played down any bad news. The new policy of segmentation and diversification called forth several evaluations of the local economy. These studies produced mixed results, but the AEC attempted to highlight the positive. When one early report did not "present an optimistic picture as regards to any substantial diversification," the AEC told Jackson that it would not be publicized. At about the same time, two versions of another AEC-GE study assessed prospects for Tri-City diversification. The first draft proved rather pessimistic, stressing Richland's "dead-end" location and the absence of a nearby four-year college. Somebody must have insisted on a happier ending, however, because the slicker final draft was significantly more upbeat. It listed no qualms about "transportation" to and from the Hanford area and advertised "extensive facilities for adult and advanced study."[109] Both the local communities and the businesses considering moving to the Tri-Cities seemed to need a rosy picture.

Despite the effort to foster a positive view of the Tri-Cities' future, studies completed in 1963–65 were not altogether confident about the prospects for diversification.[110] Although diversification implied many changes, there were some strong attachments to the status quo that proved difficult to work around. For one thing, as long as the AEC continued to produce fissionable material, it insisted on reviewing proposed new enterprises at Hanford to ensure their "compatibility with commission policies, authorities and activities." The AEC did not readily relinquish any amount of control over the Hanford area. In fact, it initially objected even to including the word "nuclear" in naming the TCNIC, suggesting that such usage "might be infringing on the prerogatives" of the AEC.[111]

All the resistance and uncertainty notwithstanding, the program of segmentation began in earnest in 1964 and 1965 when the AEC took bids from new contractors to replace GE. The commission considered only those bidders that had indicated exactly how, and how much, they would invest in the local economy. In some cases the AEC specified a particular diversification project, such as a plant to sort through Hanford wastes and recover commercially valuable isotopes left over from manufacturing. But the AEC encouraged suggestions for new enterprise from the bidders, too. In this way, Hanford began to move away from its longstanding focus on production for military purposes and tried to find a niche in the civilian economy. It also became less a satellite of the warfare state and more a creature of the welfare state.

This program of diversification by segmented contractors met with mixed results. Some ventures flourished while other proposals never reached fruition. The most successful effort was the transfer of GE's Hanford Laboratories to Battelle Memorial Institute, an Ohio-based nonprofit company devoted to industrial research. Battelle began negotiating its acquisition of the labs in 1964 and soon took over their operations. Local citizens regarded the advent of Battelle as critical because its presence promised to compensate for the university that the urban area lacked and because its research mission reduced the site's emphasis on production. Battelle–Pacific Northwest Laboratory played a significant role in developing new programs and fulfilled its contractual commitments with the AEC to help diversify the local economy.[112]

By contrast, the effort to develop an isotopes recovery plant failed. When he went to work at Hanford in 1960, botanist William Rickard recalls, people did not perceive "hazardous wastes" at the site. Rather, they assumed that the industrial by-products of plutonium manufacture could be recaptured and put to profitable use, and that this activity would help diversify the local economy. In February 1965 the AEC selected Isochem, a joint venture of Martin Marietta and U.S. Rubber, to take over processing operations at Hanford. As part of the deal, Isochem agreed to invest up to $9 million to build "a commercial plant" for recycling Hanford wastes: "Carefully packaged radioisotopes, drawn from the waste streams of Hanford's reactors, will be made available in unprecedented quantity and at sharply reduced prices. They will be used to preserve food, to sterilize medical supplies, to help manufacture chemicals, and in scores of other safe and peaceful applications. . . . This is a dividend of the Atomic Age which our nation is only beginning to cash."

Isochem started work at Hanford on January 1, 1966, but within a year the company had determined that there was no market for isotopes recovered from Hanford wastes, and decided not to build its "fission products conversion and encapsulation plant." Lee called the decision as "a very bad shock to the community," because the TCNIC had placed much hope in the idea of recycling Hanford's by-products. Still, he found solace in the thought that "if we can replace them with a better company with better diversification let's do so." The AEC felt obligated, for the sake of the integrity of the diversification program and of its standing with the TCNIC and Jackson, to replace Isochem with another contractor. By summer of 1967 the AEC had chosen the Atlantic Richfield Hanford Company to take over processing operations. The Atlantic Richfield Hanford Company also promised to invest in diversification, but the nature of its contractual commitments was telling. There was no sum devoted to a specific development of the "peaceful atom," or to any other well-defined high-tech industry. The Atlantic Richfield Hanford Company agreed instead to spend up to $3 million replacing the old Desert Inn with a "hotel-convention-resort facility" that became the Hanford House; $750,000 on a "cattle feeding yard"; $1.25 million on a meatpacking plant; $375,000 on the Center for Graduate Study; $400,000 in venture capital investments; and $300,000 on studies of the feasibility of "civilian oriented nuclear business in the Tri-City Area."[113]

The Tri-City Nuclear Industrial Council hoped for isotope recovery plants and other cutting-edge industry but ended up with feedlots and meatpacking plants instead. Boosters had insisted that their lobbying organization must be a nuclear industrial council; everyone had wanted diversification to perpetuate the local emphasis on the nuclear industry with its relatively high salaries. For communities that characterized themselves as an industrial frontier, it had been reassuring to think that they could continue to work with cutting-edge technology. But for a variety of reasons the Tri-Cities never realized this vision. Briefly, during World War II, Hanford had been at the cutting edge; thereafter it grew increasingly obsolete.

As might be expected, local reaction to the diversification program was mixed. Labor continued to protest its impact. Robert W. Gilstrap, president of HAMTC, criticized the AEC in 1968 because union members had "sustained a disproportionate number of layoffs" while "management, supervisory, and technical people have not been required to absorb their fair share of the consequences of curtailment." Furthermore, wrote Gilstrap, the new businesses started by diversification in the Tri-Cities did not generally employ the "experienced atomic workers" of HAMTC. The Atlantic Richfield Hanford Company's plans for "a cattle feed lot and hotel" hardly promised to replace the high-paying jobs lost in the course of closing down reactors. Calling union workers "the forgotten men of Hanford," Gilstrap scolded the AEC for rewarding neither their contributions to the plant's success nor the "practically irreversible commitments" they had made in order to work in the Tri-Cities.[114] The Tri-Cities business community proved more sympathetic than labor to diversification efforts. For instance, when developers planned and built the Columbia Center shopping mall in Kennewick between 1967 and 1969, they cited the AEC's diversification program as a reason for their confidence in the area.[115]

While the diversification program may have enjoyed some success, it would have been nearly impossible to satisfy the townspeople fully. The Tri-City Nuclear Industrial Council complained that the AEC was not committed enough to the program and remained too "production-minded."[116] Yet production at Hanford continued to fall off through the late 1960s as the AEC continued to shut down its reactors there. More than anything else, these closures undermined diversification efforts, as appeals from the TCNIC attested. Late in 1966 and again early in 1967, Lee asked Senator Jackson to try to postpone more reactor shutdowns until the diversification program "gets its roots down" and the business community could get its "feet on the ground." This interpretation of defense spending at Hanford as a form of local welfare payments achieved its starkest expression when Lee suggested that nobody would notice a few extra appropriations for Hanford: "With the situation with Russia, Red China, the antiballistic missile, a $73 billion defense budget, and the uncertainty in the world today, it seems reasonable that you and the Joint Committee can argue what's a few more million dollars to keep our reactors running." Lee ended his appeal with a command rather than the more usual request: "Stall any change at Hanford."[117]

Not even Henry Jackson could delay the inevitable. By 1970 all but two reactors had been closed. Then on January 26, 1971, the AEC announced that, for budgetary reasons, it would shut down the final pair of reactors, KE and N, and lay off 1,500 employees. With these closings could come the eventual end of fuel-element manufacture and, ultimately, chemical processing as well—and the loss of another 500 jobs. Once again, Hanford's reactors, and not Savannah River's, were targeted for shutdown because of their obsolete design and the fact that they produced only plutonium 239, and not enough other "isotopes of national interest such as tritium, Pu-238, Po-210, curium-244 and higher isotopes such as californium-252."[118]

This decision by the Nixon administration, made without consulting the Washington State congressional delegation, sparked a storm of protest from the Tri-Cities. The community expressed outrage that the government seemingly proposed to undermine, in one blow, all the careful work it had done in diversifying the economy. It also bitterly noted that Savannah River, which had made little effort to diversify, continued to thrive and receive new assignments while the Tri-Cities, which had toiled to diversify, were seemingly being punished. Especially shocking was word that the new N reactor, one of the few clear diversification successes at Hanford, was slated for closure. The AEC had contracts to deliver power from the N reactor to the Bonneville Power Administration, and the Northwest reportedly suffered from a shortage of kilowatts. In essence, by closing the N reactor, the Nixon administration proposed to break the bargain between the AEC and the BPA, contribute to the regional energy shortage, and nullify the progress of the diversification effort.[119] Operation of the KE reactor was halted rather quickly, but local citizens (including the local office of the Sierra Club), the state's congressional delegation, and Governor Evans fought successfully to keep the N reactor open, at first only for three more years, but eventually for much longer.[120]

In the aftermath of the fight over the KE and N reactors, observers paused to assess the impact of approximately a decade of segmentation and diversification. The overall picture remains murky, partly because local observers at the time were reluctant to declare the program either a success or a failure. On the one hand, boosters wanted to prop up con-

3.6 President Richard M. Nixon visited Hanford during September of 1971, less than one year after his administration had proposed closing down both remaining Hanford reactors, KE and N. Eventually, Nixon and the Atomic Energy Commission relented and permitted the N reactor to continue operations. Washington State's Republican governor, Daniel J. Evans, seen here on Nixon's right, had strenuously protested the decision to close the N reactor. The photo captured Evans and Nixon with a model of what would become the Fast Flux Test Facility. Photograph courtesy of the U.S. Department of Energy.

fidence in the economy among the local population; on the other hand, they did not want the AEC to withdraw resources because it got the idea that diversification had succeeded. As a result, the *Tri-City Herald* and the TCNIC alternated between applauding specific successes and calling for still more federal and contractor investment in the local economy.[121]

In historical perspective, the actual impact of segmentation and diversification on the local economy is best judged on two scales—the near term and the long term. In the short run, the program helped to cushion the blow of AEC cutbacks by generating new jobs and reducing the number of layoffs. With eight out of nine reactors shut down, Hanford was in some ways but a shell of its former self, and the plant's mission had changed dramatically. Yet employment figures did not drop drastically. At the end of 1963, prior to the shutdown of reactors and beginnings of segmentation, GE had employed 8,277 people. In May 1967, comparable contractor employment on AEC-related work, combined with employment generated by diversification efforts, stood at 8,140. This was a relatively small decrease, considering that three reactors and one processing plant had closed. By 1968, the Richland Operations Office reported, 660 new jobs had been created as a result of the commitments made by new contractors. By 1969 there were 942 "diversification employees," and in 1971 and 1972 there were about 1,100 jobs attributable to the diversification effort. A 1976 accounting of Hanford employment summarized the changes since 1964, the year cutbacks had first been announced. Total employment at Hanford, including diversification activities, stood at 9,030; the estimated number of employees in related or "diversified" jobs (those not working at Hanford for the Energy Research and Development Administration [ERDA] or under ERDA contracts) amounted to 784, or 8.7 percent of the total employment.[122] Clearly, employment figures were higher than they would have been without the programs of diversification and segmentation.

But the figures masked underlying weaknesses. Although the number of jobs in Benton and Franklin counties increased 17.2 percent between 1964 and 1973, the statewide figure for the same period was 26.1 percent. The Tri-Cities area was not keeping up with the rest of Washington. Furthermore, of the 1,220 new jobs created between 1970 and 1973, 1,075 were in construction, and many of those were temporary. The construction itself represented one kind of progress—early work on additional power-generating reactors for the Washington Public Power Supply System—but it remained unclear exactly how much long-term prosperity the WPPSS project would bring to the Tri-Cities. Moreover, economic conditions in the state in the early 1970s were bleak. Unemployment increased from 9.8 percent in early 1971 to 10.5 percent in early 1972, and the local caseload for the state Department of Social and Health Services climbed. Enrollment in Richland schools fell by 144 students between early 1971 and spring 1972, and the community voted down two school levies in 1972, forcing the district to trim its annual budget. With the closure of KE, the total number of operating employees fell by almost 1,000 between January 1971 and January 1972. Once again, new construction picked up some of the slack but did not offer the long-term stability once associated with Hanford production reactors. G. J. Keto, an AEC official in Washington, D.C., returned discouraged from a trip to Richland in late 1971 and reported that, despite the investments made and the new jobs created, "the economic outlook is not promising." Much more would be needed before the local economy became "relatively self-sustaining."[123]

Keto had put his finger on the limitations to segmentation and diversification. While government welfare programs propped up the local economy over the short term, long-term prospects for the Tri-Cities remained gloomy. Investments coerced through political channels did not create enough lasting improvement to the economy. Into the 1980s and 1990s, the Tri-Cities remained inordinately dependent on federal spending, and never developed much economic vitality apart from Hanford. Nonetheless, the ineffective program of segmentation and diversification remained securely in place. In 1985 the DOE managed to reduce the number of main contractors at Hanford from eight to four, and thereby narrowed the extent of segmentation. But when the department suggested that bidders on the new contracts not be required to invest in the local economy, it was forced by political pressure to maintain the status quo. Thus by the time that the team of Westinghouse and Boeing was selected in 1986 to become the main contractor at Hanford the following year, it had agreed to invest $10 million in the Tri-Cities economy over the five-year life of the deal. Similarly, when the DOE hired a new contractor in 1996 to head the cleanup effort at Hanford, it specified that the firm it selected would be expected to create for the Tri-Cities some 3,000 jobs—not related to work on the Hanford Site.[124] The program hastily designed by Senator Jackson and local boosters in 1963 had proven to have a very long half-life.

New Missions, New Masters: Hanford in an Era of Adversity

Problems in the local economy were not limited to the difficulties inherent in diversification programs required of contractors by the AEC and DOE. There was also the matter of bad timing and bad luck. Part of the Tri-Cities' problem lay with the fact that by staking their future on the nuclear power industry they were backing the wrong horse. After 1970, Americans became more and more suspicious of nuclear power, and as a result proved ever less supportive of the civilian nuclear energy that the communities around Hanford saw as their future. Indeed, during the 1980s the Tri-Cities were operating in a political climate that was in many ways increasingly hostile toward Hanford. Moreover, they lost their powerful champions in Washington, D.C. At the same time, the Hanford Site was no longer monopolized by the Department of Energy. By 1990 both the state of Washington and the U.S. Environmental Protection Agency possessed considerable influence over the land that had for so long been the domain of the national security state. Under the new political regime, a wide range of "stakeholders" now gained more say over operations at Hanford, and these stakeholders weakened the influence of Tri-City interests in decisions about the site.

The nuclear-power business had once promised both to integrate the Hanford Site more fully into the life of the Pacific Northwest and to save the local economy. When the N reactor started generating kilowatts for the BPA power grid in 1966, Hanford became a servant of the region as well as of the nation. As the Tri-Cities advocated diversification of the local economy, they expected that the civilian atomic industry would assume the leading role. This effort, like the broader program of diversification, entailed extensive political activity, and it took advantage of the perceived "energy crisis" of the late 1960s and early 1970s to advance its goals.

Senator Jackson, in a speech delivered in Richland during October 1972, helped set the tone for Hanford's new mission of generating electricity by comparing the situation to an earlier mobilization: "This community was born of crisis—World War II. To find an answer to bring that war to an end as fast as possible. That was 1942–1943. And now, 30 years later, we face another crisis, which a lot of people don't understand. The Energy Crisis." Hanford and the Tri-Cities had played an important role in solving the first crisis, Jackson averred, and now they stood "uniquely equipped" among American communities "to play a major role in providing a solution" to the second crisis. Because environmental and economic issues now cast doubt on the long-term future of fossil fuels, the senator explained, there was an urgent need to explore the potential of nuclear power. Recommending "a massive effort of research and development" in the area of nuclear power, Jackson urged the nation to "marshal our talents and our resources with the same kind of dedication and energy that we did in making possible the Manhattan Project in World War II."[125]

Rhetoric such as this reminds us why so many in the Tri-Cities adored Jackson. His words not only reminded them of their golden, pioneering years of the 1940s and of Hanford's previous contributions to the country, but also identified still another national crisis that the people of the Tri-Cities could help solve. Furthermore, his mention of the Manhattan Project and the need for a "massive effort of research and development" hinted that lots of federal funding would, or at least should, be forthcoming to support the Tri-Cities as they blazed new trails along the energy frontier. At a time when reactor shutdowns had lowered community morale and endangered local prosperity, Jackson not only offered hope for economic recovery but also reassured the people of the Tri-Cities that Hanford—and, by implication, they—constituted a "national asset," as the *Herald* phrased it.[126]

Jackson's words no doubt reassured the people of the Tri-Cities, but the massive federal assistance for the peaceful atom at which he had hinted was not forthcoming. Hanford's involvement in nuclear power depended much more on the local politics of the public utility districts that comprised the WPPSS. Although the WPPSS worked closely with the federal Bonneville Power Administration, although the BPA loomed large in generating political support for nuclear power by predicting energy shortages, and although electricity generated at Hanford went out over BPA lines, Hanford and the Tri-Cities would now be serving the Pacific Northwest more than the nation. In the 1970s, Hanford was becoming more a regional than a national asset, and its prosperity depended more on local than national investments.

The nuclear-power boom lasted for about a decade. Then, in the early 1980s, it expired when the WPPSS defaulted on its bonds. Only one of the five projected reactors was finished—WPPSS no. 2 at Hanford, in 1984, seven years after its projected completion date. The other four plants, including two additional reactors at Hanford, were never completed, and the WPPSS nearly collapsed beneath a mountain of debt, surplus electricity, and poor management. At Hanford, boom turned to bust. The failure of the supply system stemmed for the most part from economics, notably mistaken forecasts for energy demand and poor project management. At the same time, however, people throughout the region became less supportive of nuclear energy and less inclined to view the nuclear reservation as an asset. For the first time since the secretive era before Hiroshima, the Pacific Northwest began to show distrust of Hanford. In 1980 Washington State voted by a 3-to-1 margin in support of an initiative to ban import of nuclear waste into the state.[127] The courts overturned the initiative as improper interference with the federal government's role in interstate commerce and nuclear waste, but the vote signaled a large-scale shift in attitude toward all things nuclear.

Another sign of changing opinions toward Hanford and nuclear energy was the defeat of Governor Dixy Lee Ray in 1980. Ray was a marine biologist and associate professor at the University of Washington who had served as director of Seattle's Pacific Science Center. She saw herself as someone who could serve as a liaison between science and the general public, although a bid to have her serve on a federal science policy commission in 1966 had been blocked by Senator Magnuson, who, according to a White House memo, was "highly negative on Dixy Lee Ray's candidacy." Jackson, though, was a supporter of Ray's—he had mentored a number of female experts in federal service, especially on his office staff—and Magnuson deferred to Jackson when the latter supported Ray for a seat on the AEC, where she served as commission chair under President Nixon. Jackson also supported Ray's prospective candidacy for governor in the 1976 election. Party affiliation was not important to Ray. Because the popular Republican incumbent, Daniel J. Evans, was slow to announce that he would not run for a fourth term, Ray declared herself to be a Democrat and a candidate in March of 1976. A crowded primary field and crossover voting by Republicans in the Democratic primary helped Ray to win the party nomination, and she went on to win the general election.

This ushered in four years of turmoil as Ray found herself in conflict with the legislature, constituents, the state's news media—particularly the *Seattle Post-Intelligencer*—and Canada. Ray declared herself, as a technical expert, to be above politics; embracing a technocratic identity, she argued that the right answers were not the property of partisan groups, and indeed were best discovered by ignoring them. She championed further nuclear development in the state. Richland's Mike McCormack hired a Republican political consultant to run Ray's reelection campaign in 1980. Meanwhile Ray, assuming that Magnuson's health would not permit him to complete his term, began to offer Maggie's Senate seat to a variety of candidates, including McCormack. Not surprisingly, the 1980 Washington State Democratic Party convention in Hoquiam became the site of a pyrotechnic face-off between Magnuson and Ray. The senior senator gave a stem-winding speech, concluding, "This state is not going to be a dumping ground for nuclear waste and there are not going to be any supertankers on Puget Sound. . . ." Ray left the meeting after Magnuson's direct assault on her policies; in her absence, her rival, the quite liberal Jim McDermott, received the party's nomination for governor, and those present passed a raft of anti-Ray platform planks: against nuclear power, against a transstate oil pipeline, and against out-of-state nuclear waste storage at Hanford. Amid all the turmoil among state Democrats, and coinciding with the election of Ronald Reagan to the presidency, McDermott lost to John Spellman—the last time a Republican has won a gubernatorial election

3.7 Dixy Lee Ray visited Hanford with, as usual, her pet dogs in tow. First as head of the Atomic Energy Commission and then as governor of the state of Washington, Ray had a close association with Hanford and nuclear power. By 1980, however, both Ray and her views had become rather unpopular in the Evergreen State. She did not even receive her party's nomination for reelection to the position of governor, and in the same election of November 1980, the state's voters overwhelmingly supported an initiative to ban importation of nuclear waste into Washington. Photograph courtesy of the U.S. Department of Energy.

in Washington. In contrast to other Republican gubernatorial nominees, Spellman displayed the managerial, almost technocratic sensibility promised by Ray and delivered by Evans.

Had Ray been more willing to participate in the politics of her own party, she might have smoothed over at least some of the differences her abrasive political style had created. She might have appealed to the idea that the party platform should not repudiate outright a sitting governor who intended to run again; she might have asked for at least the appearance of party support for herself as well as McDermott. Instead, she left the convention. The result was the appearance that a rational, expert supporter of nuclear technology had been purged on behalf of a party that had seemingly committed itself to opposing the economic development of the Tri-Cities as a producer of nuclear power or a processor of nuclear waste. Before long, neither the blue-collar loggers and mill workers in Hoquiam nor the residents of the Tri-Cities felt at home in Washington State's Democratic party, which became increasingly identified with environmentalism and with hostility to nuclear power. That Seattle and other Puget Sound communities commanded so many votes in the party heightened the sense of a split between the voters of western Washington and the Tri-Cities east of the mountains.[128] A series of citizen initiatives opposing nuclear waste importation and investment in nuclear power, many of which were overturned by federal court review, tended to institutionalize elements of anti-Hanford sentiment within the state's political culture.

Another wedge between the Tri-Cities and western Washington came two years after Ray's defeat. In 1982 Seattle author Paul Loeb brought out *Nuclear Culture: Living and Working in the World's Largest Atomic Complex*. Presenting the book as the "first major work published on Hanford," Loeb had seemed a sympathetic interviewer to many in the Tri-Cities who felt it was time their stories were told, unaware of Loeb's sense of "all the reasons nuclear weapons should not have been used." He portrayed the first generation of Hanford's workers as admirable can-do figures, inventive and proud of their work. But contemporary Hanford, Loeb suggested, was a nihilistic social wasteland, with younger workers doing pointless jobs and racing to out-of-work hours of drinking and drug use. Publication of the work was suspended by a lawsuit over the portrayal of two of the Hanford workers; after an out-of-court settlement, Loeb regained the rights to the book from the original publisher. It was reprinted in 1986.[129]

Both politically and culturally, Hanford found itself increasingly isolated, less a regional asset than a growing political and environmental problem. Its outcast status grew during the early 1980s as the DOE resumed plutonium production at Hanford and considered making the site into a national, high-level, radioactive-waste repository—a mission that would have produced many new jobs and imported a great deal of waste into the state. Most Tri-Cities residents welcomed any activity that could keep their economy from stagnating, but citizens in the rest of the state—many of whom had been sympathetic to accepting more nuclear waste a decade earlier—now forcefully opposed the new proposals.[130] Washington no longer aspired to be the Nuclear Progress state.

As tensions over Hanford's future grew, it became clear that the Tri-Cities were no longer as resilient politically as they once had been. The downfall of the WPPSS, the antinuclear crusade, and the environmental movement all brought into question Hanford's various missions. Partly due to the policy of segmentation and economic diversification created to cushion the communities during the 1960s, latter-day Hanford had no single contractor in charge. It also lacked a single focus and a well-defined purpose. Such conditions made it more difficult to rise to the defense of the site. Also important was the loss of Hanford's patrons in Congress. Senators Warren G. Magnuson and Henry M. Jackson had long toiled to protect the interests of Hanford, but in the early 1980s both senators left office. Magnuson lost his bid for reelection in 1980, and Jackson died in 1983. Suddenly, Hanford no longer had two powerful senators working on its behalf. At least one of the replacements, Republican Slade Gorton, did continue to lobby hard for Hanford interests, but doing so hurt him politically in 1986 when his opponent, Democrat Brock Adams, defeated his bid for reelection in part by campaigning against Gorton's record of supporting too strongly what Adams dubbed the "bomb factory."[131]

It is not certain that Jackson and Magnuson, had they served longer in office, would have been immune to the environmentalist concerns that increasingly motivated Washington voters to criticize Hanford. As the 1986 senatorial race between Adams and Gorton demonstrated, Hanford now tended to polarize rather than unify the electorate. Yet it does seem likely that, more than their successors, Magnuson and Jackson could have deflected some antinuclear pressures while continuing to serve Hanford's interests in the nation's capital. As it was, local leaders lamented the loss of their "watchdog" in Washington, D.C. Sam Volpentest of the Tri-City Nuclear Industrial Council explained the new situation in 1985: "Things don't operate like they used to, when you could just call up Scoop or Maggie."[132]

The economic and political situations worsened in 1986 and 1987. The federal government closed down the N reactor permanently, ending the production of plutonium at Hanford. Moreover, it discontinued work on a high-level radioactive waste repository at Hanford, and turned to a site in Nevada instead, with a site for disposal of low-level waste in New Mexico. These decisions threatened to doom the local economy. One common Tri-Cities response was to look again to Washington, D.C., to see what Uncle

Sam could do for the urban area now. After canceling the waste repository and shutting down the N reactor, one local activist said, "the DOE owes us a big one."[133] Seldom had the fiscal dependence of the Tri-Cities on federal spending—based not upon national but upon local need—been expressed so starkly; seldom had the Tri-Cities sense of entitlement to "nuclear pork" been expressed so boldly. The same sentiment operated when people in the Tri-Cities accused Brock Adams, the new senator, of "betraying fundamental pork-barrel allegiances" because he did not support projects considered beneficial to Hanford.[134] Claims of "betrayal" and of unpaid debts conveyed how the Tri-Cities had come to view federal subsidies both as a kind of entitlement and as payment due in return for decades of community sacrifice.

Without continuing federal support for plutonium production and without continuing regional support for electricity production, the politics of Hanford entered a new era during the late 1980s. Outside of the Tri-Cities, the site became defined increasingly as a liability rather than an asset to the Pacific Northwest. As the federal government finally began lifting the shroud of secrecy from around the site, the precise nature of the problems it presented to the region became quite apparent. In 1986 the DOE started releasing newly declassified documents that demonstrated that Hanford's operation had secretly jeopardized the health of nearby residents and left behind a massive amount of radioactive and chemical waste that continued to threaten the environment, health, and economy of the region. The documents became the basis for sustained criticism of the federal government's actions at Hanford since the 1940s, but they also served to galvanize commitment to an expensive effort to solve the problems created by the manufacture of plutonium.

A new era at Hanford began in May 1989 with the signing of the Hanford Federal Facility Agreement and Consent Order (also known as the Tri-Party Agreement) between the Department of Energy, the Environmental Protection Agency, and the state of Washington. The agreement promised that Hanford would be cleaned up within three decades at a cost of at least $57 billion.[135] Under this framework, federal monies continued streaming into the Tri-Cities, and helped revive a local economy that had stagnated after 1986. Since 1989 the DOE has spent billions of dollars attempting to deal with the enormous waste problems at Hanford. It is widely acknowledged that, for at least the first half decade, up to a third of this money was wasted, yet DOE funds continued to flow toward the Tri-

Cities, as if they came from a broken spigot that could not be turned off.[136] And the federal spending contained provisions designed to continue the effort to diversify the local economy. Programs begun in the early 1960s to steer federal funds to the Tri-Cities and make the economy less dependent on Hanford continued to operate decades later, even though those programs had shown no broad and enduring success.

In fact, however, the Tri-Party Agreement may have spelled an end to politics as usual at Hanford, partly because it diminished local leaders' influence over the future of Hanford. Like other states, Washington negotiated with the federal government to secure a commitment to "environmental restoration" at the nuclear site. On the surface it would seem that the state's new role might perpetuate the influence of the Tri-Cities elites at Hanford. Yet during the 1980s the Tri-Cities had become somewhat estranged from the rest of Washington State. Moreover, the Tri-Party Agreement gave voice to a wide range of stakeholders—including the long-mistrusted environmentalists, urban activists from west of the Cascades, and representatives of native tribes—who did not generally share the views of Tri-Cities boosters. The new consortium in charge at Hanford also contained unfamiliar federal officials. For the first time in the history of the site, U.S. agencies other than the army, AEC, and DOE would exert significant control over Hanford. The EPA was party to the cleanup agreement and, like the state of Washington, it did not prove as responsive to Tri-Cities needs as the AEC and DOE had been. The participation of other resource management agencies, such as the National Park Service and the U.S. Fish and Wildlife Service, similarly suggested that federal offices might no longer be so sympathetic to local boosters.[137]

Even the Department of Energy had changed, out of a desperate need to reinvent itself. Largely to survive calls for its demise and to continue producing materials for nuclear weapons, the DOE consented to cleanup programs at its sites across the country, even though it may not have initially expected those programs to succeed.[138] The DOE, for its own purposes, needed to show a commitment not only to environmental restoration but also to widespread public participation in the process. These needs often put the DOE and its programs at odds with what local boosters wanted. During the 1960s and 1970s, when Magnuson and Jackson had done the heavy lifting, the economic elite of the Tri-Cities had felt assured that its concerns would be addressed. Under the new cleanup regime, it was not so clear that local needs would be accommodated. Federal funds continued to pour into the local economy, but in some ways the influence of local elites and Tri-Cities interests had diminished.

One consequence of the new political arrangements was that, over time, the local elites of Richland, Kennewick, and Pasco became more suspicious and mistrustful of the federal government. This became apparent in the debates over the future of the lands of the Wahluke Slope and the waters of the Hanford Reach, areas in the northern part of the Hanford reserve which contained some wildlife. Environmentalists from around the state (including some from the Tri-Cities) advocated some form of U.S. government control over these parcels along the Columbia, partly because they did not trust Tri-Cities business interests and political leaders to provide adequate protection for them. There had emerged over the years a pattern whereby local elites looked almost exclusively to incorporate the Hanford Reach and Wahluke Slope into plans for economic development, while neglecting environmentalist concerns for the waters and lands around the nuclear reservation. Business interests and political leaders in the Tri-Cities opposed additional federal controls over lands and waters near Hanford, proposing instead to create some sort of regulatory commission with representatives from local, state, tribal, and federal entities. Local spokespeople raised important questions of fairness; it had not been obvious that the federal government ought always to prevail over state and local interests in controversies concerning use of resources.[139]

Yet in 1999 and 2000 the local elite lost its case when President Clinton's administration decided that the Wahluke Slope and Hanford Reach would remain under federal control. In speaking out against efforts by U.S. agencies to supervise use of the resources around Hanford, Tri-Cities spokespeople joined a loud western chorus of complaint about the federal government. Since the Sagebrush Rebellion of the late 1970s one prominent feature of the states of the interior West—particularly away from largest urban areas—has been a pronounced (and, to some, a confusing) hostility toward Uncle Sam.[140] Writing about the situation at Hanford, Susan S. Fainstein, a geographer and self-identified "Easterner," notes how the sharp resentment toward U.S. agencies has been juxtaposed in the West with sizeable economic benefits from "federal largesse." "The lesson is obvious enough," she concludes, "Extreme dependence, even if it brings with it considerable material advantage, produces extreme antagonism." The frustration is worsened, she adds, "for people who feel excluded from decision-making processes."[141]

By retaining control over the Wahluke Slope and Hanford Reach, the U.S. government frustrated some Tri-Cities economic and political interests, who felt they would have little influence over a significant amount of nearby river and land. Those same interests had convinced themselves, based on a distinctive reading of history, that the federal government would almost surely mismanage the resources it had seized. On behalf of the commissioners of Benton, Franklin, and Grant counties, Adam J. Fyall wrote in 1999, "The communities adjacent to the Hanford Reach have been pushed around and dictated to by federal agencies for six decades. We have witnessed the unfulfilled promises and periodic blunders of absolute federal power, ranging from the Department of Energy's management of Hanford cleanup, to the Corps of Engineers' lands transfer debacles, to the sweeping impositions of Endangered Species mandates, to name a few. Time and again, it is the local communities that have to live with the consequences of these episodes."[142]

Fyall's statement offers an interesting version of history. There was a time, during World War II and the early Cold War, when local communities and the state of Washington were almost never consulted about developments at Hanford. Beginning in the 1950s, and then increasingly during the 1960s and 1970s, however, the communities around Hanford made sure they would not simply be "pushed around." In many instances, the communities (or, rather, local economic and political elites in those communities) had pushed U.S. government agencies back. And they had done so remarkably effectively, gaining substantially more federal investment in their communities than they had had any reason to expect. After 1950 or so, the Tri-Cities had proven repeatedly that federal power over Hanford was not "absolute," that Hanford politics were not simply a zero-sum game with one winner and one loser.

Although Uncle Sam had wielded tremendous local power and made significant mistakes at times, events of the 1990s did not demonstrate six decades of unmitigated federal omnipotence and blunder. Rather, they suggested that the federal government itself was becoming even less monolithic than before. While Fyall wanted to portray federal agencies as the chief villains, the forces seemingly arrayed against the Tri-Cities went far beyond the national government. In the 1990s there were many non-federal entities, groups, and individuals with a keen interest in Hanford, the Wahluke Slope, and the Columbia River. Perhaps the biggest difference compared to before was that these entities were now taken seriously

as stakeholders, in part because of the federal government's new commitment to large-scale public participation. The Tri-Cities political and
economic elite was also a stakeholder, of course, and had been the most
important and effective stakeholder for a long time. But by the 1980s and
1990s it had lost much of its clout in Washington, D.C. Its authority was
challenged even within the Tri-Cities, where environmentalists advocated
stronger preservation programs than the old economic and political elite
wanted. By the 1990s Tri-Cities boosters were but one of many groups lobbying the U.S. government about lands, waters, and economic opportunities on and around the nuclear reservation. Their major obstacle was not
Uncle Sam but the welter of other interests that were now more motivated
and more able than ever before to promote competing visions for the area
around Hanford. It may have seemed to some in the Tri-Cities as if they
were taking a step back to the 1940s, when local people had no say over
what transpired on or around the nuclear reservation. But the present was
not repeating the past, even though the past was often placed at the center
of political debates concerning Hanford.

By the 1990s, more or less well-articulated bodies of opinion appealed
to the past to resolve questions about how to further develop, remediate,
or provide compensation for the nuclear industry at Hanford, even while
the general public remained largely indifferent. Some critics argued that
nuclear technology, by its very nature, was undemocratic—witness the
secrecy surrounding Hanford's environmental hazards over the decades.
"One must still ask why so little postwar debate took place on the decision to develop an atomic arsenal capable of ending civilization on earth
as we know it," wrote Paul Loeb in his description of life in the Tri-Cities.
"Instead of holding a public dialogue, the matter became a military, and
therefore effectively technical choice."[143] Another view held that while
secrecy was understandable in the contexts of World War II and the Cold
War, democracy could and should be appealed to in accomplishing the
post–Cold War mission of cleaning up the site, and doing so would be a
marker of the mission's success.[144]

Could democratic processes apply to nuclear technology and its lega-
cies, or not? Absent broad participation, the question fell to those with
particularly well-defined concerns. Two groups claimed a special status
for participation in the policy debates surrounding Hanford as the 1980s
turned into the 1990s. In the Tri-Cities, many appealed to history—not to
archival evidence but to experiences of past technical expertise, direct or

indirect. Many could lay claim to one kind of lived experience or another, from the hands-on knowledge derived from working with Hanford technology to second-hand experience derived from living with a Hanford employee. This sense of commonly held community experiences stood as the source of authority for many residents of the Tri-Cities, members of the "Hanford Family." Another sort of lived experience distinguished those who saw themselves as victims of the site's operations, and so as witnesses who spoke with a special credibility. Appealing to two very different histories, both of these groups implicitly reinforced an undemocratic political culture surrounding nuclear technology by their claims for special political privileges. As plank-holders and stakeholders, respectively, their voices were loudest in the debates over Hanford's environmental legacy and its post–Cold War missions.

Four

HANFORD AND THE COLUMBIA RIVER BASIN

Economy and Ecology

BY THE EARLY 1990S, THE HANFORD SITE WAS COMMONLY IDEN-
tified as one of the most polluted places in the United States. The U.S.
Environmental Protection Agency, noting the 440 billion gallons of liquid
radioactive and chemical wastes released into the ground and distributed
about the place after 1943, ranked Hanford as "the most contaminated site
in the nation." And although production of weapons-grade plutonium
had ceased in the late 1980s, tons of plutonium remained behind, in stockpiles or mixed with other wastes held in tanks or dumped into the soil. So
polluted was Hanford that in 1989 the Department of Energy (DOE) had
planned to spend thirty years and $57 billion cleaning it up. It soon became
clear that much more time and money would be needed. And although
people initially spoke about restoring Hanford to "pristine" condition, it
soon became clear that merely containing the wastes on the site would
prove difficult enough.[1]

While the nation took stock of the ecological damage at Hanford, it
simultaneously considered some fifty miles of the Columbia River that
flowed through the nuclear reservation. During the 1930s the architects of
what would become the Columbia Basin Project had planned to convert this

201

stretch of river, known as the Hanford Reach, into a lake by building a dam near Richland. But by the early 1940s Hanford had harnessed the waters of the Columbia for cooling the reactors as well as carrying away and diluting wastes; the moving, cold river water had been one of the features of the site that commended Hanford to army and DuPont engineers. Moreover, had the river been backed up by the proposed Ben Franklin Dam, it would have disturbed the wastes deposited in the grounds around the site. Consequently, the Hanford Reach remained undammed, and during the later 1980s it gained recognition as the last "free-flowing" portion of the Columbia in Washington State. The reach's relatively undisturbed condition also made it into the major wild spawning grounds for Chinook salmon in the lower forty-eight states.[2] Because of the picturesque virtues of the Hanford Reach, beginning in 1987 Congress regularly considered naming it a "wild and scenic river" and enacting protections for both the stream and the adjacent lands.[3] State and local interests, who feared that federal management of the Hanford Reach would undermine economic opportunities, succeeded for a time in blocking protective legislation. But in 1999–2000 the Clinton administration, using executive powers, set aside the reach as a national monument and perpetuated federal protection for the adjoining lands of the Wahluke Slope by including them in a national wildlife refuge. As at other nuclear-industrial sites, this action kept some contaminated lands from development while preserving the appearance of open space.

The juxtaposition at century's end of the Hanford Site and the Hanford Reach—one the nation's most polluted spot, the other sections of river and adjacent lands preserved for their natural attributes—is emblematic of the unusual mixture of environmental conditions in the vicinity of Hanford. The kinds of land uses that bordered one another represented contradictory extremes. On the one hand there were the highly industrialized zones within the Hanford Site, with enormous pollution and the state's only commercial nuclear reactor. Industry at Hanford had helped create the highly urbanized area known as the Tri-Cities. Furthermore, both factory and metropolis have coexisted with thousands of acres of irrigated fields, treated regularly with pesticides and fertilizers. Industry, cities, and farms have all contributed to damming, polluting, and consuming portions of the Columbia and its tributaries. On the other hand, the vicinity around Hanford is studded with plots of land and stretches of water being preserved in a relatively "natural" or undeveloped state. Besides the protected segment of the Columbia River, wildlife refuges, state game preserves, county and city parks, and a state habitat management area bordering Hanford have been proposed and created. In fact, portions of the nuclear reservation have been set aside for ecological research and wildlife preservation, and other parts of the site have been recognized as historic landmarks.

The diverse forms of land use around Hanford epitomize the emergence of a hypercompartmentalized West. In the late twentieth century, the use of different resources in this region, in particular land and water, became increasingly fragmented. As more numerous and more contrasting claims were made upon the regional environs, Westerners increasingly set about dividing territory into more and more specialized parcels. Relying in particular on governments to draw boundaries on the land and to create and manage a welter of administrative districts, Westerners have come to inhabit a region consisting of a patchwork of mutually exclusive, or contradictory, uses.

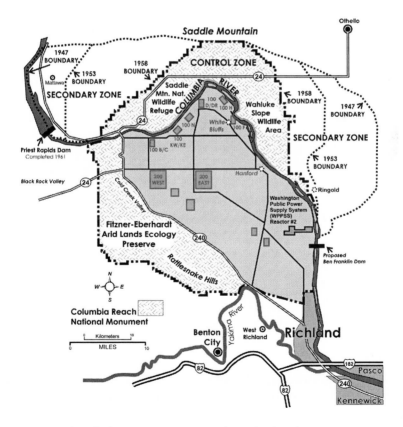

4.1 Map of Hanford Environs, c. 2000. Map by Frederick Bird.

Around Hanford and across the region, such hypercompartmentalization of natural resources was predicated on rapid expansion. At the start
of the twentieth century, many people viewed the West as a largely vacant
region. But waves of demographic and economic growth forced people to
reconsider their imagery. By the mid- and late twentieth century, Westerners had found too many uses—including preservation—for much of the
region, even in arid areas such as Hanford that had previously attracted
relatively few inhabitants. While the growth of the West's population and
economy generated more competition for the region's land and water, the
rise of environmentalism exacerbated compartmentalization by ensuring
that two different visions of western resources would come into conflict.

The powerful drive to "develop" resources, to master nature for practical gain, served as the foundation for one of the competing visions. This
drive had long been shaped by capitalism, which framed exploitation of
resources to supply markets and to benefit individuals and businesses. In
the Hanford vicinity, private enterprises included farming, food processing, transportation, and a variety of other businesses and services. But in
matters of exploiting resources, the private sector in the Columbia Basin
and around the American West was, by the middle of the twentieth century, tightly intertwined with a public sector of dams, hydroelectric and
nuclear power, irrigation canals, highways, and other government-sponsored facilities. Furthermore, the public sector in the West increasingly
included a range of activities, particularly related to national defense, that
stood at a substantial distance from capitalist markets. Plutonium production at Hanford—while overseen by such private firms as DuPont and
General Electric (GE), and while undertaken in defense of "capitalism"
and against "communism"—responded not primarily to private-sector
demands but to the needs and preferences of the national security state.
The development that occurred around Hanford reflected the drive by private as well as public interests to exploit the region's resources to the fullest
extent possible.

Over the course of the twentieth century and particularly since the
1960s, however, there has emerged in the West another impulse—to set
aside, study, and protect certain lands and waters, in their more or less
"natural" state, for aesthetic, recreational, moral, scientific, and economic
reasons. It was no accident that this preservationist impulse crested at the
same time that exploitation of resources was peaking. The rise of environmentalism was, of course, a nationwide movement during the twentieth

century, but in the American West—including at Hanford—it also represented both an outgrowth of and a response to urbanization and industrialization. By the late twentieth century the tension between exploitation and preservation became intense.

Preservationists generally used the public sector to ensure protection of lands and waters; consequently, different governmental agencies created many of the more "natural" compartments around Hanford. Indeed, the first public entity to set aside land for explicitly environmental reasons was the Atomic Energy Commission (AEC), the same body that did so much to industrialize, urbanize, and pollute the surrounding area. Local inhabitants depended upon and influenced the federal agencies in numerous ways, yet many of them immediately grew suspicious of government whenever it proposed to preserve natural resources rather than facilitate their direct economic use. Public-sector protection of lands and waters came to signify an intrusion into local affairs. And the intruders included people other than the federal government. For example, the campaign to protect the Hanford Reach received plenty of support from urbanites and fishing interests west of the Cascade Mountains and from national environmental groups such as American Rivers and the Nature Conservancy. The mostly local opponents of national protection argued that "their" river, along with the economic opportunities for which it stood, was being appropriated by outsiders. Clashes between local and nonlocal interests complicated the tension between development and preservation.

If today there seem to be too many uses for the lands and waters around Hanford, in the early twentieth century there seemed to be too few. To understand the significance of the nuclear site to the environs of the Columbia Basin, one must appreciate its relationship to the enduring idea that the interior West was an "empty" region. When the U.S. Army looked for a place to build a plutonium factory, it sought plenty of electricity and free-flowing fresh water, but it particularly wanted a largely unoccupied and isolated place where the condemnation of land and the production of fissionable material would not affect too many people. It needed to keep the plant a secret, and it needed to minimize the risks that plant operations would present to nearby populations. Similarly, the Manhattan Project had considered locating its new production reactors at Oak Ridge, Tennessee, but that site was neither isolated nor large enough.[4] So the army turned to the region that seemed to contain more "vacant" space than any other, one where land was relatively unoccupied and unencumbered.[5]

Of course, the West was never quite empty enough. There were too many people, native and nonnative, who would resist the federal takeover of land, too many people who would be exposed to the radioactive emissions and other pollutants generated at nuclear weapons facilities. The West's ecosystems could never fully dilute or discharge the wastes created while producing nuclear weapons. Even if the West *had* been truly empty, however, federal bomb making would have made sure that it did not stay that way, because nuclear weapons facilities accelerated the urbanization and the industrialization of the region. During World War II the army sought to empty relatively unoccupied places such as Hanford by evicting their inhabitants. After creating a blank slate, the army refilled selected parcels by building cities, laboratories, and factories, creating urban enclaves and attempting to run them with little consultation with local people. Adjacent tracts of land were set aside as buffer zones and waste storage areas, advancing the compartmentalization of the land. The layout of Hanford exemplified a broader regional pattern—urban and industrial nuclei surrounded by vast expanses of relatively undeveloped, and in some places highly polluted, land.

Atomic-weapons programs transformed parts of the West in several ways.[6] The new urban and industrial ways of life certainly disrupted older modes of existence, but they also served to integrate remote parts of the West into the modern patterns of growth that increasingly dominated the region in the mid-twentieth century. And because federal nuclear-weapons programs hastened economic and demographic expansion, many neighbors overlooked the disruptions they caused and the risks they presented and embraced the new opportunities to speed up development.

Westerners were generally eager to exploit the opportunities presented by the army and the AEC. While they too regarded the region as relatively empty, for them, "empty" connoted not so much a lack of population as a lack of economic growth. Most non–Native Americans and certainly most inhabitants of the West prior to the late twentieth century regarded the region as underdeveloped. The activity and people that accompanied the Manhattan Project and the AEC thus appeared as solutions to long-standing problems. Once it was no longer a secret, Hanford promised prosperity and so was welcomed by many inhabitants of the Columbia Basin.

At the same time, Hanford's neighbors did not want it to interfere unduly with other enterprises in the vicinity. In particular they hoped that the new urban and industrial activity would be contained—compartmentalized—so that it would threaten neither the established rural economy nor the projected additions to that economy. Farmers and ranchers hoped that the plutonium factory would not disrupt agriculture and stock raising. Growers eagerly awaited the benefits of the Columbia Basin Project, a proposed series of dams on the Columbia and Snake rivers, which were expected to improve irrigation, rural electrification, flood control, soil protection, and inland navigation. Hanford had benefited from this New Deal–era effort—after all, dams provided the electricity that made the site a logical place for plutonium production—but Hanford also competed against the project for access to land and water.

In sum, the advent of the Hanford Engineer Works during World War II initiated a process by which the surrounding environment became steadily more contested and compartmentalized. Here we discuss a series of episodes that illuminate Hanford's shifting relationship to the surrounding environs as well as changing attitudes toward development and preservation of resources in the Columbia Basin. The issues at play at Hanford loomed large for the West as a whole. Along with the tension between rural and urban-industrial values, there emerged the matter of federal versus local control over development. At about half the size of Rhode Island, Hanford's expansive boundaries were supposed to compartmentalize most of the waste generated on the site, holding it in storage tanks or fixed in the soil. Smaller amounts of waste were released, deliberately or by accident, into the currents of air or the water leaving the reservation. These releases began to be publicly accounted for in the late 1980s, when a vision of abundant, cheap, and environmentally friendly nuclear energy—including that associated with a proposed prototype breeder reactor embraced by the Tri-Cities' nuclear experts—ran afoul of antinuclear sentiments.

Such questions rose to the surface only gradually. World War II and the early Cold War muted rural-urban and federal-local tensions because national emergencies took clear priority over local interests. As time passed and as Hanford's mission changed, however, disputes over resource use became more contentious. The conflict became downright acrimonious after the 1960s, when it became apparent that decades of Hanford operations had produced a striking mixture of intensive industrialization, urbanization, and pollution, combined with various sorts of natural preservation—all of which seemed to jeopardize local prosperity.

This resulting pattern on the land represented a marked departure from 1930s visions of irrigated agriculture supported by federally subsidized dams. In some instances Hanford's neighbors resented the federal government appropriation of so much land and water, taken away from what they regarded as productive uses. They viewed the "preservation" of surrounding lands as the needless perpetuation of the land's emptiness. Yet the latest pattern—an urban and industrial core surrounded by large, somewhat "preserved," and in some cases immensely polluted parcels of lands—illuminated much about the American West of the late twentieth century. The Hanford area epitomized a region of expansive cities and industry, somewhat marginalized agriculture, and a patchwork of sacrifice zones and natural and cultural reserves, managed by a tangle of government agencies. It was at the same time urban and unsettled, ugly and unblemished. And it was a far cry from the relatively "empty" place that people had understood to exist earlier in the century.

Dams and Reactors: Mastering Nature in the Columbia Basin

The lands of the Hanford Site, of course, had never been empty wastes; they had been occupied for centuries. Sahaptin-speaking Indians of the mid-Columbia region ranged across the region and lived there. Hunting and gathering, they migrated seasonally as far west as the Cascade Mountains and as far east as the Blue Mountains. Along *nchi'i-wana* ("the big river") and its tributaries, families resided in semipermanent villages between fall and spring, fishing, manufacturing tools, and preserving the foods they had acquired throughout the basin. It was in these villages that the first nonnatives to enter the basin, members of the Lewis and Clark expedition, observed the mid-Columbia Indians in autumn 1805.[7]

American explorers marked the beginning of an invasion of non-Indian peoples interested in putting the region to new uses. Both European diseases and fauna—particularly horses—had preceded Lewis and Clark to the Columbia Basin, altering native societies and cultures. Following Lewis and Clark came British and American interests eager to engage Indians in the fur trade, preach Christianity, establish national claims to territory, catalogue the resources of the country, and prepare the way for settlers and their exotic species of plants and animals. The diseases introduced by immigrants reduced the Indian population and weakened its ability to interact with the newcomers on its own terms. By 1846 the United States had eliminated British claims to the territory and begun sending families to reside permanently in what was known as the Oregon Country. By 1853 the northern half of this acquisition became Washington Territory. The first territorial governor, Isaac I. Stevens, devoted his office to moving Indians out of the way of white society and onto reservations. The Yakima Treaty of June 9, 1855, helped to displace most Indians of the mid-Columbia, although some families did not sign and remained behind in traditional village sites on the river.[8]

Initially, the arid and windy lands that became the Hanford Site did not appeal to nonnative settlers. Over the last four decades of the nineteenth century, some ranchers grazed livestock there, and an assortment of trails, roads, and river routes crossed the site, taking people primarily to distant destinations. The railroad crossed the Columbia at what would become the Tri-Cities on its way to Puget Sound in the 1880s; earlier, it had turned south over the Oregon border on its way to Portland. During the first decade of the twentieth century, two small irrigation projects were built near the townsites of Hanford and Richland, allowing for the planting of orchards and inaugurating sustained agriculture in the vicinity. These efforts had a tenuous existence through the next three decades, for the price of crops did not always offset the cost of water.[9] Outside of the irrigation projects, most land remained covered with sagebrush and of little value to either farmers or ranchers.

Experiences in the Cold Creek valley, located in the northwestern sector of the future Hanford Site, demonstrated the obstacles to settling. About ten families attempted homesteads between 1906 and 1943, but most failed to prosper and eventually moved away, in large part because of the limited supply of water. Each household drilled an artesian well, but the opening of each new well tended to cause others nearby to run dry. Even wells eight miles to the west, in the Black Rock valley, were depleted by a new well drilled in Cold Creek valley.[10] Small-scale irrigation projects around Hanford and Richland enjoyed more success. Without a major effort at reclamation, however, there was simply too little water to support extensive settlement.

The New Deal offered glimpses of a solution to the problem of aridity. The Roosevelt administration embraced proposals to build federal dams on the Columbia River and develop a larger and better-financed scheme of irrigation than had been possible before. Farmers on the west side of the Columbia could not expect to benefit soon from irrigation efforts, yet proposals to dam the river offered hope of a better life ahead in such forms as protection against soil erosion, rural electrification, inland navigation,

4.2 During the first four decades of the twentieth century, irrigation permitted the development of agriculture in the vicinity of Hanford. The arrival of the Manhattan Project during the early 1940s converted farmland to urban and industrial uses. Yet some of the wartime workers at Hanford, including scientists and their families, took time out to go swimming in the canals that had delivered water to crops. The great Italian physicist Enrico Fermi appears to be sitting fifth from the right in this photograph. Fittingly enough, Fermi's Manhattan Project code name was "Farmer." Photograph courtesy of the U.S. Department of Energy.

and flood control. The proposal to build the Ben Franklin Dam just north of Richland promised to back the river up along what would become the Hanford Site.

The Columbia Basin Project, however, did not provide much help to these farmers and ranchers. Most of the dams and all of the related irrigation works were not built until after World War II, and throughout the 1930s sizeable numbers of farmers left Richland, Hanford, White Bluffs, and the Cold Creek valley because they could no longer bear up under the poverty. "Hanford had been a wonderful place to live," recalled settler Virginia Enyeart Sever in later years, "but not a good place to make a living." When the army arrived to look the place over in 1943, its observers reported an area in decline.[11]

Although the dams were mainly intended to stimulate the rural economy of the Columbia Basin, their presence in 1942 was crucial to the selection of Hanford as a plutonium-producing site. The hydroelectricity they generated attracted Colonel Franklin T. Matthias and the army, and so encouraged the development of an urbanized industrial center associated with Hanford; by the time most dams and irrigation canals were built, much of the land and water around Hanford and the Tri-Cities had been converted to purposes other than farming.

The residents and boosters of eastern Washington, who had anticipated that their needs and wishes would loom large in decisions about local development, never got much of a say over the benefits of harnessing the Columbia and its tributaries. The national government, which financed the bulk of river development, placed other interests above those of local residents. For example, during the 1930s, decisions about the distribution of hydroelectric power favored existing urban and industrial centers in the Northwest over prospective users in the Columbia Basin itself. Then after 1940 the needs of the nation, particularly during mobilization for World War II and the Cold War, took precedence over state and local desires in the development of the mid-Columbia area. Hanford ultimately took some farmland out of cultivation and prevented completion of certain planned components of the Columbia Basin Project—all in the name of national security. During the war it consumed all the electricity produced by two generators at the Grand Coulee Dam powerhouse, and it secured privileged access to Bonneville Power Administration (BPA) hydropower on an ongoing basis, at the expense of other customers.[12] Local economic interests would get to pursue their aims only when and where national considerations permitted.

A new era of government intrusion and land use at Hanford began in 1943 when the army condemned roughly 3,000 tracts of land held by roughly 2,000 owners. The army calculated that 88 percent was "sagebrush sheep range" and 11 percent was "irrigable farm land, including orchards." About one-third of the site had belonged to other government agencies before the Manhattan Project took over.[13] As underdeveloped as the land was, however, some of its occupants resented being told that their holdings were not very productive; they resented being evicted and not knowing why the condemnations took place. "We thought we were up and coming, and they called us 'sleepy,'" recalled displaced farmer Edith Hansen. "They didn't give us much money and more or less insulted us."[14]

The Manhattan Project undermined one way of life and put another in its place. Around the towns of Richland and Hanford, the government bulldozed orchards, buildings, and irrigation canals and replaced them with new manufacturing facilities and communities. Outsiders acquired an even greater ability to direct local development. From then on, the Hanford Site would operate as a cog in a bureaucratic, industrial, and scientific machine, its motion directed largely by decisions made in the nation's capital, in DuPont's Delaware headquarters, and in research laboratories across the country.

For all the disruption that Hanford caused, however, most local residents quickly decided to embrace it. True, the rural populace had little choice but to tolerate the dislocation, and those who remained had to accommodate the new industrial order. The existence of a national emergency undercut some bitterness, and the new regime also brought an unprecedented prosperity, which doubtless reduced the amount of criticism. Perhaps more importantly, though, the environmental and social conditions associated with the Manhattan Engineer District did not appear at the time to be as disruptive as they must have been in retrospect.

While the changes wrought by World War II were substantial, it is easy to overstate them. In his study of Hanford, Los Alamos, and Oak Ridge, cultural historian Peter Bacon Hales interprets the Manhattan Project as the end of American innocence by characterizing it as the dividing line between "a pastoral, individualistic, democratic past and a bureaucratic, efficient, authoritarian and technocratic future." The effort to build the bomb, he contends, stood "completely . . . at odds with the basic values of American democracy of its time, even with wartime patriotism and sacrifice taken into account." Blending together the traits of "corporate capitalism, government social management, and military codes of coercion and obedience," the Manhattan Engineer District and Atomic Energy Commission colonized American society more generally by replacing democracy with a culture of repression. Hales portrays the change as particularly harsh in less urbanized and less industrialized regions such as prewar Hanford, in which community is portrayed nostalgically as "a tight-knit mutual-aid society."[15]

But this is an exaggeration,[16] partly because Hales overestimates the project's power and influence after 1942 but also because he imagines that the United States prior to 1942 was somehow more "innocent" than it actually was. For decades before World War II, Westerners had experienced and complained about the "colonization" of their region by such big businesses as railroad companies and such government agencies as the U.S. Forest Service. But they also cultivated industrial and state actions to provide economic development that otherwise seemed lacking. In the vicinity of Hanford during the 1930s, inhabitants had become used to the idea of federal bureaucracy, big technology, and government social engineering as they witnessed the construction and operation of new dams on the Columbia River, a process as technocratic as anything associated with nuclear technology.[17] It is also important to remember that Westerners never remained passive in the face of changes imposed by outsiders. Often they had helped to bring those changes about—lobbying for the construction of both the railways and dams, for example—and they had consistently asserted their own interests against those of more powerful and more wealthy outsiders.

The Hanford Engineer Works surely disrupted life when it appeared in south central Washington in 1943, but it was hardly without precedent in the Columbia Basin or in the broader region. Some Northwesterners certainly resented its impact, but many more learned to view it as more of a blessing. Development in the vicinity was not yet viewed as a strict zero-sum game in which Hanford's growth inevitably interfered with other enterprises. Hanford did displace some farmers from the lands taken over by the army and, later, the AEC, and it called into being towns on sites where orchards and ranches had previously existed, but both the remaining rural population and the incoming urban population found much to like in the altered conditions. They witnessed a new investment in infrastructure, new markets for crops and labor, and a new atmosphere of affluence.

That Hanford struck neighbors as something that could be turned to the region's advantage was nowhere more evident than in how it came to be harnessed, at least rhetorically, to the designs of the Columbia Basin Project. A variety of economic interests, including farmers, shippers, and town boosters, quickly discovered that mobilization for national security could serve their economic aims quite well. During the early 1950s local boosters campaigned on behalf of building more federal dams along the lower Snake and Columbia rivers. The dams, which represented the completion of the Columbia Basin Project, stirred controversy because of the risks they posed to salmon populations. Commercial and sports fishing interests and Indians opposed the new construction, while most growers and shippers of produce supported development to extend inland navigation.[18]

Proponents of more dams had already pressed their case by referring to the merits of hydropower, irrigation, and inland navigation, but these arguments proved insufficient to win congressional approval. Then in the early 1950s, with the massive American military buildup, national security needs were invoked as additional justifications for the dams. When a 1951 presidential report declared that the nation needed up to 4.5 million more kilowatts of electricity for defense production, the Inland Empire Waterways Association replied that three new dams on the Snake and the Columbia could provide half the needed power. Worried about the BPA's ability to supply Hanford with electricity, the AEC manager asked Senator Magnuson whether the entire hydroelectric output of the proposed Ice Harbor Dam on the lower Snake River could be promised to Hanford. The answer was no, but the request strengthened the overall case for the dams. Responding to a constituent's concerns about the environmental effects of the dams, Senator Wayne Morse of Oregon explicitly ranked the needs of Hanford over those of salmon: "In the years ahead, we are going to need every kilowatt of electric power we can generate in this country, if we are going to maintain our superiority over Russia in the field of atomic defense."[19]

The *Columbia Basin News*, a Pasco-based newspaper that counted farmers and ranchers among its primary readership, advocated building more dams in 1950–51 by linking rural economic development to national security. The proximity of Hanford, and the concern for military preparedness that the plutonium plant symbolized, made its interpretation all the more persuasive. According to the *News*, the United States needed new dams on the Snake and the Columbia, and needed them right away because of severe international tensions. At any moment, as the war in Korea demonstrated, a major crisis might break out, requiring Americans to produce more food and generate more hydroelectricity. The nation could not afford to wait to build dams until after the crisis occurred. National-security needs also justified construction of all four proposed dams on the lower Snake River, according to the *News*, making it more difficult for the enemy to sabotage the river system. Finally, there were explicitly regional dimensions to consider. The *Columbia Basin News* urged construction on the Snake and the Columbia before other parts of the country could begin building, so that federal money would flow to the Pacific Northwest and not elsewhere. The newspaper insisted that the United States must "begin to develop every national resource at its command. The west is the last economic frontier. Its industrial and agricultural settlement is imperative."[20]

In this context, virtually nobody thought about—let alone advocated for—preservation of natural resources. For boosters, the Second World War and the Cold War only made more urgent a goal Westerners had long advocated: filling up the empty West. The *Columbia Basin News* summarized this development as the "mastery of man over Nature," an effort to achieve economic and cultural progress, affluence for the entire society, and the advance of "civilization." Boosters took note of the risk that more dams presented to salmon but argued that destroying the fisheries was a small price to pay for the expected benefits. "If the state is to be maintained as a glorified national park, then the desert will remain desert, and the productive potential must be discarded and forgotten," the *News* explained. "If full-scale industrial development is desired, it must be realized that civilization and the primitive are not compatible and some sacrifice of natural resources must be expected."[21] Local boosters defined the possibilities quite starkly. There would be either full-scale development and civilization or preservation and a primitive way of life. Few imagined a situation whereby preservation and exploitation might coexist.

The *News*'s editorial position helps explain why Hanford and its neighbors could get along so nicely. Today we look back and wonder how people could have tinkered so carelessly with natural forces that they did not understand very well. We are startled by the seeming casualness with which the army and the AEC made the river, the soil, and the air part of the industrial plant. But was this use of natural resources so dramatically different from the hydroelectric and irrigation works already harnessed to the Columbia and its tributaries? And did it differ so much from what went on in any number of company towns in the Northwest, particularly those devoted to mining and smelter work? In many respects those who made atomic bombs did not differ much from those who were willing to sacrifice species of wild salmon and stretches of wild river to the goals of prosperity and civilization. Building dams and splitting atoms just represented two different paths—both paved with massive federal investment—toward achieving fuller mastery over nature.

Living with Hanford's Risks: Debate over the Wahluke Slope

In the dominant view of the mid-twentieth-century West, it was the underutilization rather than the overutilization of resources that seemed wrong. Most people in the Pacific Northwest valued Hanford as an eco-

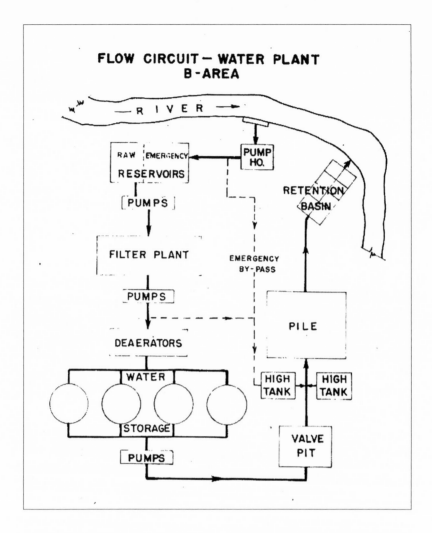

FLOW CIRCUIT – WATER PLANT
B-AREA

nomic asset that would help the region overcome its "empty" status. Boosters expected that the nearby economy could have both its agricultural and its industrial development. They assumed that their job, and the job of the government, was to ensure the maximum amount of growth by seeing to it that one kind of economic activity did not unnecessarily limit activity in other areas.

In fact, of course, the production of plutonium *did* present serious risks to the surrounding economy. While the plant did not remove commodities such as lumber, minerals, or fish, its operation did inject radioactive and chemical wastes into the air, ground, and river. Hanford contaminated the environment and thereby threatened the livelihoods (not to mention the personal health) of farmers, ranchers, and fishers. The army and the AEC, however, never revealed the full extent of Hanford's dangers. They did not

4.3 and 4.4 (left and above) Whereas most of Hanford's neighbors had regarded the Columbia River as a fount of water for irrigation, Manhattan Project plans conceived of the stream as part of the plutonium plant and essential for providing a steady flow of liquid to cool the reactors and carry away effluents. The "flow circuit" chart from World War II shows how water for the B reactor was diverted from the river, filtered, stored, circulated through the reactor, and then held in retention basins before being returned to the Columbia. Fig. 4.3 illustrates not only why Hanford's first eight piles were built alongside the river but also how much of each complex of buildings was devoted simply to moving, processing, and storing water—an ingredient absolutely essential to the operation of the site. During the Cold War, managers constantly enlarged the pump houses that piped water uphill from the Columbia to the reactors. Fig. 4.4 shows the recently expanded pump house serving the B reactor during the early 1950s. Both images courtesy of the U.S. Department of Energy.

always fully understand the risks themselves, but they also did not want to publicize those risks that they did understand, for a host of reasons. So those neighbors trying to extract wealth from the land and river had to base decisions about their livelihoods and personal well-being on the very limited knowledge they had about the effects of Hanford on the local environs. Yet it is not clear that, had the local population known more about Hanford's dangers, they would have behaved much differently. In the decades immediately following the war, Hanford's neighbors explicitly rated its environmental hazard as less important than its economic promise.

While the Columbia Basin Project and the Hanford Engineer Works shared similar assumptions about using natural resources, the two programs competed for some of the same water and land. The plutonium plant required an adequate river flow and a protected water table below the nuclear site, disturbances of which might jeopardize the storage of radioactive and chemical wastes. So it was Hanford's presence that prevented the construction of the Ben Franklin Dam. Hanford also required an uninterrupted flow of hydroelectric power, so it secured from the Bonneville Power Administration a higher priority than all other users.[22] Furthermore, the Hanford project tied up lands that had once been slated to receive water as part of either the Columbia Basin Project or some other irrigation effort. As one example, Hanford encouraged the urbanization of areas in and around the Tri-Cities, which resulted in the conversion of farmland and pasture into homesites and other urban uses. Even more emblematic of the competition, however, was Hanford's takeover of the Wahluke Slope, across the Columbia River in Franklin and Grant counties, just north and northeast of the reactors in Benton County (see figure 4.1).

Hanford's critics have claimed that the army and the AEC were not concerned enough about the threats that Hanford presented to the health and safety of its neighbors. While there may be some truth to this charge, it is important to note that the army and the AEC paid considerable attention to some threats and less attention to others. The army and the AEC took it for granted that producing plutonium would generate continuous releases of pollution to the soil, water, and atmosphere. They attempted to keep these emissions to what they regarded as a safe level, but their overriding concern was the production of plutonium rather than environmental protection. Consequently, the day-to-day release of radioactive and chemical wastes elicited relatively little concern from those who managed the plant. In contrast to their limited efforts to cope with the daily emissions from the normal operations of Hanford, the army and the AEC paid substantially more heed to the risk of catastrophic releases that might result from some sort of disaster at the plant.[23] This attitude shaped the way Hanford's managers regarded the Wahluke Slope.

Government control of much of the Wahluke Slope represented a precaution against the possibility of a major accident at the Hanford plant. In the event of an explosive release, it was believed, the prevailing winds would carry radioactive emissions north and northeast from the reactors, across the river and on to the Wahluke Slope. Consequently, in 1947 the AEC determined that it needed to prevent the occupation or development of more than 280,000 acres on the slope. Of this parcel, 88,000 acres comprised the control zone along the river and nearest the reactors, where the AEC sought "complete and permanent control" by condemning and acquiring land. The remaining acreage, mostly privately owned, was designated the secondary zone (see map in fig. 4.1 above). The AEC prevented development and settlement in this area, too, largely by getting the U.S. Bureau of Reclamation to agree not to extend irrigation to it. Through its control over the surrounding landscape, the AEC ensured a minimum of interference with Hanford production. By contrast, building new facilities, redesigning reactors, or fitting them with containment structures to protect inhabitants of the Wahluke Slope would have disrupted operations considerably. Meanwhile, the AEC held out hope that advances in reactor safety would eventually permit it to release these lands for development and habitation.[24]

The catastrophic risks posed by Hanford's reactors were the subject of several warnings from safety experts who advised the AEC. The commissioners made policy with the assistance of a Reactor Safeguards Committee, which drew attention to hazards posed to Hanford's surroundings by

4.5 During the Cold War, Hanford operators posted signs along highways adjacent to the reservation to warn motorists away in the event of an accident. Photograph courtesy of the U.S. Department of Energy.

the potential for a reactor accident, which could distribute the radioactive
fuel load and a wide variety of fission products over the landscape. The
committee had pointed out in 1949 that "the distribution of stack materials
over the Wahluke Slope and over many other areas to great distances is a
cause of much concern. Their elimination is essential for continuing safe
operations in the Hanford area." Because the recommended safety zone
around each reactor scaled up as a function of power, locating two reactors near each other would require a greatly increased buffer area. In addition, the Department of Defense wanted U.S. reactors separated by two
to two and one-half miles as a measure of protection against air attack.
Still, in the face of all this advice, in June 1952 AEC commissioner Henry
Smyth argued that "the most important consideration was time, and that
he favored the present twin-area plan rather than lose nine pile months
of production for further insurance factors." The AEC agreed to site the
reactors in pairs at Hanford, taking advantage of common cooling water
facilities. Smyth hoped to marginalize the advice of the Reactor Safeguards
Committee in favor of an emphasis on more rapid production of strategic materials, a position that also found favor with some of the industrial
contractors entering the field. This system, in which expertise on the part
of reactor safety experts flowed not into the public realm but only into
the deliberations of the AEC commissioners, manifested the "extreme cen-
tralization of decision making within the AEC," noted by historian Brian
Balogh, making safety a secondary concern.[25]

But the emphasis on centralized decision making in nuclear histories
arises in part from situating these stories so firmly within Washington,
D.C. The emphasis on production of material for nuclear weapons surely
reflected, at least in part, a broad political consensus within the Cold War
context. More to the point here, decisions about reactor safety zones—that
is, land use decisions that affected Hanford's agriculturalist neighbors—
were shaped also by the AEC's inability to ignore pressure brought to bear
by local interests in favor of opening up agricultural lands, even if to do so
entailed some risk to the users.

The AEC's preemption of the Wahluke Slope conflicted directly with
the uses proposed by the Columbia Basin Project, which had raised the
possibility of irrigating the acreage. Landowners and entrepreneurs in
the vicinity complained about this limitation on local economic development. The Bureau of Reclamation pointed out that, without the lands now
under AEC control, the development of irrigation in the vicinity would be

more expensive and less cost-efficient. At work here was the continuing assumption that the nation, the West, and the local economy all needed to maximize the utilization of resources. As the 1950s progressed, moreover, there emerged another concern, especially from Tri-Cities interests, that the local economy, too dependent on a single industry, needed to diversify. Consequently, farmers, landowners, business interests, boosters, and the Bureau of Reclamation pressured the AEC to reevaluate and, on two occasions, to reduce the size of the secondary zone. Their efforts were among the earliest occasions in which spokesmen for local interests convinced U.S. senators and representatives to lobby the AEC on their behalf.

Announcement of the first change came on January 8, 1953, when the AEC trimmed roughly 87,000 acres from the extreme southeastern and northwestern ends of the secondary zone. The commission indicated that it would also permit the temporary construction of canals and roadways through the remaining restricted areas, but continued to oppose occupation or regular work there. In publicizing this decision, the AEC carefully explained that hazards from the possibility of reactor accidents still existed. It promised to educate people who would be working or living in the newly released lands on the dangers they faced, and to provide a warning and evacuation system for emergencies. The AEC also reiterated its policy that "for safety reasons no towns or cities should be established within 25 miles of the Hanford reactor area." That a limit of this size did not apply at all atomic sites, however, suggested that Hanford posed a higher level of danger to neighbors. C. Rogers McCullogh, chairman of the AEC's Advisory Committee on Reactor Safeguards, explained in a 1958 letter to AEC Chairman Lewis L. Strauss just why the site's "unique" reactors, which lacked containment shells, presented particular risks. "The Hanford reactors have been and still are potentially dangerous facilities because of the massive escape of fission products which would occur in the event of loss of coolant. This hazard becomes progressively more serious the higher the power level at which the reactors are operated. In allowing these reactors to operate at the present power levels or at the proposed increased power levels, the Atomic Energy Commission is accepting a degree of risk which, in the opinion of the Committee, is greater than in any other existing reactor plant."[26]

The Advisory Committee on Reactor Safeguards had informed the AEC at the beginning of the Wahluke Slope controversy that a major accident at Hanford was both more likely, and bound to be more severe, because of a

20 to 33 percent increase in reactor power proposed to increase output. Yet in the early 1950s the AEC tended to control advisors rather than let them set policy or make decisions relevant to their expertise. When J. Robert Oppenheimer failed to support the development of fusion weapons in the aftermath of the first Soviet nuclear tests, and subsequently lost his security clearance in well-publicized security review board hearings held by the AEC, a broader pattern was established. Despite some enhanced power given the reactor safety experts serving the AEC by the end of the decade, advisors to the commission were generally to be subordinated to policy makers, not to act as a constraint upon them.[27]

In releasing the 87,000 acres in 1953, the AEC built expectations for additional disposal of land by promising that improvements in the "safety systems" at Hanford would soon reduce the dangers associated with the possibility of accidents. Four years later, however, the gains in reactor safety had been offset by increases in power and productivity, which meant that the AEC still regarded concession of additional acreage on the Wahluke Slope as too risky. This was a rare instance of the AEC, an agency normally quite reluctant to publicize the dangers of plant operations, coming forward with a warning—although it pursued its production program regardless.

Nonetheless, the demands from local interests to yield more land had escalated. In mid-1957, Washington State's Columbia Basin Commission, a body devoted to furthering economic growth, considered taking the AEC to court to secure the release of more of the Wahluke Slope. In October Senator Henry M. Jackson presided at informal hearings in Richland on the matter. There AEC officials explained that, in their best scientific opinion, releasing additional acreage on the Wahluke Slope remained undesirable from the point of view of safety. But Jackson, local farmers and landowners, and representatives of the Bureau of Reclamation all disputed this reasoning. In critics' minds, scientific opinion mattered less than ensuring a range of choices and opportunities within the local economy. While local residents lacked the expertise to challenge AEC opinion about the hazards of Hanford reactors, they objected to being told that they could not decide for themselves whether to live with those hazards, if they wished, in pursuing the "potential prosperity so rightfully theirs in keeping with American ideals."[28]

In rationalizing their right to expose themselves (not to mention their agricultural products and the consumers of those products) to the environmental hazards of plutonium production, Hanford's neighbors adopted a kind of nuclear fatalism. As early as the end of 1946, the Columbia Basin Commission had memorialized David Lilienthal, the first AEC chairman, noting that "this State needs and seeks a rapid growth of industrial enterprise based upon our natural advantages in basic raw materials," and thus, "the Columbia Basin Commission request[s] the Atomic Energy Commission to carry on a major portion of its experimental activities within the state."[29] In the same spirit, those who advocated opening more of the Wahluke Slope in the early 1950s explained that they were "willing to accept [the] challenge" of a possible nuclear "disaster" because in living next door to the plutonium plant they were "not in any worse position than any other spot that you pick out." A 1951 AEC report had noted, "Members of the Columbia Basin Commission have advised the Hanford Operations Office informally that, since bombing is a civilian hazard commonly faced in any locality adjacent to defense plants, or in key cities, it should not be given any added weight in appraising the hazards on the Wahluke Slope." The commissioners now claimed as well that "even a direct hit on a pile would cause only a localized disaster."[30] Hanford's neighbors naively downplayed the risks of living next door to the plutonium plant, and then claimed that if dangers did exist, they were no greater than those faced in any other locale.

Senator Jackson reiterated this spartan pragmatism in his appeal to the AEC to relax its standards. Given the stockpiling of nuclear weapons at sites across the country, the thousands of workers now handling fissionable materials, and the enemy's nuclear capabilities, he argued, "millions of Americans in this atomic age" were living with the risks inherent in "new defense systems." So, Jackson reasoned, it did not make sense to single out Hanford's hazards as any greater than those to which most of the country's population was exposed. "Life is indeed dangerous in this century of tension," and it "is becoming more hazardous every day."[31] Preventing economic development in the vicinity of Hanford was unreasonable, according to the senator, because the people living or working on the Wahluke Slope would be at risk no matter where in the United States they were.

Jackson's argument neatly ignored the fact that, as the AEC said, occupants of the Wahluke Slope would face somewhat more risk than other Americans by virtue of their being downwind from Hanford's especially worrisome reactors. Nonetheless, the political and public pressure to open additional acreage on the Wahluke Slope remained strong, and on December 30, 1958, the AEC yielded to that pressure and eliminated another 105,500 acres from the secondary zone. To justify yielding control over more of the Wahluke Slope, the AEC said that it had scheduled improvements in making reactor buildings more airtight, thereby reducing the risk of exposure to radiation "in the event of a reactor accident short of catastrophe." Clearly, some of the hazards presented by the possibility of a serious disaster remained. Nonetheless, local businessmen and landowners, elected officials, and the Bureau of Reclamation typically welcomed news of the opening of additional lands, because 62,000 of the newly available acres were deemed irrigable and thus promised an economic gain to the reputedly underdeveloped region.[32]

In changing its position concerning the Wahluke Slope, the Atomic Energy Commission may have been joining in the nuclear fatalism of the day after having learned which way the wind was blowing. Its initial anxiety about the Wahluke Slope had stemmed from meteorological information that the prevailing winds at Hanford blew toward the north and northeast of the plutonium plant. A new AEC study of wind patterns in 1958, however, revealed that the Wahluke Slope was no more vulnerable to catastrophic emissions from the plant than were the Tri-Cities to the southeast or the farmlands directly east of the plant—neither of which had been included in the primary and secondary zones.[33] In other words, the AEC had prevented occupation of the Wahluke Slope because it had seemed too risky, but now it learned that in the event of a disaster, the slope would be no more exposed than communities such as Ringold, Richland, Kennewick, and Pasco, which the AEC had overtly done little to protect. Its response was not to evacuate the Tri-Cities, of course—a practical and political impossibility—but to release more acres on the Wahluke Slope and hope that technological advances would eventually make Hanford safer.

Resisting calls to open more of the Wahluke Slope to development put the AEC in a difficult position. It had articulated its concerns that operation of the Hanford reactors posed special dangers to neighboring populations, but this admission of the public hazards of plutonium production went against the overall grain of the public relations effort. The AEC, wanting to portray Hanford as a good neighbor, emphasized the safety of the plant much more persistently than the risks it presented to surrounding areas. No doubt because of its overriding devotion to production, the AEC seemed incapable of dwelling on the special hazards that the reactors presented, and its concern for the plant's effect on the surrounding environment developed but slowly. Even when the AEC did warn of the risks of living next door to Hanford, neighbors preferred not to listen, especially if they were intent on economic growth. Boosters reinforced the pattern. Newspapers such as the *Columbia Basin News* were so committed to promoting the area that they were reluctant to find or print anything unfavorable about Hanford that might drive away investors and settlers. The *News* summarized its approach in a slogan printed on its masthead during the early 1950s: "We're So Busy Boosting, We Don't Have Much Time For Knocks."[34] Other boosters and newspapers—virtually the same thing in the Tri-Cities—followed suit.

Nature and Economic Development in the Tri-Cities

During the 1960s, much of the discussion concerning Hanford and the local environs came to be focused in the urban center of the Tri-Cities. Like farmers, shippers, and irrigators, the residents of Richland, Pasco, and Kennewick did not want Hanford to prevent economic growth; in particular, they hoped their towns could avoid becoming dependent on a single industry. Concern about a one-dimensional economy dated back at least to the late 1940s, when planners hired by GE had pondered the future of Richland Village. On the surface, broadening the economic base seemed like an easier task in Richland than it would be in Oak Ridge or Los Alamos, because Richland had never been behind a fence. Also, the continued development of the Columbia Basin Project appeared to promise an ideal counterbalance to the plutonium industry.[35]

Yet seizing new economic opportunities proved to be almost impossible for the town. The Richland Community Council took up the matter of diversification regularly during the 1950s, even creating an industrial committee to survey the issue. Hoping to develop an employment and revenue base apart from the Hanford Works, the committee's inquiries produced discouraging results. The town had too little desirable housing to accommodate new workers and probably contained too few industrial acres, municipal facilities, and available workers to attract new industry. Even more importantly, it lacked the flexibility to change its circumstances. When W. W. Birchill of the State Industrialization Utilization Committee studied prospects for diversification in 1955, he concluded that Richland could accomplish little "until Federal restrictions are removed." In light of this news, the council's industrial committee disbanded.[36] It remained difficult to exploit new economic opportunities around Hanford without meeting at least some resistance from the AEC. Even the establishment of the N reactor had to overcome opposition from within the commission.

In 1964 the need to diversify became more urgent as the AEC began shutting down production reactors at Hanford. By this time Pasco and Kennewick were nearly as dependent upon Hanford as Richland was, and all three of the Tri-Cities devoted themselves to rethinking resource use. Civic leaders defined the problem as one of attracting enough new residents and businesses to keep afloat the urban and industrial economy of the mid-Columbia area. Adhering to the venerable ethos of economic growth at virtually any cost, they pushed for a program of diversification. Their arguments helped recast ideas about nature, and they also drew on and perpetuated the idea that the Hanford vicinity was a frontier area and that the inhabitants of the Tri-Cities were pioneers.

Although Tri-Cities boosters in the 1960s hoped above all to maximize the exploitation of resources, they did not always view land and water within that narrow economic context. Rather, they tried to incorporate the Columbia Basin's natural resources into an assortment of explanations of why the region ought to be attractive to newcomers. When the army and the AEC had recruited workers to live near Hanford during the 1940s, they had counted on employees accepting the metropolis as their hometown because that was where the jobs were. Thereafter, as the Tri-Cities competed increasingly with other cities to attract new business, boosters tried to portray the surroundings more as a desirable amenity. In the 1950s and early 1960s they reasoned that the Tri-Cities area was appealing because it still had plenty of wide-open spaces. "The west is the last frontier for roomy, healthy living," the *Columbia Basin News* explained in 1951, "and we are located in the middle of it."[37] Promotional efforts highlighted assorted recreational activities in the great outdoors, trying to make the most of the "empty" expanses of land and river surrounding the towns. A corporate brochure designed for employees and customers during the mid-1960s attested, "The entire Tri-City Area is bordered by acreages devoted to country living. Lovers of the out-of-doors can find ample space to indulge rural hobbies or to test their pioneering impulses."[38]

Wide-open spaces were viewed as both amenities and assets. By the 1960s, local boosters began playing up the idea that the Basin offered open space, while other places had become too congested. They argued the logic of moving to the interior West by again terming the Tri-Cities and Pacific Northwest "the next frontier. California is crowded. The only place to go is up here."[39] Ted Van Arsdol, a local newspaperman, declared that the buildup of population and industry along the Pacific Coast, especially in California, was unwise from both an ecological and a military viewpoint. He offered the Tri-Cities as the sensible alternative. Van Arsdol's take on "the inland region"—that it "has been fighting an often losing battle against the erosion of population and the exploitation of its resources by persons whose primary interests were at other points"— summarized the hinterland's classic complaint against the metropolis.[40] It also perpetuated a blindness as to why the wide-open spaces of the interior West did not measure up to the urban and natural amenities of the coastal region. Promoters who used the natural environs of the mid-Columbia area to lure new residents and businesses to the Tri-Cities were swimming against the current.[41]

AEC sites had been chosen in large part for their remoteness, a trait that made it difficult to attract new residents and investors. A 1964 study of diversification possibilities in the Tri-Cities ranked the area's isolation "from major markets and centers of population" as the leading obstacle to be overcome.[42] Promoters might tout the Tri-Cities area as a last frontier of sorts, a place where inhabitants could enjoy the outdoor expanses and where entrepreneurs could have plenty of room, but their arguments proved mostly unpersuasive. In truth the urban area lacked the natural and cultural amenities that Americans increasingly sought.[43]

Boosters trying to lure new businesses to the Tri-Cities in the mid-1960s learned bluntly why outsiders seldom found their community and the surrounding environs attractive. Business representatives who visited the area often concluded that Richland, Pasco, and Kennewick lacked the cosmopolitan features that employees transferring from more urbane locations hoped to find. C. D. Thimsen, an executive with Computer Services Corporation, noted that the mid-Columbia region's isolation, inadequate communications facilities, lack of college or university, shortage of good restaurants, and scarcity of nice housing would make it difficult to recruit the workers necessary for his company. The Tri-Cities also lacked ample retail facilities; many residents still traveled regularly to Portland, Spokane, or Seattle for shopping.[44]

The shortcomings of the Tri-Cities became more widely discussed in 1965 and 1966, when the AEC considered Hanford as a possible home for a new 200 billion electron-volt particle accelerator. Senator Jackson and the Tri-City Nuclear Industrial Council (TCNIC) had been urging the AEC to locate some of its new or existing programs at Hanford to make up for the losses expected from reactor closures. The prospect of a national accelerator laboratory represented one of the first opportunities to see the AEC's initiative in action. Local boosters felt that Washington was over-due to win a big, federal, high-tech project. Sam Volpentest wrote Senator Magnuson, "Boston got the electronics center, Houston the space center, Florida the cape and California nearly every thing else [so] why not the A-smasher for Washington." Unfortunately for the Tri-Cities, scientists rather than politicians would make this decision—largely on the basis of which communities most appealed to the families of prospective employ-ees of the accelerator lab. Local boosters tried to argue that the Tri-Cities were "attractive to scientists, engineers, and their families" and depicted as one of their assets the absence of urban "overcrowding."[45] Local boosters continued to try to make a virtue out of the area's remoteness and absence of development.

The effort to sell Hanford as an accelerator site came to naught. The National Academy of Sciences, whom the AEC had asked to make the siting decision, left Hanford off its short list. The main reason was that the region did not appeal sufficiently to scientists or "seem desirable to [their] wives and families," the academy concluded, noting that "the cost to the project that would result from inadequate staffing would far exceed the savings that might be realized through such visible considerations as cheaper land, power, or water [at Hanford]." Rather than praising the area's wide-open spaces, scientists determined that the region was too remote and provincial. They preferred to locate the accelerator in or near a major metropolis, and did not look favorably upon having to drive four or five hours to Seattle or Portland. They also wanted the accelerator to be located near a university campus, but the Tri-Cities did not have a four-year col-lege or research university.

When the decision against Hanford was announced, opinion makers in the Northwest were bitter about having been passed over. They criticized "Space Age Pioneers" who—unlike the previous generation of atomic pioneers—were unwilling to "rough it" at Hanford.[46] Tri-Cities boosters' efforts to make "frontier" conditions a virtue failed to convince others. The experience of bidding for the accelerator made it plain that outsiders held unfavorable attitudes toward the Tri-Cities. Remoteness and lack of development—the very things that had lured the Manhattan Project to the

4.6 On-site storage of wastes began during World War II, when the army and DuPont installed underground tanks as depositories for liquid and chemical wastes. During the Cold War the number of such containers increased, as with the 1951 construction of the 241 TNX tank farm shown here. As more and more wastes were retained on the site, the tacit threat to Hanford's environs and nearby enterprises increased. Once the mission of the site became cleanup, the tanks became a major focus of attention. However, as late as the end of 2008, only 11 of Hanford's 177 tanks had had their waste retrieved for treatment. Photograph courtesy of the U.S. Department of Energy.

mid-Columbia Valley in the first place—continued to define the locale in significant ways. Boosters and residents had attempted to put these traits in the best possible light by likening their home to a frontier, but such imagery often backfired and never concealed the dearth of natural and cultural amenities.

The boosters went back to work, developing a strategy that demonstrated their genius for promotion. Told that their towns were remote and unappealing, the community leaders tried to turn these complaints to their advantage, pursuing economic development by arguing that certain types of businesses belonged in the Hanford area precisely because it was unattractive. They offered their urban area as a regional and national sacrifice zone, willing to play host to nuclear reactors and nuclear wastes that other communities did not want.

Hanford's fate, of course, was still determined by forces largely beyond local control, but the boosters' argument provided a coherent rationale for the Tri-Cities and helped to prop up a community identity. The idea of a sacrifice area reinforced residents' sense that they were performing an important mission. It reiterated the notion that Hanford and the Tri-Cities were technological and social pioneers on behalf of the nation. It took into account the environmental movement and allowed the Tri-Cities to play both sides of the street. Tri-City residents could argue that by promoting nuclear power plants they were supporting a clean source of energy. At the same time, if environmentalists opposed nuclear power plants or nuclear waste repositories *elsewhere*, Hanford could benefit by providing a home for unwanted reactors and wastes.

The notion of Hanford as home to power reactors was not new. The Tri-Cities had viewed nuclear-generated kilowatts as a logical path toward diversification since the 1950s, when they lobbied Congress to fund the dual-purpose N reactor. Then, after the AEC began closing plutonium-producing reactors in 1964, nuclear power plants were regarded as key for diversification because they would cause the fewest disruptions to the existing economy. Just as with planning for Grand Coulee Dam during the 1920s and 1930s, there appeared in eastern Washington during the 1950s and 1960s an assumption that local development of a new source of energy would attract more people and industry to the region.[47] The Tri-Cities expected to grow by developing a large, civilian, nuclear-power industry and by having the new power plants attract additional, nonnuclear businesses seeking to reduce costs by locating close to a kilowatt-producing center.

Forecasts of an imminent shortage of electricity in the Pacific Northwest appeared to make additional power plants a matter of grave importance, especially during the peak years of America's "energy crisis" in the early 1970s. It was widely claimed that the nation needed to double its power supply every decade, and that failing to meet this target would endanger its security and its economy. Congressman Chet Holifield (D-CA) of the Joint Committee on Atomic Energy, speaking at Richland in 1970, made what became a staple prediction. America's population would increase by 50 percent, from 200 million to 300 million, by the year 2000. In order to maintain an adequate standard of living, the country would need to generate seven times as much electricity as it produced in 1970 and do so in a "safe, reliable and economical" fashion. Holifield expected nuclear power plants to provide most of the new electricity. In 1970 they accounted for only 2 percent of American capacity; in 1980, he predicted, they would have to provide 50 percent.[48]

Projections such as these offered hope to the nuclear industry in general and to Hanford in particular. At both the regional and national levels, the demand for power-generating reactors grew, and for a time, communities across the country explored nuclear plants as a serious option. As part of regional power planning, nuclear reactors had been proposed for western Washington and Oregon. Over time, opposition developed to sites west of the Cascades. Yet as the tide of support for nuclear power began to turn, the argument for siting new plants at Hanford gained strength. In May 1970 voters in Eugene, Oregon, decided against continuing with a new reactor for their municipal power supply. The *Tri-City Herald*'s response was that the environs west of the mountains could be protected from the nuclear threat by building all new reactors on the east side. The mid-Columbia area, the newspaper explained, offered both a more supportive political climate and a better natural setting. People in Eugene or Seattle might well postpone construction by arguing against nuclear power and its risks, but the Tri-Cities offered what another proponent of nuclear energy deemed "the precious ingredient of 'Public Acceptence [*sic*]'" which would facilitate the rapid progress needed to avoid regional energy shortages.[49]

Moreover, the boosters contended, the relatively dry and unoccupied interior would be a better home for nuclear power plants than the wetter and more populated parts of Washington and Oregon. Recycling the idea that coastal areas were already too crowded and over-developed, one promoter wrote that siting new plants at Hanford would "help relieve current industrial and population congested areas in the Pacific Northwest through transfer of industry which, in turn, could improve their existing environment."[50] The notion of Hanford as regional sacrifice zone took shape.

In offering protection to the more populated west side of the mountains, the *Herald* conceded that nuclear power continued to present considerable environmental risks. As critics were suggesting, it may have been dangerous to locate nuclear reactors close to the large urban areas around Seattle and Portland. One alternative, the *Herald* averred, was

to place the plants east of the Cascades, away from population centers, and where prevailing winds blow from, instead of toward, the cities. Transmission costs might increase Northwest power bills fractionally but the additional

cost would be a low price to pay to protect our environment—and our people
. . . . [And,] what about the other costs? The cost of damage to marine life from
heated water dumped into Puget Sound? The cost in human misery of super-
saturating an atmosphere already dripping? The cost of "polluting" magnificent
sea and mountain scenery with transmission towers? . . . [The people] have a
right to choose between slightly higher cost of electricity and irrevocable damage to our environment.[51]

New reactors belonged "on federal land areas, particularly in the West," the
Herald concluded.[52] This argument drew upon ideas that had been advanced
since the army selected the Hanford Site for reactor construction during World
War II. The sentiment again perpetuated the view that the interior West was the
best place to put a potentially hazardous technology because it was somehow
"empty." It also built upon the growing perception that the Pacific Northwest—or
at least the western half of it—was a special place inhabited by special people who
were devoted to environmental protection.[53]

The *Tri-City Herald*'s viewpoint seemingly accepted as fact many of the
criticisms that outsiders had made and would continue to insist upon concerning the risks of nuclear power. In the newspaper's scenario, dumping
heated or radioactive water into Puget Sound and building nuclear power
plants upwind from Seattle or Tacoma was ill-advised, but diverting waste
into the Columbia River and building nuclear power plants upwind from
Walla Walla or Spokane was acceptable. The *Herald* parroted the dismissive
attitudes toward both the mid-Columbia region and nuclear energy held by
people in western Washington and Oregon and perpetuated the strategy of
tolerating the risky by-products of extractive industry to secure economic
gain by contending that it would be safer (and, of course, more profitable
for the local economy) to accept those risks in eastern Washington.

It is doubtful that citizens of the Tri-Cities fully accepted the views
expressed in the *Herald*'s editorials. Many residents had become quite
attached the area and would have defended its amenities against what
western Washington and Oregon had to offer. Furthermore, the *Herald*
seldom took environmentalist ideas very seriously and had already begun
to warn against the influence of environmental activists, an agenda that
continued through the 1970s and 1980s.[54]

Finally, residents of the Tri-Cities and advocates of power reactors
generally believed that nuclear energy was environmentally safe. Boosters might suggest that Hanford was willing to make real sacrifices to take

on more reactors, but few in the nuclear communities really thought there was much danger. The more common belief was that nuclear reactors were an ecologically sensitive method of producing new kilowatts—especially if the alternatives were hydroelectric dams and coal-fired plants. The sentiments of Glenn T. Seaborg, the "father" of plutonium, were typical. In 1968, on the twenty-fifth anniversary of Richland's occupation by the Manhattan Project, Seaborg visited Richland in his capacity as chairman of the Atomic Energy Commission. He spoke glowingly about a "nuplex" or "nuclear-powered industrial complex," and praised it not only for its economic benefits but also for its ecological advantages. Seaborg imagined an industrial park organized around nuclear power plants, a place that would allow for the healthful segregation of manufacturing from cities, making the city "once again . . . a place primarily for people" and reducing its pollution. Should the Tri-Cities collectively undertake such a development, he predicted that by 1993 they would "probably be a large metropolis thriving on its growing science-based industries. Perhaps Hanford will be its Nuplex, *able to preserve the surrounding vast and majestic area close to the way nature created it.*"[55] In sum, nuclear power promised both economic and environmental benefits, according to its advocates. Thus, by offering to "sacrifice" their region for western Washington and Oregon, Tri-Cities boosters believed they were not taking much of a chance—just as during the 1950s those who argued to open the Wahluke Slope did not see unacceptable risks in doing so.

Volunteering to play host to nuclear power plants allowed the Tri-Cities to pursue economic development within the context of a familiar self-image. While outsiders might not appreciate the significance of Hanford as a frontier, locals continued to draw on the conviction that they were pioneers.[56] This notion had two key components—the idea of sacrifice and the idea of leadership. The energy shortage projected during the 1960s and 1970s gave the people of the Tri-Cities another mission to perform on behalf of the United States. The communities had endured risk and hardship to battle fascism and communism. Now they mobilized against a different sort of enemy—threats to American affluence. Boosters occasionally allowed that nuclear power presented certain dangers—danger, after all, is what pioneers must overcome—but they also saw the trend toward nuclear power as "irreversible." "Whatever the dangers, real or imagined, of nuclear power," the *Tri-City Herald* editorialized, "the overriding reality is that it's gone beyond the point of no return."[57] And to "reject nuclear energy" was "to condemn ourselves and our children to a lower standard of living."[58]

Other American cities might waver at the prospect of nuclear power, but the towns near Hanford were prepared to lead where no communities had gone before. They proposed to explore another technological frontier, the nuclear power park, and to remain "a showcase for the nation."[59] The communities would also preserve their identification with the atom into the 1970s and 1980s. Richland continued to call itself "Atomic City," and Congressman Mike McCormack happily went by the nickname "Atomic Mike."[60] Hanford may have lost most of its original mission with the curtailment of plutonium production, but in many ways the Tri-Cities self-image remained much the same.

Another thing that had not changed was the prevailing attitude of maximizing resource use. In 1968 the *Tri-City Herald* published an article by Pat Bushey called "In a Manner Similar to Los Angeles" as part of a special issue titled "Taming of the River." The article described the Tri-Cities in the year 2000 as three sprawling cities where much of the workforce found employment in the field of generating electricity. By century's end, Bushey predicted, the "thoroughly tamed Columbia" would be no more than "a series of deep, slackwater pools impounded by dams." These dams would generate plenty of kilowatts, but together they would amount to no more than "an important secondary source to the 40 nuclear plants in the Northwest," many of which would be located on or around the Hanford reservation.[61]

This future represented an urban and industrial update of that vision of development advocated by the Columbia Basin Project three decades earlier. Through the taming of the atom and the river, the Tri-Cities imagined themselves as the future headquarters of a new industrial and technological empire. During the late 1970s, booster Sam Volpentest explained how the metropolitan area would benefit from the policies and needs of other places: "We're striving now to become a total energy center where . . . we'll have fifteen to twenty more plants, 150,000 to 200,000 people, and enough processing and storage facilities so raw uranium can come in, electricity can flow out, and that will be it. It seems Oregon doesn't want any more reactors. And maybe California with that crazy Jerry Brown doesn't either. But we have the skills and the enthusiasm. We'll be nice and sell them what we produce."[62]

Reality could never live up to aspiration. The Tri-Cities got a decent start toward their goal during the 1970s when the Washington Public Power Supply System (WPPSS) decided to build three electricity-generating reactors at Hanford (and two more at Satsop in Grays Harbor County, in western Washington). The ensuing planning and construction brought another considerable boom to Hanford and promised fulfillment of the aspiration for a diversified economy. By 1980, with three WPPSS reactors underway at Hanford, journalists claimed that diversification at Hanford had proven "highly successful," and Tri-City boosters asserted that local accomplishments would "serve as a model elsewhere."[63] The nuclear future seemed to be on track.

In retrospect, however, 1980 may be seen as the apogee of the Tri-Cities as a center for nuclear power. The WPPSS boom would soon collapse, along with ideas of a future dedicated to electricity-generating reactors. It turned out that the demand for kilowatts in the near term would not be as great as had been projected during the 1960s and 1970s; conservation was a more economical way to enhance the energy supply in the region. Nuclear power plants regarded as urgent necessities in 1972 had by 1982 become white elephants. Moreover, the problems of actually constructing and operating the proposed reactors proved overwhelming. Cost overruns, mismanagement, and labor disputes dragged the projects down until the weight of debts and delay finally proved too heavy. Collecting information for his 1982 book on Hanford and the Tri-Cities, Paul Loeb found a dispirited and cynical labor force tired of working on projects that never seemed to be completed, in contrast to the go-go successes of their fathers' times during the Cold War. In 1981 the WPPSS announced a moratorium on construction of two of its five reactors; in 1982–83 it halted work on four of the plants; in 1983 it defaulted on its bonds, the largest default on municipal bonds in financial history. The only completed reactor, WPPSS no. 2 at Hanford, began producing electricity in 1984—seven years later than initially planned.[64]

The nuclear future broke down not only because of costs and delays, however. It also broke down because people lost faith in it. Through the 1960s and 1970s the Tri-Cities remained convinced that they were blazing a trail toward America's energy future. By the 1980s most Americans had begun to reject that destiny. Although the WPPSS itself collapsed of its own weight, and not because of antinuclear activism, the public was growing much more skeptical of nuclear power. The environmental movement, the questioning of technology, and the response to the 1979 accident at the

Three Mile Island nuclear plant all contributed to this change of heart. At the same time, neither economics nor technology realized the prediction that economic development would be directly tied to energy abundance, at least for the 1970s and 1980s.

The effect of this turnabout on the Tri-Cities was harsh. The communities had realized that not all Americans shared their confidence in nuclear power; they had offered to sacrifice their own environs to house nuclear reactors and wastes that other places did not want. Even if other towns hesitated, the Tri-Cities believed that they might ride an inevitable trend toward the development of nuclear power, grasping economic and technological leadership. Yet it became more and more clear that nuclear power was hardly inevitable and that the Tri-Cities' economy and future were in grave danger. In fact, the towns were increasingly portrayed as backward oddities by outsiders. In 1979 one Seattle reporter dubbed Richland "the Ellis Island of the nuclear age. Send us your harassed and embattled reactors, says the local chamber of commerce. They are welcome here." [65]

Tri-City boosters continued to look for ways to perpetuate an orientation to high-technology industry. From the N reactor to the WPPSS, Hanford's managers, workers, and advocates had seen a trajectory leading to a stable production mission other than manufacturing plutonium for weapons. With the demise of the great ambitions for the WPPSS, many of their hopes came to focus on a reactor built between 1970 and 1978 and defended into the twenty-first century: the Fast Flux Test Reactor, or, as it came to be called, the Fast Flux Test Facility (FFTF). "WPPSS and the breeder really turned things around," the indefatigable Volpentest told writer Loeb. The FFTF was a "breeder" reactor, so-called because it would be able to produce more fuel than it consumed, in contrast to Hanford's graphite-moderated converters, for example, which produced less fissionable material than they burned. Playing host to a breeder reactor gave Hanford an unexpected role in U.S. reactor development, but in the end it proved a disappointment for those who had anticipated that it would become integral to the Tri-City economy. [66]

Responsible for an organization with the status of a stepchild within the AEC, Hanford's leaders had seen their own efforts confined overwhelmingly to production rather than invention. While the research program at Hanford had been constrained by the site's focus on production, some reactor development projects had found a home in the 300 Area, where uranium and graphite were processed for use in Hanford's piles. Small

4.7 Once it became clear that plutonium production could no longer sustain Hanford's economy, residents of the Tri-Cities envisioned the Fast Flux Test Facility as the reactor of the future for Hanford. Shown here in 1979, the FFTF seemed modern and flexible enough to meet a wide variety of commercial and scientific needs. Despite extensive lobbying by local and state politicians, however, no paying mission was ever authorized for the facility, and between 1993 and 2001 it was shut down and deactivated. This photograph (which captures FFTF in conjunction with wild flowers in the foreground, Rattlesnake Mountain in the background, and fluffy clouds in a bright sky) illustrates the juxtaposition of nuclear technology and environmental amenities that some envisioned for the vicinity. Photograph courtesy of the U.S. Department of Energy.

assemblies for the measurement of physical constants relevant to reactor design and operation, and for exploring the possibility of recycling plutonium, were constructed there during the 1950s. Atomic Energy Commission lawyers raised an objection to such development projects: given Hanford's status as a production facility, the diversion of any reactor fuel away from the main piles and, ultimately, the chemical separation plant ran counter to the commission's legal mandate to produce the greatest possible amount of weapons fuel. Still, citing Hanford's extensive operational experience as an important resource, the commissioners had permitted a small amount of fissionable material to be allocated to research projects on the Hanford Site.

Hanford seemed to lack the sort of connections within the AEC that would secure a research role to supplement the production mission that defined activities on the site. The commission's reactor development investments were directed by the Argonne National Laboratory near Chicago, which exercised management over the Idaho National Engineering Laboratory. Oak Ridge had defined itself as an important research center, presided over by scientist, public figure, and popular author Alvin Weinberg (who defined the term "Big Science" in the early 1960s). Savannah River operated within the orbit of Oak Ridge and with its own powerful patrons. But with the N reactor and plans for the WPPSS, Hanford could hope to provide the foundation for a new industry based on the production of electricity and the processing of nuclear materials from the defense economy. The Bonneville Power Administration provided the infrastructure to transport megawatts from Hanford to regional markets. In this effort, the Tri-Cities aspired to operate on the frontier by building breeder reactors—the next generation of nuclear technology—and Hanford's technical staff would provide knowledge of the cutting-edge mechanism to a worldwide market. Hopes attached to the construction of a breeder reactor, represented by investments in the FFTF, were also supported by renewed calls in the early 1980s for development of nuclear energy as a source of electrical power. Arguing for investment in new nuclear generation systems on both environmental and economic grounds, in the aftermath of the energy crisis of the 1970s, a number of authors pointed to advanced nuclear reactor designs as the rational way forward.[67]

Breeder reactors became a problematic part of the American reactor development program during the 1950s. DuPont's experts had suggested the utility of a breeder reactor by the end of World War II and had assessed a proposed Oak Ridge project with an eye to assuring the operation of a reactor that operated with a fast flux of neutrons for production and reactor development purposes in 1949. Rather than using a moderator to slow neutrons down to thermal speeds, early breeder designs proposed using a fast flux of neutrons from uranium 235 and plutonium 239 loaded as fuel, which were allowed to collide with a jacket of uranium 238 and thorium. These outer layer materials, then, would be transformed into fissionable isotopes.[68] Breeders would operate at very high temperatures; rather than water cooling, most designs proposed to use a liquid sodium compound as a coolant, a substance able to carry off a great amount of heat. As a bonus, the coolant might then be cycled through a heat exchanger in a generator facility to produce electricity. The AEC built experimental breeder reactor projects at its Idaho laboratory, and in the mid-1950s a utility consortium centered at Detroit proposed to build a breeder reactor on Lake Erie midway between Detroit and Toledo, provoking an administrative crisis among the AEC, its reactor safeguards committee, and the Congressional Joint Committee on Atomic Energy. The concept of the breeder reactor had called into question both the contradictory roles played by the AEC as promoter and regulator of nuclear energy and the political impetus for the Eisenhower-era program to promote nuclear-powered electrical generators via support to private enterprise.[69]

Interest in the breeder at Hanford dated to the mid-1950s as well, in large part because the technology appeared to be capable of meeting a wide variety of needs. General Electric representatives from Hanford who visited Argonne felt that the "fast breeder reactor" promised to accommodate GE-Hanford's "long-range goals." These reactors, by producing their own fuel, freed up resources within a national weapons complex that was always demanding fissionable material. They could be scaled up or multiplied to produce as much weapons fuel as the strategic situation seemed to call for. The technology would allow producers to control costs. Predictions called for breeders to turn out a high-quality product. The design lent itself to "the inadvertent production of high quality weapons-grade plutonium," material much less contaminated by the undesirable plutonium-240, which underwent spontaneous fission and threatened the neat assembly of an explosive designed around plutonium-239. Finally, breeders seemed to be well suited for production of steam for electricity generation, with long running lives at high temperatures and little need for refueling. The breeder concept seemed almost too good to be true. "It appears to be an article resembling in desirability Little Abner's 'Schmoos,'" able to take the shape of each observer's desires, noted the head of GE-Hanford's reactor design section.[70]

Breeders offered the possibility of producing a net surplus of fissionable material. They also produced a challenge for technological development to engage the technical staff and create a salable product to the advantage of the contractor helping to develop it. During the period of economic diversification, interest in a breeder reactor for Hanford did not wane but was passed from General Electric to Battelle–Pacific Northwest to Westinghouse. By the beginning of the 1970s, Westinghouse had vested the project in its Hanford Engineering Development Laboratory, which undertook fuel fabrication studies in support of the experimental breeder reactor program in Idaho and also began to develop specifications for fuel elements to be purchased from outside vendors and used to load the Fast Flux Test Facility on the Hanford Site. With an array of process tubes running at angles through the spherical reactor chamber, the FFTF was very different from the older, graphite-moderated piles. The FFTF did not need to be sited by the river, and so Hanford was able to inaugurate a new compartment on the site, the 400 Area, devoted to the new technology that loomed so brightly in Voltenpest's hopes by the end of the decade.[71] The Fast Flux Test Facility started initial operations in 1980 and began running at full power in 1982.

Through the efforts of Senators Magnuson and Jackson, Hanford had acquired federal funding for its breeder reactor program during the late 1960s as part of the diversification effort. Perhaps it was no coincidence that shortly after Magnuson lost his Senate seat in 1981 and Jackson died in office in 1983, the FFTF's fortunes began to decline. The reactor continued to operate until 1993, when, due to the lack of an "economically viable mission," the DOE ordered the reactor to be shut down. It went to cold-standby status in 1997, and in 2001 the government began deactivation of the facility.[72] At each step, local interests and elected officials pressed to find some research or production mission for the FFTF, but it was not to be. Some pro-nuclear commentators would identify this as another casualty of the "war against the atom." At Hanford the war had been waged incrementally; by the mid-1980s, operators at the FFTF had given up on electricity generation and fastened their hopes on using the breeder to produce isotopes to order, as well as its role as a research tool. Setting up to produce a given isotope might take fifteen months—a slow process. In a

4.8 This shot shows construction on the interior of the Fast Flux Test Facility during the late 1970s. The FFTF differed substantially in size, shape, and most other ways from Hanford's other reactors. It had no graphite core, for instance, and it did not require Columbia River water for cooling. During the 1970s and 1980s, many in the Tri-Cities embraced FFTF as a nuclear technology that could compensate for Hanford losing its Cold War mission of plutonium production. Not surprisingly, then, some in the communities expressed bitterness that the federal government would identify no research or production role for FFTF, in either the public or the private sector. Photograph courtesy of the U.S. Department of Energy.

sense, the FFTF reflected the diffuse commitments of the diversification period; in the end, the great advantage of potentially being all things to all people—the breeder technology's "Schmoo factor"—ended up with the facility never finding a strong defining mission.

Those in favor of developing nuclear power at the beginning of the 1980s emphasized the utility of the breeder design in the future, should it be developed. Not to do so, they argued, was economically irrational and environmentally unsound, dooming the United States to a dependence on coal, a more polluting, dangerous, and medically dubious option. "The antinuclear movement is without precedent in trying to make policy through a popular consensus as to certain scientific and technical facts," wrote Samuel McCracken, a neoconservative political scientist, in 1982. "The problem now facing us is that the so-called technical debate is being carried on at a high level of incompetence and mendacity and is aided by a credulous press." The problem as McCracken understood it was an almost atavistic environmental movement, which seemed to have become established orthodoxy rather than passing away after a few flower-power summers. The breeder reactor design, with its roots in a "plutonium economy" involving the broad distribution and transport of reactor and weapons fuel, was ruled out by antinuclear sentiments by the early 1980s. Hanford could have been home to a network of such generating facilities, sharing fuel and infrastructure, but there was no wider market for the vision.[73]

If the Tri-Cities seemed on the verge of becoming pariahs instead of pioneers, it was not simply because ideas about nuclear power there diverged from those held by most of the country. It was also the case that Tri-Cities ideas about nature stood at odds with how most Americans were thinking about the environment. At times it seemed as if the environmental movement, which exerted so pervasive an influence on the country after 1970, had somehow bypassed the Tri-Cities. The communities knew enough about the movement to try to exploit it economically—when Eugene, Oregon, decided against building a nuclear power plant on environmentalist grounds, the Tri-Cities offered to play host to it—but they did not embrace the goals of the movement. Rather, through the mid-1980s Richland, Kennewick, and Pasco remained rather narrowly focused on the exploitation of lands and waters, and expressed little sympathy for calls to protect resources. The extent to which this made them exceptional became apparent in debates within the state of Washington over the storage of nuclear wastes at Hanford.

For most of the 1970s the Evergreen State remained largely supportive of the nuclear industry. For example, Governors Daniel J. Evans (1965–77) and Dixy Lee Ray (1977–81) generally encouraged proposals to have other states ship their nuclear wastes to Washington for storage. This willingness to serve as a waste repository diminished at the decade's end, however. In 1980 Washington voters passed, by a three-to-one margin, an initiative to ban the import of nonmedical nuclear waste. A federal court soon ruled the initiative unconstitutional, an improper intrusion into the federally regulated sphere of interstate commerce. Yet the measure did indicate the extent to which Washington, which had once dubbed itself the "Nuclear Progress State," was changing its mind about atomic energy.[74] Washingtonians generally were not so much becoming "anti-Hanford" as they were abandoning the active support that had won the fight for the N reactor and the electrical generating facilities attached to it.

If most Washington voters objected to the idea of their state taking on more of the nation's radioactive waste, the people of the Tri-Cities generally did not. Local business leaders had envisioned waste management as an integral part of economic diversification at Hanford, and included the activity as part of their portrayals of the local nuclear future. The federal government apparently bought into the idea of Hanford as a national sacrifice zone. The Nuclear Waste Policy Act of 1982 proposed to select two places, one east and one west of the Mississippi River, as prospective repositories for 77,000 tons of high-level radioactive wastes. The Department of Energy considered ten places around the country as storage sites, but it was particularly interested in the Tri-Cities area. Through 1984 it spent more than $300 million "studying Hanford's basalt" as a potential underground facility "while virtually ignoring most of the other sites."[75]

Although strong federal interest may have stemmed in part from the perception that Washington was more receptive than other states to storing radioactive wastes, attitudes within Washington were in fact quite divided. A 1986 poll conducted by the *Tacoma News-Tribune* determined that 71.9 percent of the state's population opposed putting a national waste repository at Hanford; only 15.7 percent favored the proposal. Much of the sympathy for the idea came from the Tri-Cities, where supporters of a national waste repository outnumbered opponents by 46.5 to 32.0 percent.[76] These figures suggest the extent to which Tri-City attitudes had diverged from those of the region generally. In the same year, Democrat Brock Adams won votes (across the state, if not in the Tri-Cities) in his successful campaign to defeat U.S. Senator Slade Gorton by attacking the Republican incumbent's support of the "bomb factory" at Hanford.[77] Things nuclear could no longer be viewed as opportunities; they were either failed attempts or, more likely, problems, and debate over them increasingly set the Tri-Cities apart from the rest of the Pacific Northwest.

The effort to create a national waste repository at Hanford derailed. In 1987 the DOE terminated the Basalt Waste Isolation Project at Hanford, at a loss of 1,200 jobs, and turned its attention elsewhere in the West, particularly to Yucca Mountain, Nevada.[78] Yet in some sense Hanford did not really need to import wastes from other states; it already had plenty of its own to deal with, if it wanted to. According to a 1987 estimate the nuclear reservation contained "about two-thirds of the nation's total volume of waste involved in the production of nuclear weapons."[79] As production of weapons-grade plutonium ceased in 1986–87 and as Hanford facilities closed down, people in the Tri-Cities began to pay more attention to these wastes. But it was a particular kind of attention. Few yet viewed the wastes in *environmental* terms, as a threat to the local ecology and public health. The focus remained on economic development. Managing waste, it turned out, represented "the only long-term, full-employment solution" to periodic busts in the local economy.[80]

Cleanup was a task that the Tri-Cities could certainly undertake, but not with a great deal of enthusiasm. The communities would have much preferred employment on new reactor projects or in some other production-oriented capacity rather than at managing wastes. The attitudes resembled those from two decades earlier, when well-paid, unionized workers complained about the kinds of jobs generated through strategies of diversification. One nuclear engineer explained, "We've been asked to become janitors for the DOE when we could be more professionally satisfied being on the cutting edge of technology. . . . You can't overstate the demoralizing aspect of taking away high technology activities and asking us to become paper-pushers and janitors." Jerry White, the DOE's director of waste technology at Hanford, summed up the local view in 1985, "We're tired of managing the waste. We want to put it into a position where we can walk away and never bother with it."[81] The Fast Flux Test Facility remained the only monument to a possible future with new technology, yet its future was already clouded.

Attitudes toward cleanup would change, but in the mid-1980s the Tri-Cities were reluctant to embrace waste management as an economic mainstay. The task seemed to offer little in the way of financial or psychological reward. The Hanford workforce would not be doing work that was exciting, or over which it had much control; rather, they would become servants for the DOE. That people like Seattle environmentalists, antinuclear activists, and federal bureaucrats were strong advocates of cleanup did not increase its appeal, for the Tri-Cities had grown suspicious of such "outsiders" and their skepticism toward Hanford. The Tri-Cities felt less a part of the environmental movement than targets of it. In fact, the environmental movement was already hard at work there, in diverse forms, even if many would remain dubious about it until they discovered how well environmentalist projects could be grafted on to the better-rooted ethos of economic development.

By the end of the decade, antinuclear forces had even a stronger hand to play. Particularly after the Department of Energy made public records of large, deliberate releases of iodine-131 into the environment, resistance to new nuclear technology hardened in Washington.[82] The WPPSS had seemed to show that electricity generated from the heat produced by nuclear reactions was not feasible to produce in a regional economy defined by Columbia hydropower. Now the federal nuclear establishment provided ammunition to antinuclear groups, and Hanford was in danger of losing the opportunity to serve as the center of a green economy based on breeder reactors. The lingering controversy over whether and how to operate the FFTF, which could be used to develop the technology that would produce reactor fuel or other useful commodities from a pile that was also producing heat for electricity generation, seemed to foreclose Hanford's industrial future. It hardened the two most prominent positions among the welter of interests and voices debating the site's future in the light of its past.

The first of these voices reflected the perspective of the "plank-holders," veterans of Hanford's period at the center of the U.S. defense nuclear economy from the mid-1940s to the mid-1960s. In naval tradition, plank-holder status was accorded to the first crew of a new ship in acknowledgment of the special roles played by the vessel's initial operators, who brought the ship to life, established its operational behaviors, and set down enduring traditions for the crews that would occupy and operate the ship in the future. Because of the expansion of the navy during World War II, a ship's plank-holders often took their vessel into harm's way and established a combat record that would be attached to the ship for as long as it was part of the navy. At Hanford, the traditions were similar. Operators of the fuel-processing facilities, the reactors, and the chemical separation plants had worked out the kinks of complex mechanisms in potentially dangerous circumstances. Often coming from insular and isolated communities within the DuPont munitions production complex, these workers created similar communities for work at Hanford and for family life in Richland and Kennewick. Assertions that Hanford's costs had been greater than its benefits flew in the face of the plank-holders' sense of identity and their lived experience. In particular, it seemed unfounded and unfair for outsiders without the requisite technical expertise to weigh in on these issues.

Arrayed against this point of view was that of the "stakeholders," particularly downwinders who felt they had been damaged by airborne emissions of radioisotopes such as iodine-131. Arguably the greatest threat Hanford has posed to public health has been its contamination of the Columbia River, a pathway that particularly exposed Native Americans who practiced traditional fishing methods. (Moreover, because of the possibility of low-level wastes migrating through soils to the river or of high-level wastes in liquid form escaping from storage tanks on the site, the river pathway still poses a sizeable risk.) Nonetheless, it has been the airborne emissions from the chemical separation facilities that have drawn the greatest attention. Because of the affinity of iodine for calcium and for the thyroid, there was a clear potential route from Hanford to populations to the east and the south of the site, and children were especially vulnerable because of the uptake of iodine-131 in milk.

Because of the presumption that their concerns would be overlooked or marginalized, stakeholder status was created in the process of beginning to undertake assessments of Hanford's past impacts. These efforts had two main components: the Hanford Environmental Dose Reconstruction Project (HEDR) and the Hanford Thyroid Disease Study (HTDS). Using a retrospective mapping of the likely areas of greatest contamination by HEDR, HTDS undertook an epidemiological study beginning in January 1989 to see whether rates of thyroid disease increased with increased exposure to iodine-131 during the period of Hanford's most active plutonium production. The study's controversial result—that no statistical increase in thyroid diseases could be identified—was unacceptable to some stakeholders, who criticized the methodology of both studies and the process of interpreting their results. Some suggested that, after more than a decade of work, the studies' results should not be released until they found what the critics knew must be there: proof that Hanford was the causal center of a cancer cluster.[83]

Many environmentalists critical of Hanford's past operations have appealed to an ideal of democratic process as a counter to the power wielded by distant and secretive federal authorities, and have endorsed a strong role for stakeholders as a step in this direction. But having policy decided by either plank-holders or stakeholders seems to run counter to democratic ideals. Indeed, the category of stakeholder is a difficult one to understand in general; those with a stake in some process might be called "stake-havers," but stakeholders should be neutral parties with no interest in the outcome of a contest or wager. Both plank-holders and stakeholders found themselves in the position of being committed to historical accounts that have remained largely unproven and unprovable. For plank-holders, Hanford's operations kept the peace or at least helped to prevent a nuclear war; for stakeholders, Hanford's environmental consequences were disastrous, if largely unacknowledged.

In some sense, this confusion should come as no surprise. Studies of iodine-131 uptakes and their consequences at Hanford in the early 1950s had been inconclusive when applied to the question of whether atmospheric nuclear testing in the West had killed large numbers of sheep in Utah. Studies in the 1970s of radiological health effects on Hanford workers, for whom detailed workplace exposure records had been kept, produced contradictory results and controversy rather than clarity. By 1980 congressional advocates for downwinders began to argue that, since direct evidentiary linking of cause and effect suitable for court action could never be established, compensation for presumed victims of the nuclear industrial complex should be provided by legislation.[84]

Just as the compartmentalization of Hanford's landscape—into divisions labeled "production facility," "environmental preserve," "city," "farmland," and "wild river," for example—proved to be impossible to maintain, so have the categories "plank-holder" and "stakeholder," and "victimizer" and "victim," been difficult to establish. Interviewing some of the more senior members of the community at the end of the 1970s, Paul Loeb found that one woman mentioned that her husband had spoken "at a California church 'who'd already sent people to march against the atom,'" not realizing that those protesters might very well have been nuclear engineers from another division of General Electric who had turned against nuclear power.[85] Whistleblowers who accused the site's managers of suppressing information about radioactive releases to the environment also confused the categories of insider and outsider. Some members of the Hanford family, sons and daughters of those who spent their careers working on the site, emerged as leaders in the stakeholders' movement, while former workers—encouraged by the U.S. Energy Employees Occupational Illness Compensation Act of 2000, which recognized that workers at (but *not* neighbors of) sites involved in nuclear weapons production stood an increased risk of contracting illness from their employment—have begun in the twenty-first century to be more willing to file claims for compensation for health issues arising from workplace hazards such as chemical exposure and noise as well as radiation exposure. A common identity as residents of the Columbia Basin erodes some of the differences between the two groups, particularly on the issues concerning the environmental protection of the area on and around the reservation.[86]

The elevation of stakeholders in the process of negotiating the terms on which Hanford's past actions will be remediated both in the environment and in human populations does not necessarily represent the embrace of a democratic decision-making model. Technocracy and democracy are two political ideals never realized in practice. Focusing on the technocratic interests and power of the AEC/DOE, it is important not to forget the extent to which representative democracy, exercised via the influence of the region's citizens on their political representatives, shaped policies at Hanford that had a bearing on the public health. Similarly, the influence of stakeholders since the 1990s represents an approximation of technocracy in decision making, one based on presumptive experience, if not expertise.[87]

"Highest and Best Use": Environmentalism and Resources around Hanford

The leaders in protecting the lands and waters around Hanford were not virtuous farmers or progressive urbanites or even outside activists. They were officials of the AEC. Unlike most of the surrounding population, the commission was *not* preoccupied with economic development. In fact, it had needs as an organization that required it in some circumstances to prevent economic development in the open lands surrounding Hanford's work sites. Paradoxically, then, while the AEC was largely responsible for the pollution at Hanford, it also became a source of preservation programs and ecological awareness in the Columbia Basin.

The greening of the Atomic Energy Commission came in response to the shutdown of eight reactors at Hanford between 1964 and 1971. The agency was instructed during this time to assist local efforts at economic diversification, and it undertook that task. But it grew worried about the extent to which local interests proposed to exploit the Hanford reservation for economic gain once the production of plutonium had ceased. As the AEC closed reactors down, it was approached by people eager to develop resources under the commission's control. Farmers and ranchers wanted access to Hanford lands; business groups wanted to increase transportation and recreation on the Columbia; the Bureau of Reclamation looked anew at building Ben Franklin Dam on the river and at other measures to make the Columbia more navigable above Richland. Many acted as if the AEC were simply finished with the Hanford Site and could be expected to leave. In fact, the AEC was not finished with Hanford. It was shutting down reactors, but it continued to use the site in myriad other ways, including the storage of wastes. Moreover, it wanted to keep its options open in case it needed to produce plutonium at Hanford again or utilize the site for some other purpose. In sum, it was not ready to hand over the keys to the locals. In this context, preservation programs were quite useful to the AEC.

The agency's environmentalist career began innocently enough on the Wahluke Slope. When the Tri-Cities asked the AEC to assist in economic diversification during the mid-1960s, the commission tried to ease restrictions on some of the lands that it controlled. After studying which parcels it expected to continue using, the AEC offered to make available some of the excess acreage for "compatible" commercial, industrial, and municipal development.[88] The Wahluke Slope naturally came under further consideration. The closure of reactors through the mid- and late 1960s had reduced the risks of inhabiting and working on the slope, and TCNIC officials, local farmers, and state politicians all urged the release of additional lands from the AEC control zone.

On July 27, 1965, the commission announced it would permit "non-resident farming" on about 40,000 acres of the slope, a move viewed as a boon to local horticulture and therefore a form of diversification. The TCNIC reiterated its faith in farming as a staple for the future Tri-Cities economy. Referring to the dry lands on the perimeter of Hanford, one booster explained, "Water is all that is needed to turn these acres of sagebrush into a vast agricultural gold mine." The mining metaphor nicely summarized the extractive view of resources that prevailed in the area. However, in 1967 the Bureau of Reclamation decided not to irrigate the newly opened lands on the Wahluke Slope because of their inadequate drainage. Consequently, within a few years the AEC consented to proposals from state and federal wildlife agencies to allow some hunting, fishing, and preservation in portions of the Wahluke Slope, in conjunction with plans for the adjacent river.[89] This arrangement appealed to the commission because it limited access to resources that the agency was accustomed to controlling and at the same time it enhanced the AEC's environmentalist credentials. Variations soon proliferated around Hanford as the commission used preservation as a way to limit development of resources that it wanted to continue to control.

Creation of the Arid Lands Ecology (ALE) Reserve illustrated the new trend. In 1965 the TCNIC, the Bureau of Reclamation, and nearby farmers and ranchers approached the AEC about opening the slopes of Rattlesnake Mountain, on the southwestern edge of the reservation, to livestock grazing and irrigated agriculture. The AEC did not welcome the proposal. It feared that farming would disturb the underground water table, which in turn would disturb certain nuclear and chemical wastes, and it wanted to be able to reclaim the land instantly for future use if necessary. Before it responded officially to farmers and ranchers, however, the AEC received—and, perhaps, invited—another inquiry from scientists at Washington State University in Pullman and from biologists at Battelle–Northwest Laboratories, who wanted to study "natural" ecosystems on the same parcel of land. The AEC emphatically preferred the scientists' proposal to create a research zone at Hanford over the ranchers' and farmers' more intrusive plans. On March 29, 1967, Senator Warren Magnuson announced the creation of the Arid Lands Ecology Reserve at Hanford (for a map, see fig. 4.1 above). Ronald S. Paul of Battelle promised that the reserve would expand knowledge about man's relationship to nature and thereby help protect the environment.[90]

The AEC's decision provoked protests from local citizens. R. J. McWhorter had owned part of the property in question before 1943. He complained to Senator Jackson that the lands, if converted to an ecological study area, would no longer be used for the purposes for which the Manhattan Project had acquired them. Therefore, McWhorter argued, control ought to revert to the previous owners, who would put the acreage to economic use. A local paper, the *Prosser Record-Bulletin*, repeated the concern in a March 30, 1967, editorial. It was one thing to take the land to assist in building the atomic bomb; it was quite another thing—even "un-American"—to convert the acreage "to the study of bugs" without even holding a public hearing. Besides, the Prosser editor declared, "We have no idea what ecology is."[91] The idea that land and water ought to be utilized for practical gain still prevailed in the communities surrounding Hanford. Bernard Warby of Yakima, whose family's property the Manhattan Project had taken away, visited the site of the old farm in 1968 and lamented, "It's a pitiful thing. Nature has reclaimed the land."[92]

In originally proposing the Arid Lands Ecology Reserve, scientists had talked of their desire for a "pristine" area of study, but the reserve was hardly unblemished. In fact, one focus of research became "Hanford plant radionuclides cycled in the local environment." An apparent aim of these investigations was to compare "the effects of radionuclides" with the disturbances caused by other kinds of pollutants—and perhaps to suggest that radioactive wastes resembled other forms of pollution and were not especially worrisome. Furthermore, devotion to ecological study was projected as a means to improve Hanford's image. One AEC official explained in 1973 that the ALE had become "an important part of our plans to establish public confidence in our site ecology practices." Finally, by studying pollution, Hanford hoped to parlay its experience at environmental monitoring into new jobs.[93] There seemed to be growing awareness that environmentalism could pay dividends, although few in the Tri-Cities were yet prepared to regard the long-term cleanup of Hanford's own enormous stockpile of wastes as a profitable enterprise.

The rhetoric surrounding creation of the Arid Lands Ecology Reserve illuminated the rise of ecological thinking around Hanford. The Atomic Energy Commission and others were increasingly interested in environmentalist programs, but their motives were quite complicated. Moreover, the commission's efforts at protecting land and water from development, no matter its motives, provoked suspicion and hostility from local citizens who remained wedded to the principle of maximizing resource use. Environmentalism thus settled in rather haltingly around the Tri-Cities. Its peculiar progress was particularly well illustrated in discussions about the fate of the Columbia River—which had been going on for decades.

"Suppose now," an Oak Ridge scientist had conjectured in 1948, "that we are draining into a sewer, stream or bay, radioactive wastes sufficiently diluted that we could safely use the water. . . . Can we assume these radioactive wastes will remain so diluted?" The answer, he knew, was no. "Nature frequently shows remarkable abilities to concentrate various chemical elements especially when they are present only in trace amounts."[94] Reactor managers, he suggested, could—as Hanford's did—use water treatments to extract radioactive materials from cooling water and hold water in retention ponds to allow the most energetic radioactivity to decay before returning the water to the river. They could—as Hanford's did—study wind currents with the intention of diluting and dispersing radioactive contaminants from isotope separation plants. As did many concerned with water purification, Hanford's operators expected that sandy soil would remove and bind contaminants from some waste fluids. (In fact, the site's "thick layer of volcanic ash" seemed to provide "a discharge to nature in such a way that the activity is contained." Like many of the practical solutions to production problems on the reservation, "the solution appears indigenous to Hanford."[95]) While in the long run some of these best practices failed, Hanford did rely on best practices in assuming that the environment would mitigate the environmental effects of its operations.

The Applied Fisheries Laboratory at the University of Washington studied salmon for evidence of radioactivity in particular. Operational assumptions that releases to the atmosphere or the river would be dispersed and diluted were defeated in spots by concentrations of radioactive contaminants in the local environment, something that research projects such as those on the Arid Lands Ecology Reserve would later try to assess. When the AEC took control of Hanford from the army, academic experts from Harvard and Johns Hopkins raised questions about Hanford's "sins of emission," and particularly about the assumptions made to justify gaseous and fluid releases, as well as the practice "of constructing more and more containers for more and more objectionable material." One member of the Safety and Industrial Health Advisory Board, Herbert M. Parker, would devote his career at Hanford to assessing the significance of releases from the site.[96]

Such efforts were constrained in part by concerns over secrecy. Making public information about the rate and the content of releases from Hanford operations to allow public discussion would reveal critical information about the rate of production of fissionable material and, by extrapolation, about the rate of growth of the U.S. nuclear weapons stockpile. For example, members of the AEC's committee on reactor safeguards were told prior to an inspection trip of British facilities—and prior to giving advice on the issue of the AEC holding a safety zone on the Wahluke Slope—that such details concerning the operation of Hanford's reactors were not to be shared with British colleagues. "Topics specifically applicable to Hanford reactors or technical know-how in the production of plutonium should not be discussed," according to the ground rules laid down for the committee in 1949. "The committee will be instructed specifically not to introduce any data which would disclose Hanford production capacities, power levels, or production techniques or processes peculiar to the plant." Such constraints persisted to limit releases of information concerning releases of radioisotopes.[97]

During World War II and the first two decades of the Cold War, Hanford remained tightly compartmentalized. Federal authorities controlled not only information about its operations but also very substantial areas of land and water. For example, full-scale plutonium production had required AEC control over not only much of the Wahluke Slope but also the Hanford Reach. Beginning in the mid-1960s, though, the AEC slowly began loosening restrictions on the river in conjunction with the closure of reactors on the reservation. Prior to 1965 it had restricted access upstream from Richland; now it opened the stretch of river between Richland and Ringold. This encouraged local boosters to eye the Hanford Reach as a potential resource to be exploited. The TCNIC had already inquired about developing the river for tourism and recreation. Then in 1968 Senator Magnuson speculated that the ongoing closures of reactors might permit construction of Ben Franklin Dam. He acknowledged that the dam would "inundate the last natural stretch of the Columbia River in the United States," but he cared most for its ability to stimulate the economy. Besides, Magnuson claimed to have "received solid assurances from the Corps of Engineers that everything possible would be done to protect and enhance the fish and wildlife potential of the region." Discussion about building the dam continued in the 1970s. Proponents conceded that, if built, the structure would flood part of the nuclear reservation and that some waste burial sites would need to be exhumed, but they expressed optimism about engineering solutions to these problems.[98]

The prospect of another dam, which so pleased urban boosters in the Tri-Cities, provoked environmentalists to step forward with another vision for the Hanford Reach. In 1972 the Columbia River Conservation League urged that the river between Richland and Priest Rapids Dam, the last "free-flowing" segment of the river between Bonneville Dam and the Canadian border, become the backbone of a proposed Hanford National Recreation

Area. Hoping to preserve the Hanford Reach from further development, the league suggested federal protection as the best means to accomplish this goal. Knowing that its preservationist agenda might antagonize local communities, the league tried to accommodate the plans of local boosters. While conceding that "the Hanford reservation impinges upon the natural beauty of this stretch of the Columbia River," the league was much less concerned about the AEC site than about the proposed Ben Franklin Dam, a project sure to submerge the Hanford Reach. The league went on record as preferring nuclear plants to dams as sources of additional electricity. Furthermore, it went to considerable lengths to show how preservation efforts would not conflict with—and might even complement—development of the kind of nuclear power park advocated by Tri-City boosters. Thus preservationists offered their own version of compartmentalization:

> Establishment of a National Recreation Area would not preclude further nuclear development on the inner portion of the Hanford reservation. With careful, coordinated planning, additional power reactors, such as Hanford Number Two and the potential Hanford Number Three could be operated with minimal encroachments or impact on the Columbia River. The inner portion of the Hanford reservation could be developed into a model nuclear park, while the outer portion of the reservation could be designated a National Recreation Area or its equivalent. The combination of a National Recreation Area and Nuclear Park provides a unique opportunity to demonstrate that development can coexist with natural areas.[99]

Predictably, the preservationists' compromise approach did not win much sympathy among boosters who were more concerned about economic growth. Organizations such as the TCNIC did not oppose inviting tourists and recreationists to the Hanford Reach, but they wanted to make sure that preservation of that segment of the river did not interfere with efforts at economic development. During the early 1970s the idea of a nuclear power park loomed especially large in the boosters' estimation, not only because of its local economic importance but also because promoters assumed it would become a critical source of electricity for a region facing a short supply of energy. In separate letters, Robert F. Philip and Glenn C. Lee of the TCNIC explained to the AEC how the proposed Hanford National Recreation Area would interfere with "orderly development" along the mid-Columbia: "We want Hanford to be developed as a nuclear

4.9 The shutdown of so many of Hanford's reactors during the 1960s precipitated vigorous discussion about not only the economic prospects of the Tri-Cities but also the fate of the natural setting, and particularly the Columbia River. The scientists and engineers who laid out and operated the Hanford Site during World War II and the Cold War had regarded the river as part of the manufacturing plant. By the 1970s, other possibilities began to emerge. Some wanted to ensure that the river remained harnessed to production goals by building the Ben Franklin Dam, proposed since the 1930s for a site a few miles upstream from Richland, or by dredging the river bottom in order to facilitate barge transportation. But in the end no such development of the river occurred—partly because of growing environmentalist thinking that favored an undammed and undredged river, and partly because additional river development threatened to disturb the many wastes remaining on the Hanford Site. Photograph courtesy of the U.S. Department of Energy.

park . . . and we want to move up and down that river and in and out of that area with barges and equipment and utilize the Hanford area for its highest and best use for all of the residents of the Pacific Northwest, not just a few people who want to watch birds or catch fish."[100] As in the debate over the Arid Lands Ecology Reserve, the lines between environmentalist and economic uses of resources got drawn rather sharply. Development would allow the masses to put resources to their "highest and best use"; preservation would serve only the esoteric interests of the few.

The Conservation League's proposal initially did not go very far, but it provides a useful lens through which to view responses to environmentalism in the Tri-Cities region. The proposed recreation area was particularly interesting in two ways that it engaged the AEC. First, both proponents and opponents recognized and appealed to the AEC as a decisive authority, for both sides understood that the federal agency's support would be essential to their success. Second, the AEC's response revealed its own complex relationship to the natural resources and human communities around Hanford. For one thing, ideas like the Conservation League's proposal brought unwanted attention to AEC control over the Hanford Reach. The AEC manager at Richland, Alex G. Fremling, conceded in 1973 that "our authority to close any portion of the river is, at best, tenuous."[101] The commission did not want the limitations on its claims exposed. Even more to the point, it did not want interference with its prerogatives over the lands and waters surrounding Hanford. For this reason, preservationist programs could be helpful, provided the AEC controlled them.

The Richland office of the Atomic Energy Commission had been instructed to support efforts at diversification in the Tri-Cities. It could not afford to baldly oppose schemes for economic development that had the support of Senators Jackson and Magnuson, and it even had a limited stake in the idea of a nuclear power park. So, the AEC gave no forceful or outward support to proposals for a Hanford National Recreation Area. Yet the idea of protecting the Columbia River and adjacent lands from further economic development was in many ways compatible with AEC efforts at resisting encroachment by too many outsiders on the lands and waters around Hanford. When the AEC took no favorable action on the Conservation League's proposal, it justified the decision by saying that, through AEC agreements with the Washington State Department of Game and the U.S. Bureau of Sports Fisheries and Wildlife to manage the shoreline along the Wahluke Slope, most of the league's conservation goals had already been attained.[102] Furthermore, the next year it responded favorably to a Washington State proposal to set aside additional portions of the Hanford Reach as a wild or scenic area, partly because the proposal would "severely curtail public access and preserve the isolation desired [by the AEC] for the Hanford reservation."[103] Finally, it even went so far as to suggest the creation of a National Environmental Research Park on the Hanford Site, under its own management and that of Battelle–Pacific Northwest Laboratories. This proposal was made,

again, to help "allay public criticism of our site ecology practices," and
also to keep the resources under the management of the AEC rather than
in the less reliable hands of state or federal wildlife agencies. The proposal reached fruition in 1977, by which time the site had come under the
management of the Department of Energy.[104]

Federal managers of the nuclear site thus contributed substantially to
the hypercompartmentalization of the lands around Hanford. They did
so for a mixture of reasons. Like other Americans, AEC and DOE officials were influenced by the rise of environmentalism. They shared in
the movement's commitment to a cleaner environment; they needed for
political purposes to respond to the perceptions and pressures of a more
mobilized public; and they hoped to turn environmentalism to their own
advantage. Such motivations continued to shape federal management of
nuclear weapons facilities in the 1980s and 1990s. Symbolic of the government's shifting perspective was its commitment to the Tri-Party Agreement in 1989.

The Tri-Party Agreement was a deal struck by the U.S. Department
of Energy, the U.S. Environmental Protection Agency, and the State
of Washington. It promised that the Hanford Site would be cleaned up
within thirty years at a cost of at least $57 billion, and it put into place a
process for consulting the public.[105] Such an agreement would have been
unthinkable twenty years earlier, when the AEC still could demand and
maintain nearly exclusive control over Hanford. But voters and Congress
now pressed the DOE to change its ways. The diminution of Cold War
tensions had reduced the demand for nuclear weapons and allowed the
DOE to pay more attention to its aging and polluted facilities. The agency's
new emphasis on health, safety, and the environment was not simply the
right thing to do; it was also pragmatic. Department of Energy officials
were receptive to cleanup in large part because doing so was the only way
they could continue to operate their facilities and produce more nuclear
weapons.[106] For the DOE of the 1980s, as for the AEC in the later 1960s,
environmentalism proved convenient.

As with most things about Hanford after 1970 or so, the Tri-Party
Agreement was controversial. Before the agreement, Washington and
Oregon had been represented on the Columbia River Advisory Group,
which acknowledged that the states' public health authorities had an interest in Hanford's effect downstream. But the Tri-Party Agreement gave
states a much stronger legal position vis-à-vis the DOE, and legal trac-

tion to enforce agreed-upon environmental standards. Responding to this, the DOE allocated a great deal of money to the program during the early 1990s but received few immediate dividends. Studies of the effort reported wasted resources but little progress in cleaning up the site, and deadlines for reaching certain goals were repeatedly postponed. By the late 1990s the slow progress caused concern that Congress would cut cleanup funding severely, so the DOE and the Tri-Cities launched a public-relations campaign to reassure legislators about the progress being made.[107]

Those most satisfied with the cleanup program were local boosters, who found that after the hard times of the mid-1980s the Tri-Cities economy was booming again. The local population had initially expressed little interest in becoming "janitors and paper-pushers" for the DOE; but it turned out that they did not mind so much, provided they were paid well enough and that the work appeared to be "pioneering." In 1994 about $5 million daily was spent on the cleanup effort; among other things the money employed 18,750 workers at an average annual salary of $43,000. Once suspicious of environmentalists, local boosters now talked like converts to a new faith. Sam Volpentest, a founding member of the TCNIC who for so long had pushed for new production missions, cheered Hanford's new direction in 1992, exclaiming, "The green stuff is just raining down from heaven." He also framed cleanup activities in the customary frontier context: "All the stuff that's in the ground at Hanford I think of as a gold mine. The whole world has to be cleaned up, and this is where it could all start."[108]

The fact that Volpentest could apply the old metaphors of extractive industry and pioneering to the new cleanup orientation hinted that in some ways attitudes had not changed very much. He saw the pollution at Hanford as an opportunity, not a problem. Indeed, for some in the Tri-Cities the main measure of the cleanup program seemed to be the economic benefits it provided to the local communities, not its ecological impact. For this reason, when responding to proposals to designate the Hanford Reach of the Columbia a wild and scenic river, some boosters who had cheered the cleanup at Hanford because it lifted the economy condemned efforts to "lock up" the river for preservationist purposes.[109]

The debate over the Columbia emerged in large part for the same reason that environmentalism had taken root in the 1960s and 1970s: the decline in plutonium production moved people to look to Hanford for opportunities that could never have been developed before. In the late 1980s the Army Corps of Engineers—the agency that had built Bonneville

Dam during the Great Depression and the Hanford Engineer Works during World War II—proposed dredging the Hanford Reach to permit barge traffic as far upriver as Wenatchee. But the corps had more than dredging in mind for the mid-Columbia. As one spokesperson explained, "We are in the business of building projects. And that's the last major dam site left on the river." Constructing the proposed Ben Franklin Dam had been discussed as a form of economic diversification around the Tri-Cities since the late 1960s. "The Columbia River is not being tapped anywhere near the level it could be," said an economist in 1974, speaking for many in the Tri-Cities area. In this spirit, people continued advocating for the Ben Franklin Dam or at least a program of dredging. Agricultural, industrial, housing, and some tourist interests advocated further "taming" of the Columbia, and so did port districts, hydroelectric authorities, and irrigation agencies. Yet such proposals never got very far. As the Army Corps of Engineers admitted in 1992, "We are having a little trouble selling that dam."[110]

Proposals to "develop" the Hanford Reach provoked preservationists to demand that that stretch of river and the adjacent lands simply be left alone. Since the proposal for a Hanford National Recreation Area in 1972, there had been talk about setting the Hanford Reach aside in some way. For example, late in 1987 Washington Republican Congressman Sid Morrison suggested studying the reach for possible inclusion in the national system of wild and scenic rivers.[111] Now the idea of protecting the reach attracted increasing support from many who advocated "saving" the last undammed stretch of the Columbia within Washington State. The campaign gained strength from the fact that the Hanford Reach comprised the one healthy spawning ground remaining for Chinook salmon on the main stem of the river. As the Pacific Northwest fought to maintain its endangered species of salmon, the entire region (along with the numerous federal agencies involved with the salmon crisis) came increasingly to feel that protection of places such as the Hanford Reach—including not only the river but also the lands adjacent to it—was essential.[112] Those who advocated preservation of the Hanford Reach also appealed to economic arguments. They said that a protected river would help attract more tourists, enhance the business of recreation, and attract new industry by "deflating the image of the mid-Columbia as the United State's [sic] most polluted place."[113] Like the lands and waters themselves, environmentalism and boosterism were not always so easily compartmentalized.

Congress intervened in 1988, barring further river development while the National Park Service studied how to manage the Hanford Reach. In 1992 the Park Service recommended that the reach be named a "federal wild-and-scenic river" managed by the U.S. Fish and Wildlife Service; it also argued for an 86,000-acre national wildlife refuge abutting the riverbanks on the Wahluke Slope.[114] Congress took no immediate action on these recommendations, slowed by the resistance of local business interests—well-represented in the Senate and the House—who protested the notions of both more natural resources being protected from economic development and more federal control over local affairs. Finally, in April 1999, the Department of Energy determined that the federal government would retain control over the Wahluke Slope along the Columbia River, and in June of 2000 the Clinton administration named the Hanford Reach a national monument.[115]

The argument over the Hanford Reach was instructive. For many local boosters and for some federal agencies, little had changed since 1950: the development of natural resources for economic purposes remained the dominant motif. In another respect, however, everything had changed. Prior to 1980, few people in the Tri-Cities would have disagreed with economic boosters; by the early 1990s, many residents supported the idea of some strong protections for the Hanford Reach. Even the *Tri-City Herald*, for so long the mouthpiece of local boosters, advocated some form of preservation, and powerful federal agencies—long stymied by the AEC and DOE—made the case for protection, too. The environmental movement had made inroads after all.[116]

There remained the question, of course, of what was left for environmentalists to protect. For one thing, their focus was on not an entire body of water but a fragment of river—some fifty miles of the Columbia that, because of efforts to produce nuclear weapons, had escaped being fully incorporated into the Columbia Basin Project. For another, the stretch of river was not as pristine as some preservationists made it out to be. In recommending safeguards for the Hanford Reach, the National Park Service reported in 1992 that the river "looks the same now as it would have appeared a century ago. It is one of the last wild landscapes along the Columbia River." In fact, the reach had hardly been "wild" in 1892; recovering the river of the 1890s would have been a project in historic rather than natural preservation. In this context, "wild" seemed to mean "rural," or "underdeveloped"; it defined the appearance of the river, not its many other qualities. A proper designation for this portion of the

Columbia would have been the *relatively* wild and scenic Hanford Reach. Another claim made by preservationists was that the Hanford Reach was the last free-flowing stretch of the Columbia. It would be more accurate to say that the reach was *relatively* free flowing. While there were no dams on the stretch of the Columbia passing through the Hanford reservation, and while it did not exhibit the lakelike appearance displayed by the river for most of its length, the river's flow was still dictated by management upstream. Priest Rapids Dam, the upper boundary of the reach, "causes the river to fluctuate as much as 10 feet in a short period."[117]

In sum, while looking at a stretch of seemingly natural river, it is useful to realize that even though the river is designated "wild," "scenic," "historic," or "monumental," it may still be polluted. Radionuclides had been deposited in sediments along the Columbia's bottom, where they continued to affect life. They were one reason why dredging or damming the river would have been so problematic. And there remained the considerable risk that more radioactive and chemical wastes were moving from the lands and water table of the Hanford Site into the river itself.[118]

The continuing threats to the Hanford Reach remind us, finally, of the limitations of preservation in a hypercompartmentalized region. Americans may wish to designate the Hanford Reach as a national monument, and they may treat the Hanford Site as a national sacrifice area, but in fact the two places remain part of the same natural ecosystem and continue to affect one another profoundly, regardless of the boundaries that humans construct between them. The two places also belong to the same historical ecosystem. The site and the reach may strike us as contradictory resource uses, but they sprang from the same source. The plutonium-producing plant necessitated an undeveloped river, just as it necessitated pollution. We may talk about and label and treat the Hanford Reach and the Hanford Site as if they were entirely distinct. Naturally and culturally, however, the two places remain interdependent. The process of compartmentalization is a real factor affecting western lands and waters, but the lines being drawn in that process—marking off one resource as pristine and another as polluted—sometimes mask as much as they reveal.

As life goes on around the former plutonium factory, other types of masking occur. After 1990, eastern Washington became known for producing high-quality wine. A growing number of wineries in Richland suggested one kind of economic diversification, while numerous patches of the surrounding region gained recognition for their vineyards. Early in

4.10 The flowering of the wine industry around the Tri-Cities during and after the 1990s represented more than economic diversification. In many ways, it moved the region "back to the future," to a time when the economy of the towns revolved around agriculture. At one time Kennewick had been devoted primarily to the processing and shipment of agricultural products. In 1911–15 and 1922–33 the town had hosted annual grape festivals, a tradition it resumed in 1946–48 (this photograph is from the October 1946 event). While Hanford was the main focus of the economy during World War II and the Cold War, today the resurgence of grape growing and the rise of the wine industry have begun to characterize the area once more. Photograph courtesy of the East Benton County Historical Society, Kennewick.

2006, the United States Tobacco, Tax, and Trade Bureau recognized the Wahluke Slope as an American viticultural area. The designation, based on a nomination by a geologist at Washington State University, marked the site as a distinctive producer of wine grapes. Vintners and promoters claimed that the recognition would bring prestige and higher prices to wine made from grapes grown on the Wahluke Slope. Given what was at stake, it did not seem prudent to bring up the fact that for decades the Wahluke Slope had been off-limits to farming because of its location downwind from the plutonium plant. One Oregon winery produced a 2004 Semillon from the Desert Wind Vineyard on the Wahluke Slope. The label on the bottle included a map of the region showing the vineyard and winemaker, nearby dams and towns, and an adjacent wildlife refuge. Its text noted that "the unique soil and climate of the Wahluke Slope" were responsible for the "distinctive quality" of the "beautiful fruit" in the Semillon.[119] But no mention was made of Hanford. Indeed, in the latest efforts to develop the Tri-Cities, the future has come to look quite different from the past. Wine has become an important component of the unplanned diversification of the local economy, along with abundant recreational activities and moderate weather, all of which have attracted new residents, particularly retirees.

In 1992, reflecting on the problems of local farmers, Sam Volpentest had asked, "Can you imagine your supermarket trying to sell Hanford asparagus?"[120] Grape growers and vintners intended to avoid that problem by disassociating local produce from Hanford. They saw no reason to recall that during the 1950s and 1960s local boosters had had to fight the federal government for the right to cultivate those very farmlands that had once been in the path of radioactive wastes. While agricultural interests were content to erase Hanford's story, stakeholders and plank-holders insisted that Hanford's past be kept in mind, although that history had a different meaning for each group. As much in memory as it is on the ground, then, Hanford remained compartmentalized.

EPILOGUE

THE INTEGRATION OF THE AMERICAN WEST AND ATOMIC energy that Hanford represented drew upon broad themes in U.S. culture as well as on a particular regional history. Even before the outbreak of World War II, the country had begun imagining such mixtures of the old and the new. In 1935, for example, singing cowboy Gene Autry made his debut appearance in movie theaters around America as the hero of a twelve-part serial titled *The Phantom Empire*. Autry portrayed himself as the part-owner of Radio Ranch, a dude ranch dependent not just on the income from paying guests but also on the returns from a daily radio show broadcast from the ranch house's front yard. Radio Ranch was a technologically intensive place: in addition to the array of equipment set up to handle the remote transmission of performances by Autry and his band, the children in residence had a clandestine laboratory, hidden in the barn and accessible via a secret entrance. Moreover, Radio Ranch happened to sit atop the secret underground kingdom of Murania, "an advanced civilization through the use of radio activity" reliant on abundant stores of radium to support a high-technology culture that had been in existence since being driven underground by the last ice age. Despite all the glowing, whirring, and ray beaming associated with these advanced tools, life on the ranch still required Autry to be a superb horseman, a crack shot (with six guns and lever-action rifle), and handy with his fists. Once radium deposits of inestimable value were discovered on the property, Autry was beset by a stock character from the popular culture of the interwar period: the mad scientist driven by a lust for power and wealth. (In the serial's tenth episode, Autry is "captured by a vicious party of research scientists.") Some of the essential templates for how to understand nuclear technology—the mad scientist, the secret city, radioactive materials possessing the power of life and death—had been established before the phenomenon or the mechanisms of nuclear fission were realized and had already been juxtaposed with popular imagery of the Western frontier.[1]

Despite the machinations of greedy scientists and their henchmen, the hidden community of Murania represented the greatest threat of all to the peaceful ranchers. "There is no limit to the power of radium when it is controlled," Autry is informed when he makes his way into the underground kingdom. Illustrating popular tropes about the power of radioactivity (especially that of materials kept in hidden fastness), radium emanations in Murania could be employed to cause a slow death by a "disintegrating atom smashing machine" or used to bring someone back to life in the "radium revival chamber." Murania's most powerful secrets, though, were its weapons, including an atom-smashing machine "capable of destroying all civilization" and, perhaps, "capable of destroying the universe." When this device is set loose in the serial's final episode, the secret kingdom's ruler decides to die with her people rather than to escape and live in the uncontrolled madness on the Earth's surface. In a cascade of running photographic emulsion, Murania disappears under the disintegrating rays of its own radium machines.

It is tempting at this point to say that, unlike the fictional Murania, real communities do not just undergo slow dissolves and then disappear—except that sometimes they do. Company towns across the United States faded away when business philosophies changed and communities became the business of politicians rather than employers. Harpers Ferry, Virginia, for example, like Richland, was developed by the United States government to support a federal arsenal. Its location in the early nineteenth century was dictated in part by the presence of a river that could play a vital function in the armory's mission by ensuring reliable transportation to Washington, D.C. In addition, Harpers Ferry drew upon the gunsmiths of western

Maryland and Pennsylvania. These craftsmen were the artisanal ancestors of the machinists drawn from DuPont's Remington firearms plants who came to Hanford in 1943–44 and solved some of the basic problems of operating its nuclear reactors. Under pressure from federal authorities to increase production by giving up some of the prerogatives of their status as craftsmen, the Harpers Ferry workforce resisted, making life difficult for the arsenal's resident manager, who had to live in an isolated town with unhappy workers while being responsive to the demands of the Department of War in Washington, D.C. Like the residents of the Tri-Cities, the workers at Harpers Ferry did not always accede to the dictates of their federal sponsors. In the end, though, Harpers Ferry—indefensible during the Civil War and subject to a series of floods afterward—to a very great extent washed away and disappeared as a working community. The U.S. Army found a more tractable workforce to draw upon around Springfield, Massachusetts.[2] Still, the history and the historical significance of Harpers Ferry were not defined by federal policies alone but by the negotiations between the government and a local workforce rooted in a regional history. The same can be said for Hanford.

In assessing the significance of nuclear weapons, students of history tend to regard the atomic bombing of Hiroshima and Nagasaki as marking the end of one era and the beginning of another. They appropriately identify the period 1939–45 as the birth of the Atomic Age, and they see that brief period as the onset of a particularly problematic phase of development. But an exploration of the history of Hanford in the context of the American West suggests another way of periodizing nuclear history. Before the 1970s Westerners did not have to struggle very much to come to terms with the new reality of nuclear weapons. Rather, they easily incorporated the weapons-production complex into ways of thinking and acting that had characterized the region for decades before 1940. Initially, the bomb did not change their overriding view of the world. In contrast to Gene Autry's Murania, the West's atomic complex was not regarded as part of an ominous underworld prior to 1970 or so. Served by state and interstate highways, without a fence separating it from Kennewick or the rest of Washington, Richland was no secret city, although it remained a distinctive one.

For the latter half of the nineteenth century and most of the twentieth, Westerners were convinced that their region was underdeveloped. In that context, most Westerners viewed the production (and, for some, even the testing) of nuclear weapons as a means to advance their communities and regions. Residents of the Pacific Northwest thus saw in Hanford a means to advance their economy, to elevate the Columbia Basin so that it stood more on a par with other prosperous and industrialized areas. Yet the site represented more than factory jobs and outside investment. It attracted more people to the lightly settled territory, and the urbanization associated with the site served as another sign of progress in the underdeveloped West. Hanford also gave the peoples of the Tri-Cities and Washington State a chance to identify with scientific endeavor, advanced technology, and an important national mission. For inhabitants of a part of the country that had long been regarded as backward, an association with Hanford furnished a sense of importance and even sophistication. Richland became the Atomic City of the West, an Atomic-age Utopia, and an All-America City; the plutonium factory's workforce was said to be spearheading the nation's engagement with nuclear power.

By viewing themselves as *pioneers* for the nation and the world, Hanford and the Tri-Cities deepened their identification with the West by drawing upon American frontier rhetoric. Like the conviction that the West was underdeveloped, this rhetoric dated from the late nineteenth century. By the mid-twentieth century it had transcended the West and begun to influence how Americans understood themselves and their place in the world. In this sense Richland cast itself as Radio Ranch and Murania simultaneously, grasping ray guns and six-shooters. People working at Hanford and living in the Tri-Cities found the frontier identity especially powerful because so many of them had recently migrated to the West and because they were so frequently told (and they so frequently told others) that they were pioneers. The identity proved especially handy in lobbying the federal government for additional investment and subsidy during the 1950s and 1960s, at a time when Richland was desperate to ensure that, unlike Murania and Harpers Ferry, it would not disappear. Being associated with the frontier suggested that Hanford employees and Tri-Cities residents deserved special consideration because of the sacrifices they had endured. At this time, few locals actually believed that sacrifice came in the form of putting their health at risk by living near and working at Hanford. Rather, they spoke about the hardships encountered in moving west, settling in an arid, rural environment, and meeting construction and production goals during times of national emergency. Their hard work produced an ideal city, one associated with technological progress, from its schools to its volunteer community institutions and economic base.

Assuming an identity as pioneers, Hanford workers and Tri-Cities residents tended to embrace the idea that they had more than their share of traits associated with the American frontier. They lauded the idea that individualism loomed large in their society and culture, even though government and corporate power was especially concentrated at Hanford and the Tri-Cities. They replicated forms of entertainment associated with the bygone American West, such as the rodeo, even though the area that they inhabited had never seen much in the way of cowboys or ranches. They cherished the notion that, as defense workers, they stood between "civilization" and whatever form of savagery seemed especially menacing—German Nazism, Japanese aggression, Soviet communism, or energy shortages. As settlers in a new place, they erected landmarks to orient themselves, just as historians do when trying to understand an unfamiliar history. For those who worked and lived around Hanford and for the historians who study them, the imagery of pioneering retains its power, but here we have tried to gain a measure of perspective by placing it in a broader historical framework.[3]

Other historians have paid far more attention than we do to Hanford's environmental history, but we have tried to suggest that this broader view of the site's history is useful as well when thinking about the hazards associated with the reservation and the cleanup program that has been its mission for the past two decades. While there were hazardous materials both on the site and released into the environment, these potential threats were understood in the context of the site's mission and the effectiveness of Hanford's workers in carrying it out. Given that Hanford's production rate played a fundamental role in defining the size and rate of growth of the national nuclear stockpile, and so represented the American hole card in the arms race with the Soviet Union, secrecy was a part of life in the Tri-Cities, and this extended to environmental hazards as well. To some it has seemed remarkable that those who worked at Hanford and raised families in the Tri-Cities—as opposed to downwinders who had little or no access to information about Hanford's operations—have so little concern about radiological releases. Arguably, though, Hanford was representative of a large class of industrial communities. Company towns in more hazardous industries learned to accustom themselves to risk; in the coal mining region around St. Clair, Pennsylvania, in the late 1850s, for example, mine workers had about a fifty-fifty chance of surviving for twelve years.[4] Even taking into account the restrictions on information associated with the nuclear production facility, no comparable threat appeared to life in the Tri-Cities.

Hanford's culture of production, which contributed to the community's identity by reference to the workers' mastery of an esoteric (if not idiosyncratic) group of machines, also helps explain why locals discounted the threat of radioactive contaminants. Engineers and craftsmen at Hanford solved problems; the reliable output of plutonium from their graphite-moderated reactors was their signal achievement. It was what kept Hanford in operation despite the fact that on a number of occasions it seemed possible that the federal government would shut the facility down. While Hanford's close identification with production at times limited its mission, and so its institutional prospects, it became the key to the working community's longevity. Perceptions of risk, like successful resistance to the federal government's efforts to write Hanford's script in a one-sided fashion, have to be understood by reference to this local context of problem solving.

In short, between the 1940s and 1960s, while plutonium-plant workers, Tri-Cities residents, and Pacific Northwesterners generally grasped that the bombs being manufactured at Hanford were changing the history of the world, they tended to understand the new development in regional terms that drew on the past more than they looked to the future. For the region's citizens who thought about Hanford or supported its growth, and especially for politicians who championed its cause, the advent of Hanford may not have marked the start of the new age in world history so much as it represented the fulfillment of western ambitions for economic growth, urban expansion, and cultural sophistication. In this regard, Hanford resembled the Columbia Basin Project—another massive federally funded engineering project. Both were public works projects devoted to taming nature. A difference was that the Columbia Basin Project's service to the welfare state pervaded its career, beginning during the 1930s with the effort to create jobs, and continuing during the 1950s in an effort to prop up some version of "the family farm." Hanford made the transition from a national-security-state venture to a welfare-state project only haltingly after 1960.

In the American West, if Hanford's experience is an indication, the more significant discontinuity in history was not the dawn of the Atomic Age in the early 1940s but rather the growth of an environmentalist consciousness after the mid-1960s.[5] For decades the region had advocated growth and the development of natural resources. Both hydroelectric dams and plutonium-producing reactors represented a form of progress because they promised to reduce the region's condition of underdevelopment. During the last third of the twentieth century, however, a different way of thinking about natural resources and nuclear power emerged. This different way of thinking did not convert everyone immediately; in the Tri-Cities and around the larger West, for example, proposals to build more reactors, import more nuclear waste, and dam or dredge more rivers were hotly contested during the 1980s and 1990s. And even those who converted to cleanup only belatedly (and if only largely because of the economic benefits that it promised) continued to speak about Hanford as if it were simply another western extractive industry—"a gold mine"—capable of perpetuating local prosperity.

Nonetheless, by the 1990s it was nearly impossible to imagine construction of another reactor or another dam along the Columbia River. A region that had once been viewed as empty now seems much too "full" in too many spots. Westerners have become divided over just how much protection of natural areas is enough or just how much development is sufficient. As different groups contest the region's future, they increasingly discover that national sacrifice zones (such as the Hanford Site) and governmentally protected lands and waters (such as the wild and scenic Hanford Reach of the Columbia River) are coexisting cheek by jowl. These contests, like even more long-standing ones about the health effects of the production and testing of nuclear weapons, cannot be resolved by an appeal to the court of history; even if historians felt competent to provide accounts of "what really happened," none of the interested parties would accept them.[6] But one thing that historians can do—and that we have tried to do in this volume—is recapture the contours of past interactions between the federal presence in the West, the communities dependent on it, and the ways in which a broader regional polity variously supported, opposed, and ignored it. It is important to remember that Washingtonians, in particular, were not just passive victims of abusive federal power; both directly and indirectly, they courted federal support for Hanford in the interests of economic development and as part of the vision of modernization in the Columbia Basin.

Debates over the future uses of western lands and resources will continue. It is not clear whether nuclear power will again have much of a future in the American West. What is more certain is that the region will remain the part of the country most strongly identified with the atom. During World War II and the Cold War, the West received more than its share of America's nuclear weapons facilities. As a result, it also received more than its share of the nation's Atomic Age pollution. By trying to locate the country's major storage sites for nuclear waste in the region, the United States seems bent on ensuring that the West will remain America's most nuclear region. Like aridity, the atom is something that in many respects sets the West apart. As a result of a shift in attitudes after the mid-1960s—attitudes toward the environment, toward nuclear weapons and nuclear power, and toward economic development—the majority of the region's inhabitants don't much value their atomic identity today. But at Hanford, in the areas around the Tri-Cities, and in many other places across the region for much of the mid-twentieth century, Westerners did not regard an association with the atomic frontier as such a bad thing.

Appendix 1

Starting and Completion Dates of
Manhattan Project Facilities at Hanford, 1943–45

HANFORD FACILITY	STARTING DATE	COMPLETION DATE
100-B Processing Area (B reactor)	June 7, 1943	Sept. 15, 1944
100-D Processing Area (D reactor)	Nov. 1, 1943	Dec. 4, 1944
100-F Processing Area (F reactor)	Dec. 21, 1943	Feb. 10, 1945
200-E Processing Area (canyon)	Aug. 2, 1943	Feb. 10, 1945
200-W Processing Area (canyon)	June 22, 1943	Dec. 16, 1944
200-N Processing Area (canyon)	Nov. 17, 1943	Nov. 3, 1944
300 Area	June 10, 1943	Jan. 6, 1945
700 Administrative Area (Richland)	Sept. 14, 1943	Mar. 31, 1945
1100 Richland Village (residential)	Apr. 9, 1943	Mar. 31, 1945

E.I. du Pont de Nemours & Co., Inc., "Completion Report, Manhattan District, Hanford Engineer Works TNX Plant," Apr. 30, 1945, MEDPR, box 46.

Appendix 2

Wartime Population Increase in Hanford Area

CITY	1940 ESTIMATED POPULATION	MARCH 1945 ESTIMATED POPULATION
Benton City	200	1,000
Grandview	1,200	2,500
Kennewick	1,900	7,500
Pasco	3,900	8,500
Prosser	1,800	3,000
Sunnyside	2,000	3,500
Toppenish	3,600	5,500
"Trailer Camp Benton Co."	0	1,500
Yakima	27,000	35,000
Totals, (% increase)	41,600	68,000 (63%)
Totals, (% increase) excluding Yakima	14,600	33,000 (126%)

Source: W. A. Rothery, "Population Figures—Hanford Engineer Works," Apr. 11, 1945, MRCF-DE, box 8.

Appendix 3

Tri-Cities Demographic Data, 1950–60

	PASCO	KENNEWICK	RICHLAND
1950 population	10,228	10,106	21,809
1960 population	14,522	14,244	23,548
1950 total blacks	980	4	7
1950 % nonwhites	10.1	0.1	0.2
1960 total blacks	1,213	5	189
1960 % nonwhites	9	0.3	1.3
1950 % over 64 years	5.1	4.7	1.2
1960 % over 64 years	6.3	5.8	2.2
1950 % unemployed	12.7	11.3	3.7
1960 % unemployed	6.2	7.1	4.2

Sources: U.S. Bureau of the Census (hereafter cited as USBC), *United States Census of Population: 1950*, vol. 2, *Characteristics of the Population*, part 47, *Washington* (Washington, D.C.: U.S. Government Printing Office, 1952), 8, 43, 44, 67; USBC, *United States Census of the Population: 1960*, vol. 1, *Characteristics of the Population*, part 49, *Washington* (Washington, D.C.: U.S. Government Printing Office, 1963), 23, 45, 46, 91, 132; USBC, *United States Census of Population: 1970*, vol. 1, *Characteristics of the Population*, part 49, *Washington* (Washington, D.C.: U.S. Government Printing Office, 1973), 13.

Appendix 4

Tri-Cities Demographic Data, 1990

	PASCO	KENNEWICK	RICHLAND
Total population	20,070	42,155	32,315
Number of blacks	1,126	476	461
(% of total)	(5.6)	(1.1)	(1.4)
Number of Hispanics	8,277	3,684	983
(% of total)	(41.2)	(8.7)	(3.0)
% unemployed	11.5	7.0	5.1
Per capita income (in 1989 dollars)	$8,016.00	$12,767.00	$17,085.00
% with incomes below poverty line	33.0	13.9	7.8
Ratio of owner- to renter-occupied homes	0.90	1.13	1.65

Sources: U.S. Bureau of the Census, *1990 Census of Population: General Population Characteristics, Urbanized Areas* (Washington, D.C.: U.S. Government Printing Office, 1992), 18, 214; U.S. Bureau of the Census, *1990 Census of Population and Housing: Summary Social, Economic, and Housing Characteristics, Washington* (Washington, D.C.: U.S. Government Printing Office, 1991), 52, 54, 81, 83, 84, 85, 128, 130.

Note on Sources

Writing about the history of American nuclear-weapons programs requires understanding the different, intricate systems of storing and identifying documents generated since 1942 by the U.S. government and its contractors. The research for this book relies heavily upon declassified, and never classified, materials held by two different government entities: the United States Department of Energy (DOE) and the National Archives and Records Administration (NARA). A few words about each institution's holdings may be instructive to readers and researchers.

Over the years, production of nuclear weapons in the United States has been the responsibility of, successively, the Army (through 1946); the Atomic Energy Commission (1947–1974); the Energy Research and Development Administration (1975–76); and the Department of Energy (1977-present). Today, the Department of Energy is responsible for preserving and making available the declassified documents produced over the years by the different federal agencies and their assorted contractors. Those documents are available at more than one DOE location and are accessible in different ways. We have attempted to identify the place where we consulted or obtained each unpublished document that we cite; however, many of those documents are available in more than one location.

The Department of Energy maintains the DOE Historical Research Center within its Office of History and Heritage Resources in Germantown, Maryland. A number of collections are available there, including the Richland Diversification Files, the Atomic Energy Commission Division of Production materials, the General Manager files, and the Energy History Collection. Some of these collections have their own numbering system, which we have attempted to acknowledge consistently in our notes. Thus we identify a letter from Carroll L. Wilson to J. Robert Oppenheimer, April 16, 1948, as being from "GenMan 5580:17," that is, the collection of the General Manager Files, box 5580, folder 17. In many cases specific documents were assigned their own numbers by the agency generating them. Thus the report from the AEC Advisory Committee on Reactor Safeguards is identified as "ACRS Report on Modifications to Hanford Reactors," Feb. 4,

1958, AEC 172/22, pp. 1–2, 51–58Sec, 1284:6, "AEC 172/22" is the internal identifying number given by the commission, while 1284:6 refers to box 1284, folder 6.

At Richland, Washington, the DOE maintains the United States Department of Energy Public Reading Room (PRR) as part of its Richland Operations Office. Many documents consulted in this facility are identified with the initials PRR in the endnotes. Most (but not all) documents from this collection may also be identified with either an accession number or a document number. For example, the Hanford Works Monthly Report (HWMR) for June 1949, dated July 18, 1949, bears the PRR accession number 10373, and the document number HW 13793-Del; we have endeavored to provide one or the other. Document numbers have different provenances. The prefix HW in the above example, for instance, identifies the document as pertaining to the Hanford Works—the name for the site during the Cold War. (During World War II the site was called the Hanford Engineer Works, and many document numbers from that period commonly begin with the letters HAN.) On the other hand, some documents have identifiers that associate them with a government entity. Thus the diary of Captain F. A. Valente is associated with the Manhattan Engineer District (document numbers MED1001 through MED1004). Still other PRR documents are identified by the microfiche numbers.

The DOE has put thousands of Hanford documents and photographs on line through the Hanford Declassified Document Retrieval System (DDRS), located at http://www2.hanford.gov/declass/. Through its Advanced Search protocol, DDRS allows multiple ways of retrieving an item, e.g., through document number, accession number, author, company, title, keyword, or date. Plugging in one kind of identifier usually calls up other kinds.

Part of the mission of the DOE Office of History and Heritage Resources is to prepare DOE records for transfer to the National Archives and Records Administration. As a result, documents are constantly in motion from one agency to the other. For historians and others, this presents problems. In our research at the DOE Historical Research Center in Germantown, Maryland, we consulted the Secretariat Files of the U.S. Atomic Energy Commission, grouped then by DOE staff into the periods 1947–1951 and 1951–1958, and therefore abbreviated as 47–51Sec and 51–58Sec in our notes. Those two collections were later transferred to the NARA branch at College Park, Maryland. As part of the transfer, we have been told, the files were reboxed and renumbered; yet no inventory of the transfer, explaining how the National Archives reorganized the collection, has been found. To make matters worse, from what we can gather, a number of the files from the AEC Secretariat for 1947–1951 and 1951–1958 have been re-classified and are therefore no longer available to researchers without appropriate security clearances. After consultation and deliberation, we have decided to retain in our manuscript the information derived from research in the AEC Secretariat Files, regardless of whether the pertinent documents are now classified or declassified. Unable to develop the current provenance for these materials, we maintain the citations that we established upon doing research in the DOE Historical Research Center. In other words, in our endnotes, citations to 47–51Sec and 51–58Sec refer to erstwhile DOE holdings that may no longer be accessed using the location, box numbers, and file numbers that we provide. But the authors, titles, and dates of individual documents provide some information about our sources, and we have confirmed the accuracy of our research by checking citations carefully against the records generated in the course of our research, including photocopies of many of the pertinent documents and notes drawn from others.

The National Archives and Records Administration holds the great majority of the materials pertaining to U.S. nuclear weapons programs in Record Group 326. Many of these materials are available at National Archives II, or the National Archives at College Park, Maryland. One such collection is the David Lilienthal Papers (abbreviated in our notes as LP). However, many of the materials from Record Group 326 that we examined were located in the National Archives Center, Southeast Region, outside of Atlanta in East Point, Georgia; this branch has now relocated to Morrow, Georgia. After World War II, many papers concerning Hanford were filed together with those from the Manhattan Project site at Oak Ridge, Tennessee, and therefore ended up at NARA's facility in the southeast (rather than the northwest) region. In our notes, these materials are identified with the following abbreviations: MED-C, MEDPR, MRCF-ACC, and MRCF-DF.

Most of the illustrations in this book originated with the DOE and its predecessor agencies. We located some of them on the DDRS website. Many others were identified and obtained through Lockheed Martin, a DOE subcontractor whose offices in the Tri-Cities have assumed responsibility for many historical photographs of Hanford.

Of course, DOE and NARA have no monopoly on historical records pertaining to Hanford. Many other collections contain important materials, including the Manuscript Division of the Hagley Museum and Library, Wilmington, Delaware, which has many records pertaining to activities of the DuPont company. We have been fortunate to work next door to the University of Washington Libraries, which are particularly rich on materials concerning Hanford. Many of those materials, including the papers of U.S. Senator Henry M. Jackson (D-WA), U.S. Senator Warren G. Magnuson (D-WA), and Richland Mayor Fred Clagett, are housed in Special Collections. The Libraries' division of Government Publications has also proved important as an official repository for the hundreds of reports and publications generated by the Hanford Environmental Dose Reconstruction Project (HEDR).

Abbreviations Used in Notes

47–51SEC AND 51–58SEC

These abbreviations refer to the Secretariat Files of the Atomic Energy Commission, Record Group 326, National Archives II, College Park, Maryland. We consulted these materials in the Historical Research Center, Office of History and Heritage Resources, Department of Energy, Germantown, Maryland, where they were organized into the periods 1947–1951 and 1951–1958. Items from these collections are generally cited with the box and folder numbers used by DOE, as in 97:20.

58–66SEC

Secretariat Files, U.S. Atomic Energy Commission, 1958–1966, on file in Historical Research Center, Office of History and Heritage Resources, Department of Energy, Germantown, Maryland.

AEC

U.S. Atomic Energy Commission

CBN

Columbia Basin News

DDRS

Hanford Declassified Document Retrieval System. Online access to numerous documents and photographs concerning Hanford after 1942 via http://www2.hanford.gov/declass/.

DOE

U.S. Department of Energy

EHC

Energy History Collection. Miscellaneous materials in the Historical Research Center, Office of History and Heritage Resources, U.S. Department of Energy, Germantown, Maryland.

ERDA
U.S. Energy Research and Development Administration

FCP
Fred Clagett Papers, Accession 3543, Special Collections, University of Washington Libraries, Seattle. Materials are generally referred to by the box number, e.g., FCP4.

FTM
"Col. F[ranklin]. T. Matthias, Notes and Diary" [1942–1946]. This microfiche document was consulted at the Department of Energy Public Reading Room, Richland, Washington.

GE
General Electric

GENMAN
General Manager Files, Office of History and Heritage Resources, U.S. Department of Energy, Germantown, Maryland.

HEWMR
Hanford Engineer Works Monthly Report. These reports were consulted in the Department of Energy Public Reading Room, Richland, Washington.

HMJP
Henry M. Jackson Papers, Accession 3560, Special Collections, University of Washington Libraries, Seattle. This collection has been subdivided by period, so that, for example, Jackson's first two senatorial terms are covered mainly in Accession 3560–3. In the notes, this subdivision of the accession is given as HMJP-3. Items from the collection are generally cited with the box and folder numbers, as in 97:20.

HMLMD
Hagley Museum and Library Manuscripts Division, Wilmington, Delaware.

HWMR

Hanford Works Monthly Report. These reports were generally consulted in the Department of Energy Public Reading Room, Richland, Washington.

JCAE

Joint Committee on Atomic Energy, U.S. Congress

JGT

J. Gordon Turnbull, Inc., and Graham, Anderson, Probst, and White, Inc. (authors of *Master Plan for Richland, Washington*, 1948).

JOINTCOM

Joint Committee on Atomic Energy materials, held in Energy History Collection and consulted in the Office of History and Heritage Resources, U.S. Department of Energy, Germantown, Maryland.

LP

David Lilienthal Papers, Record Group 326, National Archives II, College Park, Maryland.

MED-C

[Manhattan Engineer District], Correspondence 1943–1949, Record Group 326, Accession 68A588, National Archives Center, Southeast Region, Morrow, Georgia.

MEDPR

Manhattan Engineer District Project Records 1943–1947, Record Group 326, Accession 4NN-326–8505, National Archives Center, Southeast Region, Morrow, Georgia.

MRCF-ACC

Mail and Records Central Files, Office Services Branch, MED & CEW; Areas and Contractors Correspondence, Jan. 1947 to June 1948, Record Group 326, Accession 67A803, National Archives Center, Southeast Region, Morrow, Georgia.

MRCF-DF

Manhattan Engineer District (CEW), Mail and Records Central Files,

284 ABBREVIATIONS USED IN NOTES

War Department, Decimal Files 1943–1948, Record Group 326, Accession
66A1405, National Archives Center, Southeast Region, Morrow, Georgia.

NYT
New York Times

PAS
Public Administration Services, Inc.

PI
Seattle Post-Intelligencer

PROD
Atomic Energy Commission Division of Production materials, held
within the Office of History and Heritage Resources, U.S. Department of
Energy, Germantown, Maryland.

PRR
U.S. Department of Energy Public Reading Room, Richland, Washington.
Documents in this collection are usually accompanied by either an accession number (assigned by PRR) or a document number (usually assigned
by the organization that generated the document).

RCC
Richland Community Council (also referred to as Richland City Council). This was the citizen advisory body consulted by the General Electric
company and the U.S. Atomic Energy Commission regarding the governance of Richland, WA, 1947–1958. Files of Richland Community Council minutes and records are located in the Fred Clagget Papers (FCP1
through 1957, FCP2 from 1958), Special Collections, University of Washington Libraries, Seattle.

RDF
Richland Diversification Files, Collection 1314, Office of History and
Heritage Resources, U.S. Department of Energy, Germantown, Maryland.
Items cited from this collection are generally given with the box and
folder numbers, as in 5:2.

RPL
Richland Public Library, Richland, WA.

RPL-CF
Richland Public Library Clippings Files (or Vertical Files), including "Historical File, Reference Office." These are files of newspaper clippings and other materials maintained by librarians at Richland Public Library. Some materials are undated or have dates handwritten onto them.

SSR
Spokane Spokesman-Review

ST
Seattle Times

SUL
Seattle Urban League Papers, Accessions 607 and 681, Special Collections, University of Washington Libraries, Seattle. Materials cited include the box and folder numbers, as in 36:6.

TCH
Tri-City Herald

TCNIC
Tri-City Nuclear Industrial Council

USBC
U.S. Bureau of the Census

UWL
University of Washington Libraries

WGMP
Warren Grant Magnuson Papers, Accession 3181, Special Collections, University of Washington Libraries, Seattle. This collection has been subdivided by period, so that Magnuson's first two senatorial terms, 1945–1956, for example, are in Accession 3181-3. In the notes, this subdivision of the accession is given as WGMP-3. Items from the collection are generally cited with the box and folder number.

Notes

Notes to Introduction

1   Mrs. J. W. Nichols to Col. F. T. Matthias, Aug. 17, 1945, box 80, MED-C.

2   Bohr cited in Richard Rhodes, *The Making of the Atomic Bomb* (New York: Simon & Schuster, 1986), 294. In 1945, upon visiting Los Alamos, New Mexico, Bohr thought back to his prediction and declared, "You have done just that" (500). Peter Galison lumps wartime Hanford under the laboratory rubric while discussing the extent to which the industrial site departed from an academic model for physics; see Galison, "Three Laboratories," in *Technology and the Rest of Culture*, ed. Arien Mack (Columbus: Ohio State University Press, 2001), 194–200, and compare to Bruce Hevly, "Hanford's Postwar Voices," epilogue to S. L. Sanger, *Working on the Bomb: An Oral History of World War II Hanford* (Portland, OR: Portland State University Continuing Education Press, 1995), 227–242.

3   Richard G. Hewlett and Oscar E. Anderson, Jr., *A History of the United States Atomic Energy Commission*, vol. 1, *The New World, 1939/1946* (University Park: Pennsylvania State University Press, 1962); Richard G. Hewlett and Francis Duncan, *A History of the United States Atomic Energy Commission*, vol. 2, *Atomic Shield, 1947/1952* (University Park: Pennsylvania State University Press, 1969); and Richard G. Hewlett and Jack M. Holl, *Atoms for Peace and War, 1953–1961: Eisenhower and the Atomic Energy Commission* (Berkeley: University of California Press, 1989).

4   The quite personal focus on second-generation studies of Hanford is epitomized by Teri Hein, *Atomic Farmgirl: Growing Up Right in the Wrong Place*, rev. ed. (New York: Houghton Mifflin, 2003). Hein writes, "I don't care that expensive scientific studies claim there is no definitive proof that Hanford caused cancers. I don't care about judges and lawyers who ping-pong evidence in and out of court—

valid one day and not admissible the next—as if it were a game of table tennis. *I just care about my father*, and he is only one of thousands of people probably, and quite tragically, affected by [Hanford] activities during the Cold War" (ix-x; emphasis added).

5   See *A Guide to the Hanford Thyroid Disease Study: Final Report* (Atlanta: Centers for Disease Control and Prevention; Seattle: Fred Hutchinson Cancer Research Center, 2002). This guide, along with the complete report and other useful sources, may be found at http://www.cdc.gov/nceh/radiation/hanford/htdsweb/index.htm. (accessed July 29, 2010). On downwinder lawsuits, see *TCH*, May 20, Nov. 24, 2005, Dec. 16, 2008. There is greater confidence that Hanford operations caused ill health for workers on the site.

6   Thomas Powers, "Downwinders: Some Casualties of the Nuclear Age," *Atlantic Monthly* 273 (March 1994): 124. For examples of titles that tend to lose sight of the national-security context behind Hanford's emissions, see Michael D'Antonio, *Atomic Harvest: Hanford and the Lethal Toll of America's Nuclear Arsenal* (New York: Crown, 1993); and Peter Bacon Hales, *Atomic Spaces: Living on the Manhattan Project* (Urbana: University of Illinois Press, 1998).

7   Michele Stenehjem Gerber, *On the Home Front: The Cold War Legacy of the Hanford Nuclear Site*, 3rd ed. (Lincoln: University of Nebraska Press, 2007).

8   An exception to this generalization is health physics and environmental research, and some reactor physics research undertaken in support of the production mission.

9   T. E. Marceau, "Historic Overview," in *Hanford Site Historic District: History of the Plutonium Production Facilities, 1943–1990* (Columbus, OH: Battelle Press, 2003), 1.5. "Weapons-grade plutonium" refers to, in this case, plutonium as the output of Hanford's reactors and chemical separation facilities that was practically all comprised of the isotope plutonium-239. In particular, plutonium tinged with a few percent of the isotope plutonium-240 (that is, plutonium with one more neutron in its nucleus) is not suitable for use in nuclear weapons, because it is subject to spontaneous fission and can lead to an early and inefficient detonation. Because the two isotopes are chemically indistinguishable, the requisite purity has to be controlled for in the operation of the production reactors.

10   Students of the American West have begun examining how defense

industries and military bases were integral to the region's economic development during the mid-twentieth century. They have paid less attention, though, to the century's later decades, when environmental concerns increasingly challenged the emphasis on economic development. Roger Lotchin, *Fortress California, 1910–1961: From Warfare to Welfare* (New York: Oxford University Press, 1992); Gerald D. Nash, *The Federal Landscape: An Economic History of the Twentieth-Century West* (Tucson: University of Arizona Press, 1999); Kevin Fernlund, ed., *The Cold War American West, 1945–1989* (Albuquerque: University of New Mexico Press, 1998). A few studies have looked at developments after 1970: Brian Gerard Casserly, "Securing the Sound: The Evolution of Civilian-Military Relations in the Puget Sound Area, 1891–1984" (Ph.D. diss., University of Washington, 2007); and Bruce Hevly and John M. Findlay, eds., *The Atomic West* (Seattle: University of Washington Press, 1998), part 3.

11   Bates McKee, *Cascadia: The Geologic Evolution of the Pacific Northwest* (New York: McGraw-Hill, 1972), 271. On the Pasco Basin, the Ringold Formation, and their relationship to the Palouse, see pp. 278–79.

12   Guy Waring, *My Pioneer Past* (Boston: Bruce Humphries, 1936). On this theme, see also D. W. Meinig, *The Great Columbia Plain: A Historical Geography, 1805–1910* (Seattle: University of Washington Press, 1968).

Notes to Chapter One

1   *The American Heritage Dictionary of the English Language*, 3rd ed., "work" entry.

2   Stanley Goldberg, "General Groves and the Atomic West: The Making and the Meaning of Hanford," in *The Atomic West*, ed. Bruce Hevly and John M. Findlay (Seattle: University of Washington Press, 1998), 39–89; Stanley Goldberg, "Groves Takes the Reins," *Bulletin of the Atomic Scientists* 48 (Dec. 1992): 32–39.

3   Richard Rhodes, *The Making of the Atomic Bomb* (New York: Simon & Schuster, 1986), 486, 449–51, 496; History Associates, Inc., *History of the Production Complex: The Methods of Site Selection* (Rockville, MD: History Associates, Inc., 1987), 7.

290 NOTES TO CHAPTER ONE

4   History Associates, *History of the Production Complex*, 7. The army
    had initially identified Stone and Webster, the contractor hired to
    build Oak Ridge, for the plutonium project work, but decided by
    Nov. 10, 1942, to put DuPont in charge of "the pile method." "D.S.M.
    Chronology," MEDPR, box 181, Nov. 11, 1942.

5   FTM, Dec. 14–16, 1942; History Associates, *History of the Produc-
    tion Complex*, 7–9; "D.S.M. Chronology," Dec. 10 and 14, 1942; E. I.
    du Pont de Nemours and Company, Inc., "Design and Procurement
    History of Hanford Engineer Works and Clinton Semi-Works," Dec.
    1945, Wilmington, Delaware, on file, PRR, IN-6263, I:18.

6   Cited in *P-I*, July 16, 1985. The army *did* test the winds at Hanford,
    after the site had already been identified. See Daniel Grossman,
    "Hanford and Its Early Radioactive Emissions," *Pacific Northwest
    Quarterly* 85 (Jan. 1994): 8–10.

7   FTM, Dec. 22, Dec. 31, 1942, Jan. 16, 1943; *TCH*, Dec. 22, 1992; S.
    L. Sanger, *Hanford and the Bomb: An Oral History of World War II*
    (Seattle: Living History Press, 1989), 6–7; DuPont, "Design and Pro-
    curement History," I:49; E. I. du Pont de Nemours and Company,
    Inc., "Construction, Hanford Engineer Works, U.S. Contract no.
    W-7412-ENG-1, DuPont Project 9536, History of the Project" (here-
    after "Construction History"), Aug. 9, 1945, Wilmington, Delaware,
    on file, PRR, HAN-10970, I:2; "D.S.M. Chronology," Dec. 14, 1942.

8   History Associates, *History of the Production Complex*, 79; *TCH*,
    Feb. 21, 1993.

9   Typical army and DuPont views of the land can be seen in DuPont,
    "Construction History," I:4, 6, 8. See also FTM, Mar. 8, 1943; Sanger,
    Hanford and the Bomb, 7–12. In a teletype from Matthias to the Dis-
    trict Engineer, U.S. Engineer's Office, Oak Ridge, TN (Oct. 25, 1944,
    in MRCF-DF), Matthias quoted a story from the *Wenatchee World*
    newspaper, Oct. 19, 1944, wherein a Justice Department attorney
    conceded that War Department personnel had made "basic errors in
    appraisal work" on farms that were condemned for Hanford.

10 DuPont, "Construction History," I:7; William L. Laurence, *Dawn
    Over Zero: The Story of the Atomic Bomb* (New York: Knopf, 1946),
    135–36; FTM, Apr. 2, 1944.

11 FTM, Sept. 15, 1943; "Monthly Field Progress Report, Hanford Engi-
    neer Works," Apr. 30, 1944, MEDPR, box 46; Peter Bacon Hales,

Atomic Spaces: Living on the Manhattan Project (Urbana: University of Illinois Press, 1997), 204–6.

12   Physicist John Marshall, quoted in Sanger, *Hanford and the Bomb*, 129; chemist Glenn T. Seaborg, Manhattan Project diary, as published in Ronald L. Kathren, Jerry B. Gough, and Gary T. Benefiel, eds., *The Plutonium Story: The Journals of Professor Glenn T. Seaborg, 1939–1946* (Columbus, OH: Batelle Press, 1994), 463.

13   DuPont, "Construction History," I:6.

14   Matthias cited in *TCH*, Mar. 21, 1993. Figures from Kevin O'Neill, "Building the Bomb," in *Atomic Audit: The Costs and Consequences of U.S. Nuclear Weapons since 1940*, ed. Stephen I. Schwartz (Washington, D.C.: Brookings Institution Press, 1998), 60.

15   Rhodes, *Making of the Atomic Bomb*, 605.

16   "Manufacturing Division, Process Group HEW, Report," Oct. 9, 1943, DuPont Atomic Energy Division Records, box 2, HMLMD. According to one participant, "In '43, when the Hanford Project came along, [DuPont Engineering] was starting to run out of things to do. The supply lines which were almost empty in '41 were all full in '43." Walter Simon, quoted in Harry Thayer, *Management of the Hanford Engineer Works in World War II: How the Corps, DuPont, and the Metallurgical Laboratory Fast-Tracked the Original Plutonium Works* (New York: ASCE Press, 1986), 187.

17   Sanger, *Hanford and the Bomb*, 13–15.

18   Rodney P. Carlisle, with Joan M. Zenzen, *Supplying the Nuclear Arsenal: American Production Reactors, 1942–1992* (Baltimore: Johns Hopkins University Press, 1996), 19–20; Rhodes, *Making of the Atomic Bomb*, 431–32; Sanger, *Hanford and the Bomb*, 15–16. For a copy of the agreement between Washington State, DuPont, and the army releasing the company from extraordinary workman's compensation claims, see DuPont Atomic Energy Division Records, box 1, HMLMD.

19   FTM, June 17, 1943; K. D. Nichols, *The Road to Trinity* (New York: Morrow, 1987), 71, 83, 84, 136–37; Herbert L. Anderson, "Assisting Fermi," in *All In Our Time: The Reminiscences of Twelve Nuclear Pioneers,* ed. Jane Wilson (Chicago: Bulletin of the Atomic Scientists, 1975), 98; Leona Marshall Libby, *Uranium People* (New York: Crane, Russak, 1979), 169–71; U.S. Atomic Energy Commission,

Report of the Safety and Industrial Health Advisory Board, AEC-10266 (Washington, D.C.: Atomic Energy Commission, 1948), 5, 7, 9, 21, 38; David A. Hounshell and John Kenly Smith, Jr., *Science and Corporate Strategy: DuPont R&D, 1902–1980* (New York: Cambridge University Press, 1988), 339–41. More discussion of tensions concerning Richland between the army and DuPont follows in Chapter 2. The best-known episode during which DuPont "saved the day" was when the company's engineers insisted upon adding one-third more tubes to the B reactor than the MetLab scientists had specified. Their industrial experience with scaling up manufacturing facilities had taught them to allow for unanticipated problems, which in the case of the Hanford reactors took the form of a contaminant that would have "poisoned" the reaction and shut down the reactor had it been built as originally designed. See David Hounshell, "DuPont and the Management of Large-Scale Research and Development," in *Big Science: The Growth of Large-Scale Research,* ed. Peter Galison and Bruce Hevly (Stanford, CA: Stanford University Press, 1992), 245–54. Another episode was when a company craftsman used production—rather than laboratory—techniques to perfect the aluminum casing on fuel slugs. See Nichols, *Road to Trinity*, 140–41; *P-I*, July 16, 1985.

20   "Revision of Ordnance Safety Manual, 1940," Bill Mackey papers, box 3, HMLMD; J. N. Tilley to Samuel Allison, Nov. 11, 1943, with topographic map of the Hanford Site, in "Manufacturing Division—Process Group HEW Report," Nov. 13, 1943, DuPont Atomic Energy Division Records, box 2, HMLMD.

21   FTM, Dec. 1, 28, 2, 1943; Michele Stenehjem Gerber, *On the Home Front: The Cold War Legacy of the Hanford Nuclear Site,* 3rd ed. (Lincoln: University of Nebraska Press, 2007), 42–43. The best account of the urgency at Hanford is Goldberg, "General Groves and the Atomic West," 51–75. See also Stanley Goldberg, "Racing to the Finish: The Decision to Bomb Hiroshima and Nagasaki," *Journal of American-East Asian Relations* 4 (Summer 1995): 117–28.

22   On dissatisfaction with the War Manpower Commission, turn to FTM, Oct. 6, 12, 1943. The figure 50–70 percent comes from the *PI*, July 16, 1985. The accounting supervisor's comments can be found in "Manufacturing Division—Process Group HEW Report," Nov. 6, 1943, DuPont Atomic Energy Division Records, box 2, HMLMD. A good overview of labor shortages can be found in "Monthly Field

Progress Reports," June 30, 1943, to March 31, 1945, in "Progress Report, HEW Enclosure" files, MEDPR, Box 46.

23   "Monthly Field Progress Reports," May–Sept., 1944; Ted Van Arsdol, *Hanford . . . The Big Secret* (Richland, WA: Columbia Basin News, 1958), 23, 50; FTM, Oct. 11, 13, 1943, Apr. 13, 1944.

24   FTM, July 19, 20, Sept. 4, 7, 8, 13, 1944.

25   On the total number of nonwhite workers (it isn't clear whether the figure includes Mexican Americans), see Richard I. Newcomb to E. B. Riley, Nov. 22, 1944; and Matthias to Gibson, July 15, 1944; both in "Diary, Book 1—H.E.W., 6/26/44 thru Dec/31/44" (a notebook of correspondence about Hanford, hereafter cited as "Diary, Book 1"), MEDPR, box 24. On Hispanic labor, see FTM, Sept. 29, 1943, Feb. 18, 26, 27, Mar. 3, 1944. The quotation on "racial conflict" comes from Nell Lewis MacGregor, "I Was at Hanford," 1969, MS on file in Nell Lewis McGregor Papers, UWL, 35. U.S. Army survey data collected during World War II suggested widespread racial prejudice among troops; among Southern soldiers who took part in a 1942 survey of white enlisted men, "only 4 percent favored equal PX privileges for their black comrades." For the Marshall quotation, see Rick Atkinson, *The Day of Battle: The War in Sicily and Italy, 1943-1944* (New York: Henry Holt, 2007), 382, citing Bernard C. McNalty, *Strength for the Fight* (New York: Free Press, 1987), 147.

26   Captain F. A. Valente, "Daily Diary, Captain Valente, 9–44 thru July 1945, 100-F Area," on file in PRR, MED-1001 through MED-1004 (hereafter cited as Valente Diary), Mar. 13, 1945.

27   Numbers and duties of women: F. T. Matthias to William S. Gibson, July 15, 1944, in "Diary, Book 1"; *TCH*, Oct. 31, 1993. Richard I. Newcomb to E. B. Riley, Nov. 22, 1944, in "Diary, Book 1," reported that 13 percent of the construction workforce was female. MacGregor's quotation comes from her "I Was at Hanford," 42. Seaborg's observation on the jail population is found in Kathren, Gough, and Benefiel, *Journals of Professor Glenn T. Seaborg*, 467. The data on ages and draft status of men in the Hanford workforce come from Van Arsdol, *Hanford . . . The Big Secret*, 26.

28   Matthias's diary contains many entries dealing with labor issues. See, for instance, FTM, Oct. 11, 13, 1943, Apr. 13, July 22, 23, 24, 26, 1944.

29   FTM, Aug. 23, 24, 28, 1944. See also the pointed correspondence between DuPont and the army: Roger Williams to Groves, Aug. 24,

1944; Groves to Williams, Aug. 26, 1944; and Williams to Groves, Aug. 30, Sept. 1, 1944; all in MEDPR, box 52. See also the September 1944 entries in Valente Diary. See FTM, July 12, 1944, for Matthias's decision to inform only a limited number of employees about the "hazard" of working at Hanford.

30   FTM, Apr. 11, 15, May 24, June 12, July 22, 24, Sept. 7, 8, 13, 1944, Apr. 23, June 21, 1945.

31   James W. Parker memoir, n.d., Accession 2110, HMLMD, 1–16 passim.

32   Parker memoir, 15–16.

33   Richard G. Hewlett and Oscar E. Anderson, Jr., *A History of the United States Atomic Energy Commission*, vol. 1, *The New World, 1939/1946* (University Park: Pennsylvania State University Press, 1962), chaps. 4, 6.

34   The preceding paragraphs are based on Roy E. Gephart, *Hanford: A Conversation about Nuclear Waste and Cleanup* (Columbus, OH: Battelle Press, 2003), 1.14–1.23.

35   DuPont, "Construction History," 22.

36   Valente Diary, Mar. 28, 1945.

37   FTM, Sept. 4, 1944, May 8, 1945.

38   Lillian Hoddeson, "Mission Change in the Large Laboratory: The Los Alamos Implosion Program, 1943–1945," in Galison and Hevly, *Big Science*, 265–89.

39   Sanger, *Hanford and the Bomb*, 159; Goldberg, "General Groves and the Atomic West," 61–65; FTM, Feb. 2, 26, May 5, 9, 18, June 8, 10, July 30, 1945; Capt. F. A. Valente, 200 Area Daily Logs, July 4, 1945, HAN-45761, PRR (hereafter cited as Valente 200 Area Daily Logs). Ferenc Morton Szasz, *The Day the Sun Rose Twice: The Story of the Trinity Site Nuclear Explosion July 16, 1945* (Albuquerque: University of New Mexico Press, 1984), 26, points out that Groves made the important decision to hold the Trinity test (rather than simply using the weapon without testing it) because he recognized that Hanford had begun to produce enough plutonium to make such a test possible and that another plutonium bomb would be ready shortly thereafter. The bomb used at Hiroshima on August 6 had used as fuel the uranium that Oak Ridge had produced via isotope separation.

40   On the end of the war, see, for example, Hewlett and Anderson, *New World*, 401–7. Henry DeWolf Smyth, *Atomic Energy for Military Pur-*

poses (Princeton: Princeton University Press, 1945); FTM, Apr. 19, 1943, Aug. 6–17, 1945.

41   For one account of possible espionage involving Hanford's reactor designs, see Richard Rhodes, *Dark Sun: The Making of the Hydrogen Bomb* (New York: Simon & Schuster, 1995), 267–69 and illustrations 38 and 39.

42   HEWMR, Aug. 1945, HW-7-2361-Del, p. 85; Barton C. Hacker, *The Dragon's Tail: Radiation Safety in the Manhattan Project, 1942–1946* (Berkeley: University of California Press, 1987).

43   "D.S.M. Chronology," Nov. 10, 1942, Jan. 23, 1943; Grossman, "Hanford and Its Early Radioactive Emissions," 8–9.

44   Lt. Col. K. D. Nichols, memo to files, Feb. 2, 1943, MRCF-DF, box 10.

45   Army engineer F. A. Valente, promoted to major by the end of the war, noted in November of 1945 that "current dissolving practice specifies that dissolving be discontinued during periods of rain," but that a research project would be undertaken to attempt isotope separation during the next showery period and then to collect rainwater on the site for tests. Valente 200 Area Daily Logs, Nov. 27, 1945, PRR. Note that experience at Hanford gave DuPont basic understandings it applied to the design of exhaust stacks at Oak Ridge as well. "Oak Ridge Stack Height and Design" file, DuPont Atomic Energy Division Records, box 1, HMLMD.

46   Grossman, "Hanford and Its Early Atmospheric Releases," 9–12; U.S. Atomic Energy Commission, *Report of the Safety and Industrial Health Advisory Board*, 67; D. H. Denham, E. I. Mart, and R. K. Woodruff, *Notes from Key Former Hanford Employees Workshop on Vegetation Data Biases and Uncertainties 1944 to 1948, September 19-21, 1988* (Richland: Battelle Memorial Institute—Pacific Northwest Laboratory, 1988), 4–5.

47   Grossman, "Hanford and Its Early Atmospheric Releases," 10–11; Technical Steering Panel of the Hanford Environmental Dose Reconstruction Project, *Summary: Radiation Dose Estimates for Hanford Radioactive Material Releases to the Air and the Columbia River* (n.p.: Technical Steering Panel, 1994), 13, 59.

48   Technical Steering Panel, *Summary*, 39–48, 60.

49   Clarke cited in *TCH*, Mar. 24, 1989. Workweek data come from FTM, Sept. 11, 1945; HEWMR, Sept. 1945, HW-7-2548-Del, PRR, 3; Technical Steering Panel, *Summary*, 11.

50   Gerber, *On the Home Front*, 36; G. C. Houston, "Memorandum for the File; Village Administrative Experience—January through August 1946," in E. I. du Pont de Nemours and Company, Inc., "Memorandum for the File; Village Operation, Part I, Hanford Engineer Works, Richland, Washington," PRR microfiche 3097 (Richland, Sept. 10, 1946), 21, Supp. 9; HEWMR, Oct. 1946, HW 7-5362-Del, 106; "Plan for Transfer of Responsibilities and Functions of the Manhattan District to the Atomic Energy Commission," [Nov. 1946], MEDPR, p. A-10.

51   "History of Expansion of AEC Production Facilities," Aug. 16, 1963, AEC 1140, EHC.

52   *SSR*, Nov. 13, 1994; *TCH*, Oct. 9, 1994; William J. Weida, "The Economic Implications of Nuclear Weapons and Nuclear Deterrence," in Schwartz, *Atomic Audit*, 535n49, 602.

53   O'Neill, "Building the Bomb," 64.

54   Technical Steering Panel, *Summary*, 11, 41.

55   "The Great Inquiry: Testimony at AEC Hearings," *Bulletin of Atomic Scientists* 5 (Aug.–Sept. 1949): 240.

56   Walter J. Williams to manager, Hanford Directed Operations, "Subj: Industrial Mobilization Plan," Oct. 14, 1947; T. J. Hargrave to David Lilienthal, Sept. 17, 1947; and Carroll L. Wilson to Williams, Sept. 17, 1947, all in PRR accession 8921.

57   David Lilienthal and James Forrestal to Truman, Mar. 16, 1948; and Patterson, Forrestal, Lilienthal, and Leahy to Truman, Apr. 2, 1947, 47–51Sec 4927:9; AEC minutes, meeting of Aug. 30, 1949, 47–51Sec 1234:28.

58   David Lilienthal and Louis Johnson to Truman, Apr. 6, 1949, 47–51Sec, 4927:9.

59   History Associates, *History of the Production Complex*, 11–12, 21–23; Carlisle with Zenzen, *Supplying the Nuclear Arsenal*, 73; Mary Beth Reed et al., *Savannah River Site at Fifty* (Stone Mountain, GA: New South Associates, 2002), 61.

60   History Associates, *History of the Production Complex*, 11–12; M. W. Boyer to Robert LeBaron, June 18, 1952, in 51–58Sec, 1282:10. Boyer, an AEC general manager, summarized the commission's "primary objective" as "providing new productive capacity at the earliest possible date and at the lowest reasonable cost."

61   C. M. A. Stine to DuPont executive committee, Feb. 12, 1945, "Research in Nuclear Physics," TNX project file, W. S. Carpenter, Jr.,

Papers, box 830, HMLMD. See also Hounshell and Smith, *Science and Corporate Strategy*, 339, 342–45; R. M. Evans to K. D. Nichols, Jan. 16, 1946, MRCF-DF, box 96. DuPont soon reconsidered its position, apparently, because the company built and operated the Savannah River reactors between 1950 and 1989.

62 Nichols, *Road to Trinity*, 231; Hounshell and Smith, *Science and Corporate Strategy*, 244.

63 Hewlett and Anderson, *New World*, chaps. 13–14, 17; Daniel J. Kevles, *The Physicists* (New York: Knopf, 1977), chaps. 21–22.

64 "Report of the Advisory Board on Relationships of the [AEC] with its Contractors," June 12, 1947, box 4, LP.

65 Paul John Deutschmann, "Federal City: A Study of the Administration of Richland, Washington, Atomic Energy Commission Community" (master's thesis, University of Oregon, 1952), 7; Carlisle with Zenzen, *Supplying the Nuclear Arsenal*, 56–57; Gerber, *On the Home Front*, 37–39. See also Chapter 2 on Richland.

66 Gerber, *On the Home Front*, 31, 33; Richard G. Hewlett and Francis Duncan, *A History of the United States Atomic Energy Commission*, vol. 2, *Atomic Shield, 1947/1952* (University Park: Pennsylvania State University Press, 1969), 146; Gephart, *Hanford*, 1.9, 1.24–1.26; *TCH*, Jan. 9, 1964.

67 HEWMR, Jan. 1947, HW-7-5802-Del, p. 110; RCC, "Richland Census Data—August 1958" (Aug. 25, 1958), FCP2.

68 Lt. Col. Frederick J. Clarke to District Engineer, Oak Ridge, Nov. 21, 1946, MRCF-DF, box 108; J. E. Travis to Director of Operations, AEC, Oak Ridge, Apr. 3, 1947; Col. P. F. Kromer, Jr., to Carroll L. Wilson, Apr. 29, 1947, MRCF-ACC, box 219; J. E. Travis to Director of Operations, Oak Ridge, May 19, 1947, MRCF-ACC, box 219.

69 L. L. Wise, "The Richland Story—Part I: Development and Management," *Engineering News-Record* 143 (Sept. 8, 1949): 16.

70 "Great Inquiry," 242.

71 "Great Inquiry," 229.

72 "Design and Construction History, Project C-165-A, Pile Area 'H,'" July 1952, PRR, HW-24800-2, pp. 7–8.

73 "Atomic Energy Commission: Materials Testing Reactor," AEC 149/2, June 20, 1949, 47–51Sec 1207:13; "Nucleonic Events: Bacher Outlines AEC Nuclear Reactor Plans," *Nucleonics* 4 (Mar. 1949): 72–73; Hewlett and Duncan, *Atomic Shield*, 495–96, 515; "Atomic

Energy Commission Relations with the General Electric Company,"
AEC 196/1, Apr. 4, 1949, 47–51Sec 1207:3, pp. 4, 5.

74   Vannevar Bush, *Science—The Endless Frontier* (Washington, D.C.:
U.S. Government Printing Office, 1945).

75   Frederick J. Clarke to P. F. Kromer, Mar. 24, 1947, MRCF-ACC, box
45; D. S. Lewis, "Operating the Hanford Reactors," *Electrical Engineering* 76 (Nov. 1957): 951. During its time as contractor, GE operated a relatively small research laboratory at Hanford. This facility
was turned over to Battelle Memorial Institute in 1965, and in 1995—
years after plutonium production had ended at Hanford—it was
renamed the Pacific Northwest National Laboratory, one of ten such
facilities designated by the U.S. Department of Energy.

76   "Prof[.] Personnel, Hanford Engineer Works," July 25, 1947, MRCF-
ACC, box 45, "Experts" file.

77   "Nucleonics Editorial," *Nucleonics* 2 (June 1948): 2.

78   L. B. Borst, "Engineering Opportunities in Nuclear Energy," *Nucleonics* 5 (Dec. 1949): 67.

79   Parker cited in *TCH*, Aug. 6, 1965, clipping found in "Hanford Works"
newspaper clipping file, Penrose Library, Whitman College, Walla
Walla, Washington. Workers at Los Alamos, also bearing down on
the job of readying a plutonium device for the Trinity Test and fabricating uranium and plutonium bombs for use against Japan, likewise
tended not to be interested in the petitions.

80   "Hearings: Reaction to Russian Atomic Development," Sept. 28,
1949, JointCom 5312:5, pp. 42–43. DuPont contracted with the AEC
to develop better techniques to recover fissionable materials during
chemical separation and from Hanford's waste tanks to help meet
the demands of American weapons policy after the Soviet nuclear
test. See handwritten notes (n.d.), "Atomic Projects Committee Survey, 12/17/48–3/10/50" file, box 2, DuPont Atomic Energy Division
Records, HMLMD.

81   Ibid., 45.

82   Clark Goodman, "Nuclear Principles of Nuclear Reactors," *Nucleonics* 1 (Nov. 1947): 24–25.

83   Versions of this artifact have been preserved at Hanford and were on
display at the B reactor. See also Paul Loeb, *Nuclear Culture: Living
and Working in the World's Largest Atomic Complex*, rev. ed. (Philadelphia: New Society Publishers, 1986), 37–38.

84 "Summary, from Standpoint of Control, of the Graphite Situation," June 21, 1948, 49–51Sec 1225:6, pp. 6–7.

85 H. E. Hanthorn, "Hanford History, Technology, Expansion and Present Efforts," June 24, 1957, HW 51188, PRR.

86 Most writers root these arguments in particular contexts and then generalize them. See, for example, Edwin Layton, "Mirror Image Twins: The Communities of Science and Technology in 19th-Century America," *Technology and Culture* 12 (Oct. 1971): 562–80. For more general discussions, see George Basalla, *The Evolution of Technology* (Cambridge: Cambridge University Press, 1988), chap. 2, and John M. Staudenmaier, *Technology's Storytellers: Reweaving the Human Fabric* (Cambridge, MA: Society for the History of Technology; MIT Press, 1985), chap. 3.

87 George Wise, "Science and Technology," *Osiris* n.s. 1 (1985): 229–46.

88 Fred C. Schlemmer to David Lilienthal, n.d. (c. 1948), box 6, LP.

89 HWMR, June 1949, HW-13793, pp. 3–6; HWMR, Dec. 1950, HW 19842, p. 9; *TCH*, Dec. 12, 1958.

90 William J. Satterfield, Jr., to Roger Harris, Jan. 30, 1947, MRCF-ACC, box 219; HEWMR, Jan. 1947, p. 87.

91 Jack Curts to Walter J. Williams, Aug. 21, 1947, pp. 1, 3, 5, MRCF-ACC, box 43; Curts to Williams, Aug. 21, 1947, MRCF-ACC, box 45.

92 Curts to Williams, Aug. 21, 1947, MRCF-ACC, box 45.

93 Oscar Smith, "Letter to Be Sent to GE—Dictated from Richland," Mar. 8, 1949 (draft), 47–51Sec, 1234:23, pp. 1–5.

94 Lewis, "Operating the Hanford Reactors," 951; H. E. Hanthorn, "Hanford History, Technology, Expansion and Present Efforts, June 24, 1957," address in Richland, on file in PRR, HW-51188, p. 9; Hanson W. Baldwin, "New Atomic Capital," *New York Times Magazine*, July 30, 1950, p. 19. The notion of nuclear production as a rather unexceptional kind of industry was reiterated in a Jan. 6, 1961, column in the *Richland Villager* by Ted Van Arsdol, a Tri-City newspaperman who covered Hanford. Criticizing the sensational coverage of a nuclear accident that killed three people at Arco, Idaho, Van Arsdol lamented the fact that "atomic energy public relations men haven't been able to 'sell' outsiders on the idea of atomic workers going about their work routinely in an industry tending toward a more normalized condition." Ted Van Arsdol, *Dateline—Hanford* (Vancouver, WA: n.p., 1964), 15. Van Arsdol worried that Amer-

icans remained more afraid of things nuclear than they needed to be.

95   U.S. Atomic Energy Commission, *Report of the Safety and Industrial Health Advisory Board,* 44–45, 81.

96   "ARCS Report on Modifications to Hanford Reactors," Feb. 4, 1958, AEC 172/22, 51–58Sec, 1284:6, pp. 1–2; quotation cited in Gerber, *On the Home Front,* 102–3.

97   U.S. Atomic Energy Commission, *Report of the Safety and Industrial Health Advisory Board,* 63, 74. See also Carlisle with Zenzen, *Supplying the Nuclear Arsenal,* 100.

98   The Green Run also released radioactive xenon to the atmosphere. Gerber, *On the Home Front,* 90–92. Gerber relies on an estimate of 7,780 curies released; Technical Steering Panel, *Summary,* 13, offers the span of 7,000 to 9,000 curies. A more careful estimate appears in Maurice A. Robkin, "Experimental Release of ^{131}I: The Green Run," Technical Steering Panel of the Hanford Environmental Dose Reconstruction Project, *The Green Run Source Term Study: Special Report* (Olympia, WA: Department of Ecology, 1995), 487–95. Robkin suggests the range of 8,000 to 14,000, or 11,000 curies plus or minus 3,000 curies (11±3kCi) or 11,000 curies plus or minus 30 percent (11kCi±30%).

99   Technical Steering Panel, *Summary,* 39–51; quotation from a worker interviewed by Brian Freer, "Atomic Pioneers and Environmental Legacy at the Hanford Site," *Canadian Review of Sociology and Anthropology* 31 (Aug. 1994): 305–23.

100   Rafe Gibbs, "The World's Hottest 'Garbage,'" *Popular Mechanics* 103 (Apr. 1955): 127.

101   H. M. Parker, "Control of Ground Contamination" (memo to D. F. Shaw), Aug. 19, 1954, PRR, HW-32808.

102   H. M. Parker, "Status of Ground Contamination Problem," Sept. 15, 1954, PRR, HW-33068, p. 6. Parker's report contained some striking characterizations of the problem. He was not certain about just how dangerous the ruthenium-106 was; at one point he wrote that particles of the isotope "can give skin contact doses well above conventional safe limits" (p. 7). Then he spent many words suggesting that nobody on or off site was really in any danger. "As a graphic illustration of the severity of the current deposition about off-reservation locations, one can picture the entire population of Richland

lying unclothed on the ground for one day. There would be about 25 identifiable particles in contact with skin; not more than three would be in an activity type range that could produce a significant effect; not more than one would probably produce an effect." (Parker's data said that small towns to the north of Richland would have received heavier doses.) Later, Parker compared the effect of a particle of ruthenium-106 to "plunging a lighted match head on to the skin." Then he switched from the metaphor of the nudist colony to one of barnyard animals: "Pig skin and human skin are sufficiently alike that if a pig can wear a 400 mrad/hr. particle for five days, I would be willing to wear one for one day" (pp. 3, 5). Much of this effort by Parker was aimed at persuading his readers that the dangers from the emission of ruthenium-106 did not warrant public discussion. "It is suggested that the probability of a significant injury to personnel off-site is so low that a demonstrable public health hazard does not exist. On the reservation there appears to be the potential for uncomfortable superficial injury to tissue" (p. 12). Without a "demonstrable hazard to the public" (p. 10), Parker concluded that "nothing is to be gained by informing the public of a risk that, off-site, is probably non-existent" (p. 9). Although he intended to say nothing, Parker was not confident that secrecy would be maintained. "There is a definite probability that information, or rather misinformation, on the off-site condition will leak to the public in the near future. Not all the residents will be as relaxed as the one who was recently quoted as saying, 'Living in Richland is ideal because we breathe only tested air.' To prepare for adverse questions, a suitable press release is being developed to be held in readiness" (p. 10).

103   Quotation from AEC General Manager's Bulletin no. 63, Dec. 15, 1947, quoted in Deutschmann, "Federal City," 57–58.

104   Carlisle with Zensen, *Supplying the Nuclear Arsenal*, 159.

105   *TCH*, Sept. 27, 1963.

106   Carlisle with Zensen, *Supplying the Nuclear Arsenal*, 31.

107   Ibid., 76.

108   History Associates, *Building the Production Complex*, 32; Carlisle with Zenzen, *Supplying the Nuclear Arsenal*, 76; O'Neill, "Building the Bomb," 69–71. The half-life of tritium is slightly more than twelve years, which means that perhaps 5 percent of the national stockpile decays annually.

109   Carlisle with Zenzen, *Supplying the Nuclear Arsenal*, 148.

110   "Techniques and Ideas for Minimizing Effects of Cutbacks," July 22, 1965, anonymous draft memo, in RDF 5:2, p. 1; History Associates, *Building the Production Complex*, 32–34; Carlisle with Zenzen, *Supplying the Nuclear Arsenal*, 147, 150–51.

111   Alex J. Fremling to G. J. Keto, Mar. 12, Aug. 25, 1976, RDF 4:6.

112   In separate off-the-record conversations, two well-educated former employees of Hanford suggested that after about 1970 Hanford did not really do—or produce—anything. The implication was that the productive tasks ceased and the employees did make-work jobs.

113   *ST*, Mar. 12, 1982; Technical Steering Panel, *Summary*, 47–48.

114   John C. Ryan to F. B. Michaels, Mar. 18, 1965, RDF 4:3; Glenn T. Seaborg, "Large-Scale Alchemy—25th Anniversary at Hanford-Richland," AEC press release, June 7, 1968, FCP3, pp. 11–12.

115   *TCH*, Oct. 13, 15, 1981; *ST*, Mar. 12, 1982, Sept. 18, 1983; *P-I*, July 16, 1985.

116   *TCH*, Mar. 3, 1985; Terry McDermott, "Atomic City," *Pacific Magazine*, Sunday supplement to *ST*, July 28, 1985; *P-I*, July 16, 1985.

117   B. J. Williams, "The Decline of the Nuclear Family," *Pacific Northwest* 22 (Aug. 1988): 36; *TCH*, Sept. 16, Dec. 18, 1987; Michael D'Antonio, *Atomic Harvest: Hanford and the Lethal Toll of America's Nuclear Arsenal* (New York: Crown, 1993), 223. Debate over the shutdown of the N reactor was discussed by Eliot Marshall, "End Game for the N Reactor?" *Science* 235 (Jan. 2, 1987): 17–18.

118   *ST*, Jan. 12, Feb. 4, 1993.

119   *NYT*, May 29, 1986; Barry S. Shanoff, "Tons of Nuclear Waste May Go West," *World Wastes* 28 (Jan. 1985): 21; *TCH*, Oct. 8, 1986, Feb. 25, 1985 (on polls of attitudes in Washington State).

120   *ST*, Oct. 20, 1991, Mar. 29, 1992; *Wall Street Journal*, Sept. 17, 1992. Hanford had enjoyed early cooperative efforts with fish biologists at the University of Washington. Beginning in 1964 it became the site of the Joint Center for Graduate Study, by which a consortium of faculty from University of Washington, Washington State University, and Oregon State University offered technical and management courses. And in 1989 the Tri-City campus of Washington State University was established. But these interactions were all relatively limited.

121   *NYT*, July 12, 1990; *PI*, Jan. 29, 2000; Gerber, *On the Home Front*, ch. 8, 270–73. The announcement that workers had suffered threatening

exposure was greeted with skepticism by some former employees at Hanford, who doubted that the dangers had been great. For example, H. P. Smith argued that people who were truly not familiar with Hanford exaggerated the problem. "Ninety percent of the people out there thought this was the best job they ever had." *PI*, Jan. 31, 2000.

122   *TCH,* Feb. 28, Mar. 19, Oct. 1, 1986.

123   *TCH*, Oct. 2, 1986, Feb. 6, Sept. 7, 1987, Feb. 4, 1990.

124   D'Antonio, *Atomic Harvest*, chap. 10; Gerber, *On the Home Front*, 210–11; *Denver Post*, Mar. 1, 1994; *ST*, Feb. 4, 1990, Aug. 23, 1992.

125   See the special weeklong series called "Wasteland" in *SSR*, Nov. 13–17, 1994.

126   *ST*, June 1, Aug. 23, 1992, Nov. 1, 1993.

Notes to Chapter Two

1   George Bernard Shaw, *Major Barbara*, in *Six Plays by Bernard Shaw, with Prefaces* (New York: Dodd, Mead & Co., 1948), 420–23.

2   Henry G. Alsberg, ed., *The American Guide: A Sourcebook and Complete Travel Guide for the United States* (New York: Hastings House, [1949]).

3   On company towns as regional phenomena, see James B. Allen, *The Company Town in the American West* (Norman: University of Oklahoma Press, 1966); and Linda Carlson, *Company Towns of the Pacific Northwest* (Seattle: University of Washington Press, 1997). See also Peter Galison on the "trading zone" in Peter Galison, *Image and Logic: A Material Culture of Microphysics* (Chicago: University of Chicago Press, 1997), 803–84.

4   Carl Abbott, "Building the Atomic Cities: Richland, Los Alamos, and the American Planning Language," in *The Atomic West*, ed. Bruce Hevly and John M. Findlay (Seattle: University of Washington Press), 96–99, argues that wartime Richland reflected two aspects of pre-1940 federal thinking about towns—a communitarian and a public-works orientation.

5   Peter Bacon Hales, *Atomic Spaces: Living on the Manhattan Project* (Urbana: University of Illinois Press, 1997), takes the army's rhetoric too much at face value and overstates its success at controlling residents of Oak Ridge, Los Alamos, and Richland. Life on the Manhattan Project was never as antidemocratic as Hales claims.

6   McHale cited in S. L. Sanger, *Hanford and the Bomb: An Oral History of World War II* (Seattle: Living History Press, 1989), 86; FTM, July–Aug., 1943; Nell Lewis MacGregor, "I Was at Hanford" (1969), MS on file, Nell Lewis MacGregor Papers, UWL, 7–9.

7   E. I. du Pont de Nemours and Company, Inc., "Construction, Hanford Engineer Works, U.S. Contract no. W-7412-ENG-1, DuPont Project 9536, History of the Project" (hereafter cited as DuPont, "Construction, Hanford Engineer Works"), MS vols. 1 & 2, Aug. 9, 1945, Wilmington, Delaware, on file, PRR, HAN-10970, I:42. Ted Van Arsdol, *Hanford . . . The Big Secret* (Richland: Columbia Basin News, 1958), 23, states that Hanford attained its peak wartime payroll of 45,096 on June 21, 1944.

8   DuPont, "Construction, Hanford Engineer Works," I:43; MacGregor, "I Was at Hanford," 5.

9   It is worth noting that "the village" is not commonly used in the American West to denote towns. The term strikes one as an import from back east—perhaps via DuPont.

10   MacGregor, "I Was at Hanford," 70.

11   FTM, May 21, June 23, Dec. 20, 1943.

12   Franklin T. Matthias to District Engineer, Oak Ridge, Tenn., Aug. 4, 1945, MRCF-DF, box 91.

13   L. L. Wise, "The Richland Story—Part I: Town Development and Management," *Engineering News-Record* 143 (Sept. 9, 1949): 16.

14   W. A. Rothery, "Population Figures—Hanford Engineer Works," Apr. 11, 1945, MRCF-DF, box 8.

15   Cited in Delbert W. Meyer, "A Study of the Development of the Tri-Cities: Pasco, Kennewick and Richland, Washington," July 24, 1959, TS (by Carole Van Arsdol, Apr. 1980), on file in East Benton County Historical Society, Kennewick, 39–40, 23.

16   On attitudes toward housing minorities and on assigning Mexican American workers to housing in Pasco, see FTM, Sept. 29, 1943, Feb. 18, 26, Mar. 3, 1944; MacGregor, "I Was at Hanford," 35. On Pasco conditions, see FTM, Dec. 21, 1943, May 10, 1944.

17   Matthias to District Engineer, Oak Ridge, Aug. 4, 1945; F. T. Matthias, "Conference Notes," Wilmington, Delaware, Apr. 1, 1943, MEDPR, box 182; Office of G. Albin Pehrson, Architect-Engineer, "Hanford Engineer Works Village, Richland, Washington," Nov. 1943 (hereafter cited as Pehrson, "Hanford Engineer Works Village"), 6–8.

18   The quotation comes from MacGregor, "I Was at Hanford," 15. On the fate of pre-existing houses, see M. T. Binns, G. C. Houston, T. B. Mitchell, and H. B. Price, "Memorandum for the File: Village—Housing Experience to July 1, 1945," Aug. 3, 1945, Richland, Washington, in E. I. du Pont de Nemours and Company, Inc., "Memorandum for the File: Village Operation, Part I, Hanford Engineer Works, Richland Washington," 1945, Richland, Washington (PRR microfiche 3097), 1–3.

19   Architect G. Albin Pehrson recalled in 1958 that part of the town was meant to endure only five years, because a shortage of building materials meant that some structures would last no longer, but most of the houses "should be good for at least 50" years. *CBN*, Dec. 12, 1958.

20   Pehrson, "Hanford Engineer Works Village," 6; FTM, Jan. 19, 1944.

21   Quotation from AEC depiction of wartime housing, cited in Paul John Deutschmann, "Federal City: A Study of the Administration of Richland, Washington, Atomic Energy Commission Community" (master's thesis, University of Oregon, 1952), 22, 22n2.

22   "Here's Hanford" brochure, Hanford Engineer Works, 1944, copy in Dr. George W. Swickard Papers, Accession 2003, HMLMD.

23   Groves cited in FTM, Apr. 21, 1943; Mary Day Winn, "Out of This World," article MS for *SSR*, Jan. 20, 1946, in RPL-R-History file, p. 4. On Hanford as an example of Groves running the Manhattan Project so as to avoid negative scrutiny by Congress, turn to Stanley Goldberg, "General Groves and the Atomic West: The Making and the Meaning of Hanford," in Hevly and Findlay, *Atomic West*, 39–89. Groves's idea of proper town planning was summarized in his description of Oak Ridge: "The townsite would be laid out on a utility basis and not for beauty." "D.S.M. Chronology," Oct. 17, 1942, MEDPR, box 181.

24   FTM, Mar. 2, May 21, June 23, 1943; Roger Williams to Lt. Col. K. D. Nichols, Apr. 29, 1943, MRCF-DF, box 9.

25   G. C. Houston, "Memorandum for the File, Village—Administrative Organization and Control to July 1, 1945," July 23, 1945, Richland, Washington, in E. I. du Pont de Nemours and Company, Inc., "Memoranda for the File; Village Operations, Part I, Hanford Engineer Works, Richland, Washington" (Richland, 1945; PRR microfiche 3097), p. 1; FTM, June 28, 1943.

26 FTM, Apr. 16, 21, June 10, 24, 28, July 23, 1943.

27 FTM, Feb 7, 9, 1944.

28 Deutschmann, "Federal City," 112–14. See also the modifications proposed in 1947 by those responsible for planning for the Cold War expansion of the town: J. E. Travis to Director of Operations, AEC, Oak Ridge, Apr. 10, May 19, 1947, MRCF-ACC, box 219.

29 Kenneth D. Nichols, *The Road to Trinity* (New York: Morrow, 1987), 59.

30 Pehrson, "Hanford Engineer Works Village," 27, 42.

31 Groves's ideas were conveyed in FTM, June 10, July 23, 1943. On the resulting segregation by housing type, see David W. Harvey and Katheryn H. Krafft, "A Nuclear Community: The Establishment of the Hanford Engineer Works Village (Richland, Washington)," paper presented at Pacific Northwest History Conference, Richland, Washington, Mar. 25, 1995, p. 6.

32 M. T. Binns, G. C. Houston, T. B. Mitchell, and H. B. Price, "Memorandum for the File: Village—Housing Experience to July 1, 1945," Aug. 3, 1945, Richland, Washington, in DuPont, "Memoranda for the File; Village Operations, Part I," 4–6, 9–10.

33 FTM, Mar. 2, Sept. 7, 1943; Matthias to District Engineer, Oak Ridge, Aug. 4, 1945.

34 Robert Jungk, *Brighter than a Thousand Suns: A Personal History of Atomic Scientists*, trans. James Cleugh (New York: Harcourt Brace, 1958), 116; FTM, Dec. 20, 1944.

35 FTM, Sept. 11, Oct. 18, 23–28, 1944; W. O. Simon to J. N. Tilley, Sept. 13, 1944, and F. T. Matthias to K. D. Nichols, Sept. 18, 1944, both in MRCF-DF, box 9.

36 On women in the workforce and the childcare center, see Williams to Nichols, Apr. 29, 1943; C. F. Barnes and J. A. Ricker, "Memorandum for the File; Richland Village—Schools—Experience to July 1, 1945," July 21, 1945, Richland, Washington, in DuPont, "Memoranda for the File; Village Operations, Part I," 10–11. On Villagers, Inc., see Simon to Tilley, Sept. 13, 1944, and Matthias to Nichols, Sept. 18, 1944. The purposes of Villagers were detailed by J. A. Ricker, "Memorandum to the File; Village—Villagers, Inc.—Experience to July 1, 1945," July 10, 1945, Richland, Washington, in DuPont, "Memoranda for the File; Village Operations, Part I," 2–3. The term "veto power" comes from FTM, Oct. 23–28, 1944.

37 G. C. Houston, "Memorandum for the File; Village—Administrative

Organization and Control to July 1, 1945," July 23, 1945, Richland, Washington, in DuPont, "Memoranda for the File; Village Operations, Part I," 4.

38 Asher A. White, "Census Tabulation of Richland, Washington," June 26, 1945, MRCF-DF, box 8; G. C. Houston, "Memorandum for the File: Village Administration Experience—January through August 1946," Sept. 10, 1946, Richland, Washington, in DuPont, "Memoranda for the File; Village Operations, Part I," 21, and Supplement, 9.

39 HEWMR, Jan. 1947, HW-7-5802, p. 110; RCC, "Richland Census Data—August 1958" (Aug. 25, 1958), FCP2; *Richland Day Souvenir Program*, Sept. 1, 1947, RPL-CF-Richland-Description, Guidebooks file, p. 3.

40 Memorandum of a meeting with General Leslie Groves, General Dwight Eisenhower, Secretary of War Robert Patterson, Apr. 11, 1946, in TNX project file, box 830, W. S. Carpenter, Jr., Papers, HMLMD.

41 Minutes, executive session, JCAE, Mar. 10, 1949, p. 3, in 47–51Sec, 5312:2; Carroll L. Wilson to J. Robert Oppenheimer, Apr. 16, 1948, GenMan 5580:17.

42 Minutes of the General Advisory Committee second meeting, Feb. 2, 1947, p. 3, EHC; J. Robert Oppenheimer to David Lilienthal, Apr. 26, 1948, attached to minutes of the General Advisory Committee ninth meeting, Apr. 23–25, 1948, EHC.

43 "Division of Production—Monthly Status and Progress Report, December 1948," p. 3, Prod 74:18; minutes of executive session of JCAE, Mar. 10, 1949, p. 4, in 47–51Sec, 5312:12; "Operation of Hanford Piles at Higher Power Levels," attached to Carleton Shugg to Brien McMahon, July 25, 1950, 47–51Sec, 4944: 412.14.

44 H. E. Hanthorn, "Hanford History, Technology, Expansion and Present Efforts," June 24, 1957, HW 51188, PRR.

45 J. Gordon Turnbull, Inc., and Graham, Anderson, Probst & White, Inc., *Master Plan for Richland, Washington* (Cleveland: J. Gordon Turnbull, Inc.; Graham, Anderson, Probst & White, Inc., 1948) (hereafter cited as JGT, *Master Plan*). When one of the authors visited the Richland municipal offices in the early 1990s to examine the plan, he was allowed to look at but not borrow it, because it was the only extant copy and was still consulted.

46 JGT, *Master Plan*, 70, 72; HWMR, Nov. 1949, HW 15267, p. 6.

47 JGT, *Master Plan*, 8.

48   Travis to Director of Operations, Oak Ridge, Mar. 20, 1947, MRCF-ACC, box 50; Travis to Director of Operations, Oak Ridge, Apr. 10, May 19, 1947; [Frederick J.] Clarke to S. R. Sapirie, teletype, n.d. [May 1947], MRCF-ACC, box 219.

49   AEC, *Report of the Safety and Industrial Health Advisory Board* (Washington, D.C.: AEC, 1948; AEC-10266), 11. William C. Bequette, a journalist who had once tried to cover "the Richland beat," in 1993 recalled the lack of news about the town, which he attributed to its governance: "The new Richland was a company town. GE ran it with autocratic firmness." *TCH*, Oct. 3, 1993.

50   Deutschmann, "Federal City," 146–47; "The Great Inquiry: Testimony at AEC Hearings," *Bulletin of Atomic Scientists* 5 (Aug.–Sept. 1949), 229–30; Wise, "Richland Story—Part I," 17; Donald O. Carlson, "Atomic Workers at Hanford Enjoy Modern Bus System," *Bus Transportation* 30 (Apr. 1951): 41. During the war, one Hanford veteran recalled that "busses existed in incomprehensible numbers in the Hanford area." See James W. Parker memoir (n.d.), Accession 2110, HMLMD, 5.

51   Dean cited in Deutschmann, "Federal City," 96–97, 275.

52   Fred Clagett, "Richland Is like Washington, D.C.," undated MS of remarks on "Council Talks," a 1952 Richland radio forum on disposal and incorporation, FCP1; Collins cited by Ted Var Arsdol in Vancouver *Columbian*, Mar. 9, 1993.

53   Pehrson, "Hanford Engineer Works Village," 8.

54   See, for example, the recollections of Louise Cease in Sanger, *Hanford and the Bomb*, 139–40.

55   For fur traders' accounts, turn to Ted Van Arsdol, *Tri-Cities: The Mid-Columbia Hub* (Chatsworth, CA: Windsor Publications, 1990), 13; and George Simpson, "George Simpson's Remarks connected with the Fur Trade &c. in the course of a Voyage from York Factory Hudson's Bay to Fort George Columbia River and Back to York Factory 1824/25" (TS, Hudson's Bay Company Archives, Provincial Archives of Manitoba, Winnipeg), 77–78, 179. Wartime conditions are documented in MacGregor, "I Was at Hanford," 29; recollections of physicist John Marshall in Sanger, *Hanford and the Bomb*, 129; *TCH*, Feb. 1, 1968, RPL-R-History File. For discussion of the dust problem and the planting of a shelter belt during the immediate postwar years, turn to R. L. Brown, "Recommendation Report

No. 78; General Planting Program—Richland," Dec. 23, 1946, in MRCF-ACC, box 49; A. Tammaro to Carroll L. Wilson, Mar. 5, 1947, MRCF-ACC, box 219; J. E. Travis to AEC, Oak Ridge, Mar. 7, 1947, MRCF-ACC, box 219; HEWMR, June 1947, HEW-7096, p. 122; HWMR, Jan. 1948, HW-8931, p. 159; JGT, *Master Plan*, 21, 36, 38, 56. A vertical file, "Dust Storms," in RPL contains newspaper clippings concerning the problem after 1960 or so. See also *ST*, Nov. 19, 1991.

56   In the middle of 1943, DuPont recruited to Hanford operating personnel from its ordnance works around the country. The largest single group came from Alabama. See "HEW Report for Week ending Nov. 20, 1943," box 2, DuPont Atomic Energy Division Records, HMLMD. Of the 1,532 employees moved by DuPont to Richland Village between February 1944 and January 1945, 858 came from states east of the Missouri River and 674 from the West. Ricker, "Memorandum to the File: Village—Villagers, Inc.—Experience to July 1, 1945," 3–4; DuPont, "Monthly Report—February 1944," MEDPR, box 182; FTM, Mar. 9, 1944.

57   Sanger, *Hanford and the Bomb*, 140.

58   Collins cited in Vancouver *Columbian*, Mar. 9, 1993.

59   Winn, "Out of this World," 2.

60   Wilbur Zelinsky, *The Cultural Geography of the United States* (Englewood Cliffs, NJ: Prentice Hall, 1973), 135. See also John M. Findlay, *Magic Lands: Western Cityscapes and American Culture after 1940* (Berkeley: University of California Press, 1992), 270–71.

61   Fred Clagett, remarks at "Home Show Opening," Apr. 18, 1958, FCP2. Visiting in the early 1980s, the writer Paul Loeb had this insight: "In some ways the sensibilities of Hanford's founders resembled that of the healthy white Westerners Wallace Stegner described in a 1964 essay titled 'Born a Square' . . . . Richlanders viewed themselves in these frontier terms, and like Stegner's pioneers and Hanford's original farmers, their labor did make a new land their own." While we see this idea as reflecting a deep commitment on the part of community members, we see it also as arising from the need to justify claims for ongoing federal support. See Paul Loeb, *Nuclear Culture: Living and Working in the World's Largest Atomic Complex* (New York: Coward, McCann and Geohegan, 1982), 83.

62   Richland *Villager*, Sept. 3, 1945. See also John M. Findlay, "Atomic

Frontier Days: Richland, Washington, and the Modern American West," *Journal of the West* 34 (July 1995): 32–41.

63 *Richland Day Souvenir Program*, Sept. 1, 1947.

64 On the pervasiveness of frontier imagery in the modern United States, see Patricia Nelson Limerick, "The Adventures of the Frontier in the Twentieth Century," in *The Frontier in American Culture,* ed. James R. Grossman (Berkeley: University of California Press, 1994), 67–102; and Richard Slotkin, *Gunfighter Nation: The Myth of the Frontier in Twentieth-Century America* (New York: Atheneum, 1992).

65 Frederick Jackson Turner, "The Significance of the Frontier in American History," *Annual Report of the American Historical Association for the Year 1893* (Washington, D.C.: American Historical Association, 1894), 200.

66 John W. Ward, "The Meaning of Lindbergh's Flight," *American Quarterly* 10 (Spring 1958): 6, 9.

67 *Atomic Frontier Days: A New Light on the Old Frontier, Richland, Washington, Sept. 4-5-6, 1948* (Richland: Junior Chamber of Commerce, 1948); *Richland—Richland, Washington: A Key City of the Atomic Age* (n.p., n.d. [mid-1950s?]; RPL-Richland-Annual Reports file); *2nd Annual Atomic Frontier Days Program* (Richland: Junior Chamber of Commerce, 1949), 9, cover.

68 Bekins Moving & Storage Co., *Map of the Tri-Cities, Kennewick, Pasco, Richland, Washington* (Kennewick: Inland Map Publishers, [1958?]); FCP2.

69 *CBN*, July 14, 20, Aug. 8, 1950.

70 Bekins Moving & Storage Co., *Map of the Tri-Cities; Richland Welcomes You! The Atomic City* (Richland: n.p., 1958; RPL—Richland—Description—Guide Books); column by Ted Van Arsdol in *CBN*, Feb. 12, 1957, reprinted in Ted Van Arsdol, *Dateline—Hanford* (Vancouver, WA: n.p., 1964), 6; *2nd Annual Atomic Frontier Days Program*, 3.

71 On Fission Chips, see HWMR, June 1952, HW 24928, p. L-1; on PEP, see *The Columbian*, Columbia High School 1951 Annual.

72 U.S. Congress, Joint Committee on Atomic Energy, *Disposal of Government-Owned Community of Richland, Wash.: Hearings before a Subcommittee on Disposal of Government-Owned Communities of the Joint Committee on Atomic Energy, Congress of the United States, Eighty-Fourth Congress, First Session, on Disposal of Government-Owned Community at Richland, Wash., June 18 and 19, 1954* (here-

after cited as JCAE, *Disposal of Government-Owned Community*) (Washington, D.C.: Government Printing Office, 1955), 25.

73 Minutes of RCC, May 11, 1953, FCP1. Minutes of RCC meetings through 1957 are located in FCP1; minutes of meetings from 1958 on are located in FCP2.

74 JCAE, *Disposal of Government-Owned Community*, 59–60, 89–90.

75 Cited in Deutschmann, "Federal City," 274.

76 Minutes of RCC, Aug. 17, 1953, July 6, Nov. 22, 1954; "Atomic Cities Boom," *Business Week*, Dec. 18, 1948, p. 66.

77 Deutschmann, "Federal City," 260–61.

78 "Great Inquiry," 245; Deutschmann, "Federal City," 57–58, 140–41, 275; Lyman S. Moore to Carroll L. Wilson, Feb. 8, 1947, 47–51Sec, 1231:15; minutes of RCC, Sept. 7, 1954.

79 Deutschmann, "Federal City," 283, 295–97; *TCH*, Oct. 3, 1993.

80 Minutes of RCC, Apr. 9, 1951; Deutschmann, "Federal City," 268–69.

81 "Atomic Cities Boom," 65–66; "Model City," *Time* 54 (Dec. 12, 1949), 21.

82 Deutschmann, "Federal City," 263–66; AEC, *Report of the Safety and Industrial Health Advisory Board*, 10–11, 92, 96; *SSR*, July 24, 27, 1949; "Great Inquiry," 237.

83 Stephen I. Schwartz, "Congressional Oversight of the Bomb," in *Atomic Audit: The Costs and Consequences of U.S. Nuclear Weapons since 1940*, ed. Stephen I. Schwartz (Washington, D.C.: Brookings Institution Press, 1998), 485–518; Deutschmann, "Federal City," 54–57.

84 For an example of one senator's complaints, see Gordon Dean to Joseph C. Mahoney, June 6, 1950, in AEC 198/25, 47–51Sec. Dean not surprisingly claimed in 1950 that the AEC's towns had generated "the largest number of headaches for us" (cited in Deutschmann, "Federal City," 96).

85 See, for example, AEC, "Draft Terms of Reference for Guidance of a Panel on AEC Community Operations," May 16, 1950, AEC 87/5, 47–51Sec, 1; remarks of Carroll Wilson cited in Marjorie Bell Chambers, "Technically Sweet Los Alamos: The Development of a Federally Sponsored Scientific Community" (Ph.D. diss., University of New Mexico, 1974), 298; Deutschmann, "Federal City," 92–96.

86 Deutschmann, "Federal City," 86, 192, 125; GE, *"Atomic Test" . . . at Hanford Works, Richland, Washington* (n.p.: n.p., n.d.; RPL-GE-Hanford Works file).

87 Chambers, "Technically Sweet Los Alamos," 361–62; AEC, "Compar-

ison of Community Disposal Legislation at Los Alamos, Oak Ridge and Richland," AEC 87/85, Mar. 17, 1962, in "Community Management-1, General Policy, Vol. 1" folder, 1329: 4, 58–66Sec.

88   Deutschmann, "Federal City," 85; Fred Clagett cited in *TCH*, Mar. 15, 1961; J. E. Travis, quoted in "Town Meeting: 'Buying Your House,'" TS of public meeting, Richland, Washington, Feb. 12, 1953, FCP2, p. 5. As contractor, GE did not set policy, but it tended to view disposal and incorporation as disruptive to its operations. See Deutschmann, "Federal City," 85.

89   AEC, "Management of Atomic Energy Commission Towns," Apr. 29, 1948, AEC 87, 47–51Sec, p. 5.

90   The different steps toward Richland's independence were laid out in R. G. Scurry, Frederick M. Babcock, George E. Bean, and George Grove, "Report and Recommendations of the Panel on Community Operations on Oak Ridge and Richland, to the U.S. Atomic Energy Commission" (hereafter cited as Scurry Report), Washington, D.C., Aug. 11, 1951, AEC 87/14, 51–58Sec.

91   "Information sheet," n.d., RPL-CF-Richland-Incorporation.

92   Some of the key phases of government decision making were listed in "Town Meeting: 'Buying Your House,'" 1–2.

93   Lyman S. Moore to Carroll L. Wilson, Feb. 8, 1947, 1947–51Sec, 1231:15, "Memorandum on Community Problems Submitted for Consideration of General Manager"; AEC, "Management of Atomic Energy Commission Towns," 3; U.S. Congress, Joint Committee on Atomic Energy, *Proposed Legislation to Effect Disposal of Government-Owned Communities at Oak Ridge, Tenn., and Richland, Wash., and Other Pertinent Documents* (hereafter cited as JCAE, *Proposed Legislation*) (Washington, D.C.: Government Printing Office, 1954), 3.

94   AEC, "Management of Atomic Energy Commission Towns"; AEC, "Management of Atomic Energy Commission Towns—Conference of Managers and Washington Staff on May 24, 1948," Aug. 9, 1948, AEC 87/1, 47–51Sec; Public Administration Service, *A Report on the Feasibility of Municipal Incorporation and Real Estate Disposition in Richland, Washington, Prepared for the U.S. Atomic Energy Commission* (hereafter cited as PAS, *Report*) (Chicago: Public Administration Service, 1950).

95   Walter J. Williams to David F. Shaw, Jan. 2, 1951, in "Rentals of Hous-

114   Charles B. Shattuck, "Appraisal Review Report; Appraisal Atomic Energy Community, Richland, Washington," Jan. 2, 1957, FCP2; K. E. Fields to Albert M. Cole, Jan. 25, 1957, in 51–58Sec, 1243:3; minutes of RCC, Feb. 21, 1957; "Goodbye to All That," *Time* 72 (Dec. 22, 1958): 18.

115   *CBN*, Dec. 12, 13, 1958; *Richland Commencement Day Souvenir Program* (Richland: n.p., 1958; RPL-R-Commencement Day File).

116   "All America Cities," *Look* 25 (Apr. 11, 1961): 95; *TCH*, Mar. 15, 1961.

117   B. B. Field, *The Tri-City Area Handbook on Economic and Human Resources* (Richland, WA: GE Hanford Atomic Products Operation, 1965), 11; Fred Clagett, "Richland Growth Since Incorporation," July 12, 1965, FCP2; Fred Clagett, "Richland General Government Comparison, First Full Year of Operation [1960] and Latest Full Year of Operation [1964]," July 12, 1965, FCP2.

118   *TCH*, Apr. 9, 1961; Field, *Tri-City Area Handbook*, 4–5; "Plutonium Town," *Newsweek* 63 (Apr. 20, 1964): 82. On San Diego, see Abraham J. Shragge, "'I Like the Cut of Your Jib': Cultures of Accommodation between the U.S. Navy and Citizens of San Diego, California, 1900–1951," *Journal of San Diego History* 48 (Summer 2002): 230–55.

119   D. G. Williams to John C. Ryan, Sept. 15, 1966, RDF 5:7; A. M. Waggoner to Sherman B. Boivin, Aug. 18, 1969, RDF 9:5.

120   Deutschmann, "Federal City," 260–61.

121   JGT, *Master Plan*, 8, 18; Abbott, "Building the Atomic Cities," 103.

122   *2nd Annual Atomic Frontier Days Program*, 9.

123   AEC, *Report of the Safety and Industrial Health Advisory Board*, 51, 56; Frances Taylor Pugnetti, *"Tiger by the Tail": Twenty-Five Years with the Stormy Tri-City Herald* (Pasco: Tri-City Herald, 1975), 100; HWMR June 1950, HW-15843-Del, pp. 245, 258.

124   "G. E. Moves into Hanford," *Business Week*, Aug. 31, 1946, p. 18; AEC official cited in Deutschmann, "Federal City," 280.

125   Abbott, "Building the Atomic Cities," 100–102; David Kaiser, "The Postwar Suburbanization of American Physics," *American Quarterly* 56 (Dec. 2004): 851–88.

126   Minutes of RCC, Apr. 13, 24, 1950. See also Columbia Engineers Services, Inc., "Summary of Questionnaire on Richland Citizens' Attitudes toward City" (hereafter cited as CES, "Summary") (n.d. [mid-1960s?]; RDF 11:1), 3.

127   For a sampling of praise for Richland, turn to PAS, *Report*, 66; Deutschmann, "Federal City," 21; U.S. Housing and Home Finance

Agency, Community Disposition Office, Richland, Washington, "A Commercial and Industrial Survey of Richland, Benton County, Washington," draft, Richland, 1957, FCP4, pp. 15–16; *Richland—Richland, Washington: A Key City of the Atomic Age* (n.p.: n.d.; copy in RPL-CF-Richland-Annual Reports); *Richland: The Atomic City* (Richland: GE Community Operations, 1958).

128   AEC, *Report of the Safety and Industrial Health Advisory Board*, 20.

129   D. S. Lewis, "Operating the Hanford Reactors," *Electrical Engineering* 76 (Nov. 1957): 952; U.S. Bureau of the Census (hereafter cited as USBC), *United States Census of Population: 1950*, vol. 2, *Characteristics of the Population* (hereafter cited as *1950 Characteristics of the Population*), part 47, *Washington* (Washington, D.C.: U.S. Government Printing Office, 1952), 44, table 11; USBC, *United States Census of Population: 1960*, vol. 1, *Characteristics of the Population* (hereafter cited as *1960 Characteristics of the Population*), Part 49, *Washington* (Washington, D.C.: U.S. Government Printing Office, 1963), 91, table 32; EBS Management Consultants, Inc., *Comprehensive Plans for the Urbanizing Areas of Benton County, Washington, Prepared for the Benton Regional Planning Commission* (San Francisco: EBS Management Consultants, Inc., 1965), 106; Mayor's College Committee, "Expansion of Opportunities for Higher Education in the Tri-Cities Area," Richland, Washington, 1962, FCP2, p. 19.

130   EBS, Inc., *Comprehensive Plans*, 107; Deutschmann, "Federal City," 169–73. Deutschmann (173) cited three different studies of Richland income from the early 1950s, each measuring a slightly different set of people. In 1948 the average family income in Richland was $4,400, a fact that suggests that the town's middle-class families took in sums closer to the $5,000 than to the $2,000 level. "Atomic Cities' Boom," 72.

131   JGT, *Master Plan*, 4; Deutschmann, "Federal City," 135–38; *Richland: The Atomic City*.

132   EBS, Inc., *Comprehensive Plans*, 105; "Interview with Betty Jacobs, Class of 1943," by Peggy Corley, Nov. 5, 1990, MS on file, Penrose Library, Whitman College, Walla Walla, Washington, p. 22; minutes of RCC, May 27, Nov. 4, 1957.

133   JGT, *Master Plan*, 4, 21; Greater Richland Chamber of Commerce, *A Commercial and Industrial Survey of Richland, Benton County, Washington, Atomic Energy City of the Tri-Cities Area* (n.p.: n.p., 1951), 10.

134 Deutschmann, "Federal City," 303–4; JGT, *Master Plan*, 34.

135 "Interview with Melvin Jacobs, Class of 1939," by Peggy Corley, Mar. 19, 1990, MS on file at Penrose Library, Whitman College, p. 17.

136 For evidence of disgruntlement in Pasco between old-timers and newcomers, see *CBN*, July 29, Dec. 2, 7, 1950.

137 Pugnetti, *Tiger by the Tail*, 3–4; General Electric Hanford Atomic Products Operation, Richland, Washington, to HAPO Recipients of Community Leader Mail Survey, Oct. 28, 1957, FCP2; R. A. Stafford, "The Reality of an Atomic Utopia: The Town without a Past," *Intellect* 104 (Dec. 1975): 228.

138 USBC, *1950 Characteristics of the Population*, part 47, *Washington*, 43, 71; Hanson W. Baldwin, "New Atomic Capital," *New York Times Magazine*, July 30, 1950, p. 19. General Electric medical authorities in Richland blamed construction workers for increases in venereal disease (HWMR, Jan. 1948, p. 208), and residents of Richland similarly distrusted soldiers stationed at Camp Hanford, who, it was alleged, "pick[ed] up teen-age girls on the streets of Richland" (minutes of RCC, May 11, 1953).

139 When Congress held hearings in Richland during 1954 on proposals to dispose of the town, the Hanford Atomic Trade Metals Council was not represented; C. C. Ohlke memo to files, June 30, 1954, in AEC, "Report of Congressional Hearings at Richland on Community Disposal Bill," July 2, 1954, AEC 87/51, 51–58Sec, p. 2. Similarly, contrast the participants from Oak Ridge with those from Richland during the 1956 hearings on disposal before Congress. The Oak Ridge delegation included three or four people from organized labor who testified before the Joint Committee on Atomic Energy; the Richland delegation included one spokesman for organized labor—the union's attorney; JCAE-SC, "Stenographic Transcript," vol. 1 (June 11, 1956) and vol. 2 (June 19, 1956). Similarly, when Richland elected a board of freeholders to write a city charter in 1958, only a small minority of candidates came from blue-collar ranks. *TCH*, July 14, 1958, clipping in RPL-Richland-Freeholders file. Disregard—even hostility—for labor became even more pronounced during Richland's diversification efforts during the 1960s.

140 CES, "Summary," 3.

141 *TCH*, Feb. 19, 1950. See also Van Arsdol, *Tri-Cities*, chap. 5.

142 "Diversification at Hanford Means Planning and Finding New Busi-

nesses Today for Tomorrow's Needs," *General Electric News*, Hanford Atomic Products Operations, Nov. 8, 1963, pp. 4–5; Robert F. Steadman, Memorandum for U.S. Secretary of Defense, July 23, 1963, copy in HMJP4, 85:4; EBS, *Comprehensive Plans*, 43.

143   USBC, *1950 Characteristics of the Population*, part 47, *Washington*, 43, 44, 74; USBC, *1960 Characteristics of the Population*, part 49, *Washington*, 91, 92, 134.

144   Warren M. Banner, "Review of the Economic and Cultural Problems of the Tri-City Area, as They Relate to Minority People, Conducted for the Tri-City Committee on Human Relations, Pasco, Washington," 1951, on file in Special Collections, UWL, p. 6; quotation from Meyer, "Study of the Development of the Tri-Cities," 23.

145   Dorothy B. Fassitt to R. R. M. Carpenter, June 24, 1944, Carpenter Papers, HMLMD.

146   Banner, "Review of Problems," 20–25, 32–36; Walter A. Oberst, *Railroads, Reclamation and the River: A History of Pasco* (Pasco, WA: Franklin County Historical Society, 1978), 163; *TCH* Sept. 26, 1993. For more on segregation of African Americans, see Robert Bauman, "Jim Crow in the Tri-Cities, 1943–1950," *Pacific Northwest Quarterly* 96 (Summer 2005): 124–31.

147   William Borden to Henry M. Jackson, May 2, 1950, copy in SUL, 36:6; Charles P. Larrowe, "Memo on Status of Negroes in the Hanford, Washington, Area," Apr. 1949, box 6, LP, pp. 3–4.

148   *TCH*, Sept. 26, 1993.

149   Quintard Taylor, *The Forging of a Black Community: Seattle Central District from 1870 through the Civil Rights Era* (Seattle: University of Washington Press, 1994), chap. 6; Quintard Taylor, "The Great Migration: The Afro-American Communities of Seattle and Portland during the 1940s," *Arizona and the West* 23 (Summer 1981): 109–25.

150   See correspondence and memoranda concerning discrimination in employment and living conditions against African Americans at Hanford and the Tri-Cities in SUL 34:2, 34:31, 34:33, 35:22, 36:6; Banner, "Review of Problems," 1, 6, 9–12, 17, 53, 57, 64, 68.

151   Cited in Hales, *Atomic Spaces*, 195.

152   Larrowe, "Memo on Status of Negroes," 3–4; *TCH* Sept. 26, 1993. Note the common western explanation of racism as imported from the South. This was repeated to me in an interview on March 24, 1995,

with Fred Clagett, who suggested that a lot of DuPont's employees during the war came from the South and brought southern racism with them.

153 H. C. Woodland, summary, Mar., 1950, in SUL 34:33. As another example of AEC complacency, note that in 1951 the commission expressed no interest in assisting African American and Puerto Rican troops stationed in segregated conditions at Camp Hanford, the army installation designed to protect the plutonium facility. The troops fell under the army's jurisdiction, not the AEC's. Lewis G. Watts to David F. Shaw, Aug. 17, 1951, and Shaw to Watts, n.d., in SUL, 36:6.

154 H. C. Woodland, memoranda and correspondence, Oct. 1949–Apr. 1950, SUL 34:33; HWMR, Feb. 1950, HW 17056, p. 247; AEC, "Negro Relations in the Atomic Energy Program," AEC 412, Mar. 7, 1951, 1209:18, 47–51Sec, 1, 2, 4.

155 R. W. Cook and Oscar S. Smith to W. W. Boyer, Apr. 22, 1953, in "Community Management—Racial Relations" file, 1243:4, 51–58Sec.

156 *TCH*, May 19, 1966, newspaper clipping on file, Penrose Library, Whitman College.

157 *TCH*, Sept. 26, 1993, Mar. 3, 1969 (RPL-R-Police); *ST*, May 19, 1963.

158 USBC, *United States Census of Population: 1970*, vol. 1, *Characteristics of the Population* [hereafter cited as *1970 Characteristics of the Population*], part 49, *Washington* (Washington, D.C.: U.S. Government Printing Office, 1973), 69, 254, 263, 269; USBC, *United States Census of Population: 1980*, vol. 1, *Characteristics of the Population* (hereafter cited as *1980 Characteristics of the Population*), chapter C, part 49, *Washington* (Washington, D.C.: U.S. Government Printing Office, 1982), 17, 33, 133; USBC, *1970 Census of Housing*, vol. 1, *Housing Characteristics for States, Cities, and Counties*, part 49, *Washington* (Washington, D.C.: U.S. Government Printing Office, 1972), 30; USBC, *1980 Census of Housing*, vol. 1, *Characteristics of Housing Units*, chapter A, *General Housing Characteristics*, part 49, *Washington* (Washington, D.C.: U.S. Government Printing Office, 1982), 45.

159 Elouise Schumacher, "Weathering the Storm," *Pacific Magazine*, Sunday supplement to *ST*, June 21, 1987, p. 21.

160 *P-I*, July 16, 1985; Schumacher, "Weathering the Storm," 21.

161 USBC, *1980 Characteristics of the Population*, chapter C, part 49, *Washington*, 23, 140–41; *TCH*, July 31, 1969.

162   USBC, *1970 Characteristics of the Population*, part 49, *Washington*, 77, 249, 255, 263; USBC, *1980 Characteristics of the Population*, chapter C, part 49, *Washington*, 17, 24, 33, 113, 133; USBC, *1980 Census of Housing*, vol. 1, *Characteristics of Housing Units*, chapter B, *Detailed Housing Characteristics*, part 49, *Washington* (Washington, D.C.: U.S. Government Printing Office, 1983), 55.

163   Glenn C. Lee to Carl Downing (Senator Magnuson's press secretary), May 13, 1968, WGMP4, 232:26.

164   *P-I*, Sept. 13, 1979; *TCH*, Feb. 12, Mar. 20, 31, 1985; *ST*, Oct. 22, 1989.

165   Terry McDermott, "Atomic City," *Pacific Magazine*, Sunday supplement to *ST*, July 28, 1985, p. 12; *TCH*, Nov. 16, 1990, Jan. 14, 1991.

166   Michele Stenehjem Gerber, "Historical Truth and Rebirth at the Hanford Nuclear Reservation," *Columbia: the Magazine of Northwest History* 4 (Winter 1990–91): 3.

Notes to Chapter Three

1   *Congressional Record*, Oct. 9, 1951, of the 82nd Congress, First Session, vol. 97, part 10, p. 12868.

2   Bailie quoted in *NYT*, July 12, 1990. More background on the farmer is provided by Michael d'Antonio, *Atomic Harvest: Hanford and the Lethal Toll of America's Nuclear Arsenal* (New York: Crown, 1993).

3   On interpretations of downwinders, Thomas Powers writes, "You can persuade some Americans that the downwinders were callously abused for no reason at all, and you can persuade some Americans that a resolute military posture and a willingness to confront Soviet expansionism won the Cold War at a modest price, all things considered. But each of the two groups seems unable to grasp what the other is getting at." "Downwinders," *Atlantic Monthly* 273 (Mar. 1994): 124.

4   William J. Weida, "The Economic Implications of Nuclear Weapons and Nuclear Deterrence," in *Atomic Audit: The Costs and Consequences of U.S. Nuclear Weapons since 1940*, ed. Stephen I. Schwartz (Washington, D.C.: Brookings Institution Press, 1998), 524. "In fact, some special interest groups and regional political forces have advanced their short-term economic interests even when those interests did not coincide with the good of the nation as a whole [and] sometimes even when their own health and safety might be

threatened." Weida also argues (522–23) that, while purchases of nuclear weapons buy some kind of "security" for the United States, they contribute less to the economy than most other types of expenditures. "They are end items, representing a nonrecoverable sunk cost."

5   *ST*, Mar. 12, 1982.

6   Figures come from "Manhattan District History, Book IV—Pile Project, X-10, Volume 4—Land Acquisition, Hanford Engineer Works," Dec. 1947, PRR 10965.

7   M. Grace Merrick to Col. K. D. Nichols, Aug. 28, 1943, and accompanying correspondence, in MRCF-DF, box 10. See also *ST*, Feb. 14, 1993. On the numbers of inhabitants uprooted, see Michele Stenehjem Gerber, *On the Home Front: The Cold War Legacy of the Hanford Nuclear Site,* 3rd ed. (Lincoln: University of Nebraska Press, 2007), 22.

8   Leslie R. Groves, *Now It Can Be Told: The Story of the Manhattan Project* (New York: Harper, 1962), 75–77, contended that the arguments of the displaced landowners were groundless.

9   *ST*, May 2, 1943. The article claimed that there would be one new city for blacks and another separate city for whites. Colonel Matthias, in his reaction to the article (FTM, May 3, 1943), confirmed that the racial aspects of the story added to local tensions: "It appears that if the *Seattle Times'* principal object is to see what damage they can do to this project, they could not pick a better method of accomplishing that result."

10   *ST*, Feb. 14, 1993.

11   FTM, Aug. 18, Mar. 5, 1943.

12   FTM, Apr. 21, Aug. 18, 1943, Feb. 25, Nov. 6, 1944.

13   "Future Assured," *Business Week*, Feb. 3, 1945, p. 41. See the reaction of Matthias in FTM, Jan. 31, Feb. 27, 1945, and his handling of other inquiries in FTM, May 16, June 13, 1945.

14   Lewis B. Schwellenbach to Harry S. Truman, July 1, 1943, and Truman to Schwellenbach [copy], July 15, 1943, S.V.P. File, Papers of Harry S. Truman, Harry S. Truman Library, Independence, Missouri (text exactly as it appears in the original). For this information we are indebted to Alonzo S. Hamby, who told us about the letters and then provided photocopies. The letters hint at the kind of security risks that worried the Manhattan Project. Both were ostensibly sent via the U.S. mail, and both were typed by secretaries.

15   FTM, Dec. 7, 1943.

16   Paul Boyer, *By the Bomb's Early Light: American Thought and Culture at the Dawn of the Atomic Age*, rev. ed. (Chapel Hill: University of North Carolina Press, 1994), chap. 1.

17   *ST*, Aug. 10, 1945; Mrs. J. W. Nichols to Col. F. T. Matthias, Aug. 17, 1945, MED-C, box 80.

18   William L. Laurence, *Dawn over Zero: The Story of the Atomic Bomb* (New York: Knopf, 1946), 163.

19   FTM, Jan. 4, 1944.

20   On Seattle's political mobilization on behalf of Boeing, see Richard S. Kirkendall, "The Boeing Company and the Military-Metropolitan-Industrial Complex, 1945–1953," *Pacific Northwest Quarterly* 85 (Oct. 1994): 137–49. Roger L. Lotchin, *Fortress California, 1910–1961: From Warfare to Welfare* (New York: Oxford University Press, 1992), develops the pertinent concept of the "military-metropolitan-industrial complex" and traces its development in the Golden State over five decades.

21   George W. Wickstead, "Planned Expansion for Richland, Washington: A.E.C. Development Embraces All Phases of Land Planning," *Landscape Architecture* 39 (July 1949): 167.

22   "Richland Is like Washington, D.C.," undated MS of Fred Clagett's remarks on "Council Talks," 1952 radio show, FCP4.

23   Huck cited in Paul John Deutschmann, "Federal City: A Study of the Administration of Richland, Washington, Atomic Energy Commission Community" (master's thesis, University of Oregon, 1952), 125.

24   On the meddlesome "Congressman Thomas of Texas," see TS of Gordon Dean office diary, Oct. 12, 1951, EHC.

25   Sample criticisms of the AEC's management of towns can be found in Atomic Energy Commission, "Report of the Safety and Industrial Health Advisory Board" (Washington, D.C., 1948; AEC 10266, on file in PRR), 10, 11, 92, 96; *SSR*, July 24, 27, 1949; "The Great Inquiry: Testimony at AEC Hearings," *Bulletin of Atomic Scientists* 5 (Aug.–Sept. 1949): 237. On O'Mahoney's proposal to sever ties between the AEC and Richland, see Gordon Dean to Senator Joseph C. O'Mahoney, June 6, 1950, in AEC 198/25, 47–51Sec.

The *SSR*, July 25, 1949, reported the comments of Glenn C. Lee, the "youthful publisher of the Tri City Daily Herald," a "vehement critic of AEC and G.E. policies" at Hanford, and the source of some congressional criticisms of Hanford operations. Lee claimed that the AEC and GE were running Richland "101 per cent with their plutocratic socialistic monopoly."

26 AEC, "Draft Terms of Reference for Guidance of a Panel on AEC Community Operations," May 16, 1950, AEC 87/5, 1231:15, 47–51Sec, 1; Wilson cited in Marjorie Bell Chambers, "Technically Sweet Los Alamos: The Development of a Federally Sponsored Scientific Community" (Ph.D. diss., University of New Mexico, 1974), 298; Deutschmann, "Federal City," 92–96, 6–7; Jack M. Holl, "The National Reactor Testing Station: The Atomic Energy Commission in Idaho, 1949–1962," *Pacific Northwest Quarterly* 85 (Jan. 1994): 16; M. Mead Smith, "Labor and the Savannah River AEC Project: Part III," *Monthly Labor Review* 75 (Aug. 1952): 150–51.

27 Carroll L. Wilson, "Instruction GM-63, Community Management: Determination of Community Policies," Dec. 15, 1947, 47–51Sec, 1231:15; AEC, "Management of Atomic Energy Commission Towns," Apr. 29, 1948, AEC 87, 47–51Sec, 3.

28 Deutschmann, "Federal City," 274; *TCH*, Dec. 11, 1959, clippings on file in HMJP-3, 52:10.

29 On the self-silencing of GE employees, see the recollections of former reporter William C. Bequette in *TCH*, Oct. 3, 1993; Merrill paraphrased in *TCH*, July 18, 1993.

30 See U.S. Congress, Joint Committee on Atomic Energy, *Proposed Legislation to Effect Disposal of Government-Owned Communities at Oak Ridge, Tenn., and Richland, Wash., and Other Pertinent Documents* (Washington, D.C.: Government Printing Office, 1954), 4.

31 The conclusion summarizes our reading of RCC minutes for the period 1950–56, FCP1.

32 The analysis here draws upon the model laid out in Lotchin, *Fortress California*. By focusing on changes to the state after 1940, we do not mean to say that Washington had not sought federal defense projects previously, for it had. For instance, see Brian Gerard Casserly, "Securing the Sound: The Evolution of Civilian-Military Relations in the Puget Sound Area, 1891–1984" (Ph.D. diss., University of Washington, 2007).

33 Howard W. Allen and Erik W. Austin, "From the Populist Era to the New Deal: A Study of Partisan Realignment in Washington State, 1889–1950," *Social Science History* 3 (Winter 1979): 115–43; Paul

Kleppner, "Politics without Parties: The Western States, 1900–1984," in *The Twentieth Century West: Historical Interpretations,* ed. Gerald D. Nash and Richard W. Etulain (Albuquerque: University of New Mexico Press, 1989), 317–24.

34   See P. G. Holsted to Clarence C. Ohlke, Apr. 16, 1965, in "Richland — Joint Center for Graduate Study 1965–1971" file, RDF:7. The WSU–Tri-Cities campus was created in 1989. Prior to that time, there had been a consortium of universities providing a limited amount of higher education to the urban area.

35   From the 1920s until the early 1970s, the Fourth District consisted roughly of the southern half of Washington east of the Cascade crest. After a realignment following the 1970 census, the Fourth District became the central tier of counties running eastward from the Cascade range roughly to the Columbia River.

36   Quoted in *P-I,* May 21, 1989, column about Magnuson, and reprinted in *Memorial Services Held in the Congress of the U.S., Together with Tributes Presented in Eulogy of Warren G. Magnuson, Late a Senator from Washington* (Washington, D.C.: Government Printing Office, 1990).

37   *CBN,* Mar. 3, 7, 1950; HWMR, Nov. 1950, HW19622, p. 281.

38   Frances Taylor Pugnetti, *"Tiger by the Tail": Twenty-Five Years with the Stormy Tri-City Herald* (Pasco: Tri-City Herald, 1975), 258. Of the hawkish Henry M. Jackson, Senator Eugene McCarthy once quipped, "You can't get enough security for Henry. If he had his way the sky would be black with supersonic planes, preferably Boeings, of course." Quoted in Carlos A. Schwantes, *The Pacific Northwest: An Interpretive History* (Lincoln: University of Nebraska Press, 1989), 359.

39   "Testimony Taken at Public Hearing Concerning Wahluke Slope, Senator Henry M. Jackson, Presiding, Hearing Held at Community House, Richland, Washington, on October 1, 1957, 10:00 A.M.," in Ted Van Arsdol, "The Wahluke Slope and Hanford," MS, 1980, UWL, 1–2, 4, 6–9, 17.

40   John S. Graham to W. B. McCool, Apr. 7, 1958, 51–58Sec 1280:4.

41   Joint Committee on Atomic Energy, Subcommittee on Communities, "Stenographic Transcript of Hearings Before the Joint Committee on Atomic Energy, Congress of the United States," Washington, D.C., June 20 and 21, 1956, on file in FCP2.

42   The background for and introduction of Jackson's legislation are

covered in Bonnie Baack Pendergrass, "Public Power, Politics, and Technology in the Eisenhower and Kennedy Years: The Hanford Dual-Purpose Reactor Controversy, 1956–1962" (Ph.D. diss., University of Washington, 1974), chap. 3.

43 Fremont Ellsworth Kast, "Major Manufacturing Industries in Washington State: Changes in Their Relative Importance and Causes of Changes" (D.B.A. thesis, University of Washington, 1956), 326. We are grateful to Moran Tompkins for bringing this source to our attention.

44 Kast, "Major Manufacturing Industries," 331, 333, 335.

45 U.S. Senator Homer T. Bone to Jackson, Apr. 21, 1941, HMJP-2, 30:3.

46 Press release from office of Senator Hugh B. Mitchell, Dec. 20, 1945, quoting Mitchell and Jackson, copy in HMJP-2, 12:1.

47 Statement to House Committee on Public Works, June 20, 1949, copy in HMJP-2, 44:4.

48 On groups supporting the CVA idea, see Henry M. Jackson to Hugh J. Tudor, Feb. 16, 1950, HMJP-2, 30:11. On shifting policies concerning who had access to farms to be irrigated with federally provided water, see Kathryn Louise Utter, "In the End the Land: Settlement on the Columbia Basin Project" (Ph.D. diss., University of Washington, 2004).

49 Henry M. Jackson, "Columbia River Frontier," *AMVET Magazine*, Sept. 1946, on file in HMJP-2, 30:7b; G. W. Lineweaver, U.S. Bureau of Reclamation, to Jackson, Mar. 26, 1952, HMJP-2, 22:21. The following paragraphs rely on correspondence in HMJP-2, 12, 22, 30.

50 Robert G. Kaufman, *Henry M. Jackson: A Life in Politics* (Seattle: University of Washington Press, 2000), 47, 67.

51 Glenn C. Lee to Jackson, Nov. 10, 1949, with enclosed map of the "*Tri-City Herald* Trading Zone," HMJP-2, 51:23.

52 HR 4287, 81st Congress, Sec. 2, b. 4, copy in HMJP-2, 44:3.

53 Jackson's opponents in his campaign for reelection were making these charges at the time, attacking his supposed socialist tendencies rather than the CVA or the BPA explicitly. See HMJP-2, 48:10 ("Literature—Anti Jackson 1950"). For Jackson's attack on *Readers Digest*, see HMJP-2, 48:22.

54 Lon D. Leeper to *Reader's Digest*, July 26, 1950, copy in HMJP-2, 48:22.

55 John H. Stevens, president ILWU Local 9, Seattle, statement of Feb. 15, 1949, copy in HMJP-2, 51:20.

56 Wenatchee *Daily World*, June 12, 1962.

326 NOTES TO CHAPTER THREE

57   C. M. A. Stine to DuPont executive committee, Feb. 12, 1945, "Research in Nuclear Physics," TNX project file, box 830, W. S. Carpenter, Jr., Papers, HMLMD.

58   General Electric Company, *1953 Annual Report to Stockholders* (Schenectady, NY: General Electric, 1954), 10, cited in Kast, "Major Manufacturing Industries," 338–39.

59   For doubts about the simplicity of going from plutonium to electricity production, see Ward F. Davidson, "Some Design Problems of Nuclear Power Plants," *Nucleonics* 5 (Nov. 1949): 4–15. Early calls for a dual-purpose design, which cited Hanford as a model site, include J. R. Menke, "Reactor Designs for Commercial Power," *Nucleonics* 12 (Jan. 1954): 66–68; "U.S. Power Reactor Program . . . Goal: Economic Power in 10 Years," *Nucleonics* 12 (July 1954): 48–51.

60   Davidson, "Some Design Problems," 11.

61   Charles Allen Thomas, "What Are the Prospects for Industrial Nuclear Power? Good—Here's a Plan to Consider," *Nucleonics* 6 (Feb. 1950): 73, 77–78. Also on dual-purpose reactors from the point of view of industry, see C. G. Suits, "Power from the Atom: An Appraisal," *Nucleonics* 8 (Feb. 1951): 3–9. On legal and economic restrictions, see Bennett Baskey, "The Atomic Energy Act and the Power Question," *Nucleonics* 10 (Oct. 1952): 10–13. On the broader policy argument, see Chauncey Starr, "The Role of Multipurpose Reactors," *Nucleonics* 11 (Jan. 1953): 62–64; Carroll A Hockwalt and Philip N. Powers, "Dual-Purpose Reactors: First Step in Industrial Nuclear Power Development," *Nucleonics* 11 (Feb. 1953): 10–13.

62   "The Editors Hear . . . ," *Nucleonics* 10 (Aug. 1952): 73. Cf. minutes, JCAE hearings, "Reaction to Russian Atomic Development," Sept. 28, 1949, Joint Com 5312:5, p. 50.

63   On pressure from Gore and Jackson, and on Zinn's reaction, see "Late News and Commentary," *Nucleonics* 14 (June 1956): 17–18; on the reduced cost of plutonium under a cogeneration plan, see minutes, 29th GAC meeting, Feb. 15–17, 1952, EHC; "Dual-Purpose Production Reactors," *Nucleonics* 16 (May 1958): 105–7, 146.

64   Rod Fowler, "Dual-Purpose Politics: The Origins of Hanford's N Reactor," 1993, MS in authors' possession, pp. 6–9; Richard G. Hewlett and Jack M. Holl, *Atoms for Peace and War, 1953–1961: Eisenhower and the Atomic Energy Commission* (Berkeley: Univer-

sity of California Press, 1989), 409–14, 424–29; Pendergrass, "Public Power, Politics, and Technology," 77–82; "Roundup: Pu Dual Reactor Okayed," *Nucleonics* 16 (Oct. 1958): 32; Irving Gabel and George Zelensky, "Nuclear Power Plant at Hanford, Nation's Largest, Ready This Year," *Electrical World* 165 (Feb. 21, 1966): 80.

65 Ward Bowden, secretary of Washington Senate, to Henry M. Jackson, Apr. 16, 1959, and John A. McCone, AEC chairman, to Henry M. Jackson, Apr. 22, 1959, HMJP-3, 43:19; Russ Holt to J. M. McClelland, Jr., n.d. [1959], HMJP-3, 43:20.

66 Hugh B. Mitchell to David Lilienthal, Oct. 29, 1947, box 7, LP; "Colorado, Coast Groups Want Nuclear Power," *Nucleonics* 12 (Sept. 1954): 80–82. For a collection of clippings documenting editorial support for the DPR, see HMJP-3, 62:7, 233:46.

67 "Hanford Dual-Purpose Reactor: Statement of Basic Assumptions and Objectives," May 1959, in HMJP-3, 43:19; Owen W. Hurd to Henry M. Jackson, May 5, 1959 (regarding meetings between WPPSS and AEC), HMJP-3, 43:19.

68 Pendergrass, "Public Power, Politics, and Technology," passim, especially p. 165.

69 "Briefing Data—Richland, for Chairman Seaborg's Mar. 29, 1967, Visit," Mar. 1967, RDF 4:10.

70 Pendergrass, "Public Power, Politics, and Technology," 175; TCH, Sept. 27, 1963.

71 Glenn T. Seaborg, "Statement," Jan. 8, 1964, AEC press release, RDF 4:1.

72 President Kennedy initiated the reduction in plutonium production in 1962. "It would have been logical on economic grounds to shut down all of the operations at one of the two complexes, preferably Hanford, since most of its reactors were older than those at Savannah River." But the AEC wanted to prevent economic collapse at any one site and also wished to retain the ability to resume full production if needed. History Associates, Inc., *History of the Production Complex: The Methods of Site Selection* (Rockville, MD: History Associates, Inc., 1987), 32–34.

73 William H. Slaton to Clarence C. Ohlke, Feb. 19, 1964, RDF 4:1.

74 Robert F. Steadman, memo for Secretary of Defense, July 23, 1963, in 1969 file concerning reactor shutdowns, HMJP-4, 85:4 (emphasis added).

75   Rodney P. Carlisle, with Joan M. Zenzen, *Supplying the Nuclear Arsenal: American Production Reactors, 1942-1992* (Baltimore: Johns Hopkins University Press, 1996), 164.

76   Walter A. Oberst, *Railroads, Reclamation and the River, A History of Pasco* (Pasco: Franklin County Historical Society, 1978), 183. Evidence of the *Columbia Basin News* orientation toward agriculture can be found in an editorial of March 27, 1951, which viewed farming as the economic base for the mid-Columbia region and Hanford as a kind of industrial supplement.

77   Pugnetti, *Tiger by the Tail*, 13–14, 334–36.

78   Ted Van Arsdol, *Tri-Cities: The Mid-Columbia Hub* (Chatsworth, CA: Windsor Publications, 1990), 50–51; Pugnetti, *Tiger by the Tail*, 2, 13, 15, 17.

79   On the standard metropolitan statistical area, see Glenn Lee to Carl Downing, May 13, 1968, WGM-4, 232:26.

80   *TCH* editorial reprinted in Pugnetti, *Tiger by the Tail*, 287–88.

81   Ted Van Arsdol, "Richland Diversification," MS of statement to Industrial Review Committee of the Atomic Energy Commission, Richland, Washington, June 11, 1962, UWL; Fred Clagett, "AEC Hearings," June 9, 1962, FCP2; Murrey W. Fuller, "Presentation to the A.E.C. Committee on Behalf of the City of Richland," June 17, 1962, FCP2.

82   AEC, "Report on AEC Cooperation in Industrial Development Efforts of Communities Such as Richland, Washington, and Oak Ridge, Tennessee," Nov. 1962 (hereafter cited as Slaton Report), 58–66Sec 1329:13, pp. 1–3; J. E. Travis (on "moral obligation") cited in Jerry E. Bishop, "Unique Plan Launched to Convert Big A-Bomb Plant to Peaceful Use," *Wall Street Journal*, Oct. 1, 1964, p. 10; "Diversification through Segmentation," *Hanford Project News*, Dec. 30, 1965, FCP7.

83   AEC, Slaton Report, 8, 10–11, 13–14.

84   John T. Conway to Henry M. Jackson, Nov. 5, 1962, HMJP-3, 88:5; Glenn C. Lee to Jackson, Nov. 28, 1962, HMJP-3, 88:5; Glenn C. Lee cited in Thomas P. Murphy, *Science, Geopolitics, and Federal Spending* (Lexington, MA: Lexington Books, 1971), 448. Chapter 14 in Murphy's book, "Closing Down Hanford's Reactors," examines the origins of diversification and segmentation.

85   The Henry M. Jackson Papers and Warren G. Magnuson Papers contain much often quite detailed correspondence between the

senators and Tri-Cities business leaders that evinces a pattern of considerable knowledge of and influence on AEC policies and practices at Hanford.

86   P. G. Holsted and F. W. Albaugh, *The Potential for Diversification of the Hanford Area and the Tri-Cities* (Richland: n.p., 1964), 1; Robert F. Philip to Glenn T. Seaborg, Mar. 11, 1963, 58–66Sec, 1329:13; *TCH*, Feb. 5, 1963, Oct. 1, 1964.

87   Carlisle with Zenzen, *Supplying the Nuclear Arsenal*, 160. Carlisle with Zenzen also suggest that proponents of diversification at Hanford were "grassroots community leaders" (162).

88   On the influence of TCNIC, see Murphy, "Closing Down Hanford's Reactors," 18–19; memo to Files (author unknown), Feb. 25, 1970, "Subject: Telephone Interview with Mr. Luther Carter, *Science Magazine*, on Feb. 24, 1970," RDF 3:10, p. 2; Edward Bauser, "Remarks by Captain Bauser at the Annual Meeting of the Tri-City Nuclear Industrial Council at Pasco, Washington," Jan. 15, 1975, RDF 11:2, p.3. The Tri-City Nuclear Industrial Council was modeled after the Wenatchee Industrial Council as a nonprofit entity, a form of organization that allowed not only businesses but also government agencies to join and contribute money. Ports, pubic utility districts, and cities all became members of TCNIC. Pugnetti, *Tiger by the Tail*, 292–94.

89   Christian Calmeyer Fleischer, "The Tri-City Nuclear Industrial Council and the Economic Diversification of the Tri-Cities, Washington, 1963–1974" (master's thesis, Washington State University, 1974), 142, 46. On the *Herald*'s booster role, see Cassandra Tate, "Letter from 'The Atomic Capital of the Nation,'" *Columbia Journalism Review* 21 (May/June 1982): 31–35.

90   Pugnetti, *Tiger by the Tail*, 298, 295; Fleischer, "TCNIC and Economic Diversification," 42–43. On Jackson's July 1963, visit to Richland, see Paul G. Holsted to Brian Corcoran, July 15, 1963, RDF 4:4; James T. Ramey to Glenn T. Seaborg, July 18, 1963, RDF 4:4.

91   *TCH*, Feb. 5, 1963; Sam Volpentest to Warren G. Magnuson, May 8, 1964, WGMP-4, 232:23.

92   On the Richland federal building, see Shelby Scates, *Warren G. Magnuson and the Shaping of Twentieth-Century America* (Seattle: University of Washington Press, 1997), 200; Volpentest to Magnuson, May 8, 1964. Campaign contributions are covered by Fleischer, "TCNIC and Economic Diversification," 93; Luther J. Carter, "Swords

into Ploughshares: Hanford Makes the Switch," *Science* 167 (Mar. 6, 1970): 1361.

The senators' staffs paid fairly close attention to Tri-Cities coverage of Jackson and Magnuson. Glenn Lee sent two clippings from the *TCH* to Magnuson's office in late 1965. One fairly long article featured snapshots of Jackson, Seaborg, and the presidents of companies newly selected to be Hanford contractors, but it had no picture of Magnuson. One of Magnuson's assistants wrote, "These boys are always after the boss and staff here about something, but I don't see Senator's pic in the leaflet." "JR" to Carl [Downing], memo, n.d. [around late December, 1965], WGMP-4, 241:37.

93 Carlisle with Zenzen, *Supplying the Nuclear Arsenal*, 161.

94 *TCH*, June 6, 1969; memo to files, "Subject: Telephone Interview with Mr. Luther Carter, *Science Magazine*, on February 24, 1970," 1.

95 Richland Operations Office, AEC (hereafter cited as ROO-AEC), "Report on Segmentation and Diversification Program," Nov. 20, 1968, p. 3; Seaborg, "Statement," Jan. 8, 1964.

96 Murphy, "Closing Down Hanford's Reactors," 15; Fleischer, "TCNIC and Economic Diversification," 26; constituent to Jackson, July 8, 1967, HMJP-4, 56:1; Lee to Jackson, Nov. 28, 1962. See also Pugnetti, *Tiger by the Tail*, 288–89. As early as 1950, the AEC complained that reactor development efforts at Knolls deprived Hanford of "GE talent," where it was more needed. TS of Gordon Dean office diary, Mar. 17, 1950, EHC.

97 Glenn C. Lee to Ralph D. Bennett, June 12, 14, 1963; Ralph D. Bennett to Glenn C. Lee, June 17, 1963; Glenn C. Lee to Glenn T. Seaborg, June 27, July 17, 1963; all in 58–66Sec, 1329:13. Frederick H. Warren, "Prime Factors in Aiding Economic Diversification at Hanford," Sept. 4, 1963, RDF 4:4.

98 W. E. Johnson, "Memorandum: Hanford Diversification," Jan. 21, 1964, RDF 4:1; Leonard F. Perkins, "Impact on an Operations Office of Changing Operating Contractor, or Segmentation at Hanford with DT's," talk at Idaho Falls, Oct. 6, 1965, RDF 4:3, pp. 3–4; ROO-AEC, "Report on Segmentation and Diversification Program," 4; E. J. Bloch, "Status of Segmentation and Diversification Activities at Hanford," memo, Mar. 31, 1964, RDF 4:1, p. 2; Pugnetti, *Tiger by the Tail*, 298.

99 Donald Fielding Koch, "Master Plan: A Program for the Development of the State of Washington in the Nuclear and Space-Age," Oct.

1964, RDF 9:4; *TCH*, Sept. 10, 1964, in RPL-Richland-Industries file; Joseph L. McCarthy, "The Evolution of Enabling Legislation for the Thermal Power Plant Site Evaluation Council," Sept. 16, 1971, transmitted by Lawrence B. Bradley to James R. Schlesinger, Oct. 4, 1971, RDF 9:4; *TCH*, Jan. 13, 1965. On Washington's self-designation, see *Nuclear Energy in Washington State: The Nuclear Progress State*, undated brochure on file at Hanford Photography, Lockheed Martin Co., Richland, Washington. The brochure was probably produced around 1967 by the State Office of Nuclear Energy Development.

100   *TCH*, Oct. 1, 1964; Warren, "Prime Factors in Aiding Economic Diversification at Hanford." On TCNIC's better access to news, see the complaint from a Yakima newsman to Senator Jackson that the *Tri-City Herald* received preferential treatment from Washington's congressional delegation in receiving word of new developments at Hanford: Thomas Bostic to Jackson, Sept. 7, 1967, HMJP-4, 56:10.

101   *TCH*, Jan. 9, 1964.

102   Glenn C. Lee to Clarence C. Ohlke, Dec. 7, 1964, in Ohlke memo, Dec. 14, 1964, RDF 4:2.

103   In July of 1963, months before the announcement of cutbacks, Glenn C. Lee wrote AEC chairman Glenn T. Seaborg to explain that, while TCNIC did not agree with all the commission's decisions, the council and the *Herald* would both speak positively in the community about the "wonderful cooperation" received from the AEC and GE. "As we continue to write a distinct and firm record of effort and cooperation in this respect," Lee continued, "no one can complain, if some cut-backs do occur at Hanford—because everyone tried hard and cooperatively to change the total picture in the meantime." Lee seemed to imply that the AEC and GE ought to live up to the favorable image being manufactured by TCNIC and the *Herald*. Lee to Seaborg, July 17, 1963, 58–66Sec, 1329:13.

104   On "peace-mongers," see Glenn C. Lee, "Remarks by Glenn C. Lee—Governor's Council—Richland, February 2, 1966," RDF 4:3, p. 1. A *TCH* editorial of Mar. 31, 1967, criticized environmentalists by disputing claims that Hanford's thermal discharges harmed Columbia River salmon by comparing the claims to the "emotional . . . hysteria" surrounding the damage done by dams to the salmon runs. In a 1992 book, a Hanford chemist, Robert Moore, complained about "Northwest anti-nukes, many of whom masquerade as 'environmentalists.'"

Robert L. Moore, *As I Knew Him: Recollections of a Shy, Timid Soul Who Led a Quite Life Spanning Some Interesting Times, An Autobiography* (Richland: R. L. Moore, 1992), cited in Brian Freer, "Atomic Pioneers and Environmental Legacy at the Hanford Site," *Canadian Review of Sociology and Anthropology* 31 (Aug. 1994): 315.

105 When Senator Jackson visited Hanford in July 1963 to discuss the details of diversification, he assumed that the unions would eventually cooperate but urged the AEC and TCNIC not to defer any actions because of questions or objections from labor. P. G. Holsted to Brian Corcoran, July 15, 1963; Ramey to Seaborg, July 18, 1963.

106 *TCH*, May 19, 1964; Glenn C. Lee to Clarence C. Ohlke, June 26, 1964, in Ohlke to E. J. Bloch, July 9, 1964, RDF 4:2; Lee to Ohlke, Dec. 7, 1964, in Ohlke memo to files, Dec. 14, 1964, RDF 4:2; Glenn C. Lee to Henry M. Jackson, Dec. 13, 1963, HMJP-3, 88:13.

107 A. M. Waggoner to Frank Thomas, Oct. 2, 1969, RDF 5:7; "F-Reactor as a Historic Landmark," in "Briefing Data—Richland, for Chairman Seaborg's Mar. 29, 1967 visit," Mar. 1967, RDF 4:10.

108 General manager to AEC, Jan. 14, 1964, RDF 4:1; AEC general manager memo to files, Jan. 14, 1964, RDF 4:1.

109 E. J. Bloch to Henry M. Jackson, Apr. 23, 1964, RDF 4:7. Compare Holsted and Albaugh, *Potential for Diversification*, to P. G. Holsted and F. W. Albaugh, *Hanford Capabilities* (Richland: AEC-GE, 1964).

110 "Diversification at Hanford Means Planning and Finding New Businesses Today for Tomorrow's Needs," *General Electric News* [HAPO], Nov. 8, 1963, pp. 4–5; Holsted and Albaugh, *Potential for Diversification*; Holsted and Albaugh, *Hanford Capabilities*; Battelle Memorial Institute (hereafter BMI), *Summary Report on Economic Analysis of the Tri-County Area* (Columbus, OH: BMI, 1964); EBS Management Consultants, Inc. (hereafter cited as EBS), *Comprehensive Plans for the Urbanizing Areas of Benton County, Washington* (San Francisco: EBS, 1965).

111 Addendum stapled to Holsted and Albaugh, *Hanford Capabilities*, front inside cover; Pugnetti, *Tiger by the Tail*, 296.

112 BMI, *Summary Report on Economic Analysis of the Tri-County Area*, 1; *TCH*, Jan. 1, 1965; Fleischer, "TCNIC and Economic Diversification," 65–66.

113 *TCH*, Feb. 9, 1965; William Rickard, phone conversation with John M. Findlay, Sept. 1, 1999; Isochem, Inc., *Isochem at Hanford* (Richland:

n.p., n.d.; RPL-CF-Richland—Businesses 1940–1970 file), 6; *TCH*, Jan. 19, 1967, in RPL-R-Businesses 1940–1970 file; Glenn C. Lee to Brian Corcoran, Jan. 23, 1967, HMJP-4, 56:8; ROO-AEC, "Report on Segmentation and Diversification Program," pp. 5–6, appendix p. 4. The entry of the Atlantic Richfield Hanford Company into the feedlot and meatpacking businesses elicited protest from local businessmen, who disapproved of what they perceived as government subsidy to a large corporate competitor in the region. The protests did no good. See correspondence in HMJP-4, 56:8, 10.

114 Robert W. Gilstrap to Donald Williams, Jan. 26, 1968, in D. G. Williams to H. T. Herrick, Feb. 8, 1968, RDF 4:5. Gilstrap's remarks echoed the complaints of T. J. Deen, president of Local 369 of the International Chemical Workers Union, about "the lack of jobs being provided for bargaining unit people through the diversification efforts of the successful bidders in the Hanford Atomic Products Operations Diversification Program." Deen argued that workers' long service and loyalty to the Hanford Atomic Works "entitle[d] them to something better than an unconditional layoff." Deen to Jackson, Feb. 27, 1967, HMJP-4, 56:8. The AEC conceded that union members had suffered more than nonunion employees, at least in absolute numbers. Between 1964 and 1968 HAMTC members had lost 880 jobs, while non-HAMTC employees had lost 380 jobs. In the former Hanford Laboratories, HAMTC workers had gained 110 jobs, while non-HAMTC workers had gained 825 jobs. The AEC also agreed that the jobs created by new contractors were concentrated in noncraft positions. It may have been able to force its contractors to invest in the Tri-City economy, but it could not compel them to invest in particular types of businesses. Williams to Gilstrap, Feb. 7, 1968, in Williams to Herrick, Feb. 8, 1968.

115 *TCH*, Apr. 4, 1967, Oct. 14, 1969.

116 Glenn C. Lee to Henry M. Jackson, Dec. 21, 1966, HMJP-4, 56:8.

117 Lee to Jackson, Dec. 21, 1966, Jan. 23, 1967, and Glenn C. Lee to Brian Corcoran, Jan. 23, 1967, in HMJP-4, 56:8.

118 "AEC to Shut Down Last Two Production Reactors at Hanford," AEC press release, Jan. 26, 1971, RDF 7:3; "Questions and Answers on Hanford," included in "AEC to Shut Down Last Two Production Reactors," 12. The Tri-City Nuclear Industrial Council and Washington's senators fought with some success to postpone reactor closings.

See, for example, Henry M. Jackson to Richard M. Nixon, Feb. 28, 1969, Glenn T. Seaborg to Henry M. Jackson, Apr. 14, 1969, and Sam Volpentest to Henry M. Jackson, Apr. 15, 1969, all in HMJP-4, 85:4.

119   John C. Ryan to R. E. Hollingsworth, Feb. 12, 1971, RDF 7:3; *TCH*, Jan. 31, 1971; P. M. Boffey, "Hanford's Reactors Down but Not Out," *Science* 171 (Feb. 12, 1971): 555; D. A. Snyder to Glenn T. Seaborg, Jan. 30, 1971, RDF 6:9. Although the N reactor was still fairly new in 1971, it was regarded by some as obsolete. An official in the Nixon administration contended that the N reactor was not very safe. It did not meet AEC standards of reliability and safety for commercial reactors, and it did not have to because it was operated by the AEC on AEC land. The official called N "a sloppy engineering job" and characterized it as subject to too many breakdowns. Presciently, he also suggested a more important priority at Hanford than the dual-purpose reactor—80 million gallons of "high-level radioactive liquid waste" that needed to be "managed." *NYT*, Feb. 7, 1971.

120   "Hanford KE and N Reactor Shutdown" files, RDF 7:1–2. The Sierra Club was "greatly concerned about the environmental effects of proposed new reactors" but opposed "the closure of existing ones, where they are producing electric power in quantities such as that produced by the one at Hanford." Better an existing reactor at Hanford, the thinking went, than another dam on the Columbia River. See Brock Evans to Henry M. Jackson, Feb. 12, 1971, HMJP-4, 114:14.

121   Memo to files, Feb. 25, 1970, "Subject: Telephone Interview with Mr. Luther Carter, *Science Magazine*," 1; Robert F. Philip to James R. Schlesinger, Nov. 10, 1971, RDF 4:6. Editorials in the *Tri-City Herald* and letters from TCNIC to the AEC alternated between positive and negative.

122   E. W. Johnson to Henry M. Jackson, July 12, 1967, HMJP-4, 56:11; ROO-AEC, "Report on Segmentation and Diversification Program," 9–10; memo to files, Feb. 25, 1970, "Subject: Telephone Interview with Mr. Luther Carter, *Science Magazine*," 2; AEC, "Staff Report: Economic Conditions in Richland, Washington," Apr. 26, 1972, RDF 9:4, p. 1; Alex Fremling to G. J. Keto, Mar. 12, Aug. 25, 1976, RDF 4:6.

123   Marvin Clement et al., "Study and Forecast of Tri-City Economic Activity and Its Related Impact on Gasoline Needs and Housing" (Richland: Pacific Northwest Laboratories, May 1974; on file in RDF 11:1), 31, 36; *TCH*, Apr. 11, 1972; AEC, "Staff Report: Eco-

nomic Conditions in Richland, Washington," Apr. 26, 1972, RDF 9:4, pp. 1–3; G. J. Keto to William O. Doub, Dec. 3, 1971, RDF 6:1; G. J. Keto, "Report on Trip to Richland Operations Office," draft, Jan. 3, 1972, RDF 6:1. It should be noted that during the early 1970s the entire state of Washington fell into a steep local depression, with enormous layoffs by the Boeing Company. These conditions compounded the problems around Hanford and flavored the reactions of elected officials. When writing to President Nixon to complain about the cutbacks at Hanford, Governor Daniel J. Evans implored the president to postpone additional reductions at the plant, "at least until the economy is in a much stronger position to absorb the impact." Daniel J. Evans to Richard M. Nixon, June 23, 1970, copy on file in RDF 5:2.

124 *TCH*, Mar. 13, Aug. 23, 1985, Apr. 3, Oct. 9, Dec. 12–18, 1986; Elouise Schumacher, "Weathering the Storm," *Pacific Magazine*, Sunday supplement to *ST*, June 21, 1987, p. 25; *ST*, Aug. 7, 1996. Boeing eventually dropped out of Hanford operations. In 1988 a plan for the City of Richland spoke as if diversification—elusive since 1948, at least— lay just around the corner: "A more stable and gradually increasing population base should be expected as economic diversification of the area is achieved." City of Richland, Washington, *Parks and Recreation, Comprehensive Plan: 1988* (Richland: City of Richland, 1988; RPL), 10.

125 "Remarks by U.S. Senator Henry M. Jackson at an Appreciation Luncheon Given for Him by the Tri-City Nuclear Industrial Council (Verbatim from Transcript)," Oct. 24, 1972, RDF 6:1, pp. 1, 9–10, 7. It is important to note that Jackson made these remarks roughly one year before Arab members of OPEC suspended sales to the U.S. in the aftermath of the 1973 Arab-Israeli War, precipitating the "Energy Crisis" as it is generally understood by historians.

126 *TCH*, Oct. 25, 1970.

127 On the economic collapse of the WPPSS, see Daniel Pope, *Nuclear Implosions: The Rise and Fall of the Washington Public Power Supply System* (New York: Cambridge University Press, 2008). On the 1980 initiative regarding nuclear waste, see *Wall Street Journal*, Jan. 17, 1980, June 29, 1981; *TCH*, June 26, 1981.

128 On Ray, see Erik Ellis, "Dixy Lee Ray, Marine Biology and the Public Understanding of Science in the United States (1930–1970)" (Ph.D.

diss., Oregon State University, 2006); Kurt Kim Schaefer, "Right in the Eye: The Political Style of Dixy Lee Ray," *Pacific Northwest Quarterly* 93 (Spring 2002): 81–93; Scates, *Warren G. Magnuson*, 296–308.

129   Paul Loeb, *Nuclear Culture: Living and Working in the World's Largest Atomic Complex* (New York: Coward, McCann and Geohegan, 1982), 253, 199; *ST*, July 6, 1986.

130   *TCH*, Feb. 25, 1985, Oct. 8, 9, 1986. An editorial published by the *Tacoma News-Tribune* on February 9, 1975, had advocated making Hanford a repository for nuclear wastes from other states and had noted Governor Evans's support for the idea. Much had changed over a decade.

131   The role of Hanford in the Adams-Gorton contest was chronicled by the *TCH*. See, e.g., issues of May 2, Aug. 17, Sept. 20, 29, and Oct. 8, 1986.

132   Volpentest cited in Schumacher, "Weathering the Storm," 22; *PI*, July 16, 1985.

133   B. J. Williams, "The Decline of the Nuclear Family," *Pacific Northwest* 22 (Aug. 1988): 70.

134   *ST*, Aug. 4, 1988; Williams, "Decline of the Nuclear Family," 70.

135   Gerber, *On the Home Front*, 210–11. The agreement has since been renegotiated and modified.

136   On the waste of cleanup funds, see *ST*, Apr. 26, 1994; Jim Lynch and Karen Dorn Steele, "River of Money," five-part series of articles on waste at Hanford, *SSR*, Nov. 13–17, 1994.

137   The National Park Service was charged with studying the issue of wild-and-scenic-river status for the Columbia River; the U.S. Fish and Wildlife Service began managing the Arid Lands Ecology Reserve on the Hanford Site in 1997.

138   Upon leaving the employment of the Department of Energy, President Bush's undersecretary of energy, John Tuck, said that the DOE knew upon signing the agreements that there wouldn't be enough funds for cleanup and that environmental efforts were doomed to fail. "We got into the compliance agreements, in my view, because we had to stay in production to produce the requirements for the military." *Denver Post*, Mar. 1, 1994. The redefinition of the Department of Energy's mission to emphasize cleanup is discussed in d'Antonio, *Atomic Harvest*; *ST*, Feb. 4, 1990, Aug. 23, 1992.

139   Richard Morrill, "Inequalities of Power, Costs and Benefits across Geographic Scales: The Future Uses of the Hanford Reservation," *Political Geography* 18 (Jan. 1999): 1–23.

140   This western animosity toward the U.S. government has been depicted by, among others, Richard White, "The Current Weirdness in the West," *Western Historical Quarterly* 28 (Spring 1997): 5–16; and Timothy Egan, *Lasso the Wind: Away to the New West* (New York: Knopf, 1998).

141   Susan S. Fainstein, "Power and Geographic Scale: Comments on Morrill," *Political Geography* 18 (Jan. 1999): 43.

142   Adam J. Fyall, "Shared Management for Hanford Reach," *ST*, Nov. 16, 1999.

143   Loeb, *Nuclear Culture,* 199.

144   Michele Stenehjem Gerber, "Historical Truth and Rebirth at the Hanford Nuclear Reservation," *Columbia: the Magazine of Northwest History* 4 (Winter 1990–91): 3.

Notes to Chapter Four

1   *TCH*, Dec. 20, 1992; *SSR*, Nov. 13, 1994; *ST*, July 28, 1996. The amount of radioactive material remaining at Hanford is subject to constant revision, because of DOE efforts to remove stockpiles, because of cleanup efforts, and because of changing calculations of the amount of waste on the site. For recent estimates of the amount of plutonium on the reservation, see *NYT*, July 10, 2010; *TCH,* July 13, 2010. By many accounts the most polluted place in the world was the vicinity around Chelyabinsk-65, one of the former Soviet Union's counterparts to Hanford. See William J. Weida, "The Economic Implications of Nuclear Weapons and Nuclear Deterrence," in *Atomic Audit: The Costs and Consequences of U.S. Nuclear Weapons since 1940,* ed. Stephen I. Schwartz (Washington, D.C.: Brookings Institution Press, 1998), 535n49.

2   The importance of the Hanford Reach as salmon habitat is discussed in *TCH*, Oct. 9, 1994; Portland *Oregonian*, Sept. 28, 1994. In different contexts, the length of the Hanford Reach has been given as fifty-seven and as fifty-one miles.

3   *ST*, Aug. 27, 1995.

4   History Associates, Inc., *History of the Production Complex: The*

Methods of Site Selection (Rockville, MD: History Associates, Inc., 1987), 7.

5   On the use of the West by U.S. nuclear weapons programs, see Bruce Hevly and John M. Findlay, "The Atomic West: Region and Nation, 1942–1992," in *The Atomic West,* ed. Bruce Hevly and John M. Findlay (Seattle: University of Washington Press, 1998), 3–9; Tad Bartimus and Scott McCartney, *Trinity's Children: Living Along America's Nuclear Highway* (1991; repr., Albuquerque: University of New Mexico Press, 1993).

6   Valerie L. Kuletz, *The Tainted Desert: Environmental and Social Ruin in the American West* (New York: Routledge, 1998), reminds us that Indians were important occupants of lands adjoining nuclear reservations as well as other military sites. She focuses largely on the Southwest.

7   Eugene S. Hunn, with James Selam and Family, *Nch'i-Wána, "The Big River": Mid-Columbia Indians and Their Land* (Seattle: University of Washington Press, 1990), chap. 4; Gary E. Moulton, ed., *The Journals of the Lewis and Clark Expedition,* vol. 5, *July 28–November 1, 1805* (Lincoln: University of Nebraska Press, 1988), 276–99.

8   Hunn, *Nch'i-Wána,* chap. 8. See also Alexandra Harmon, ed., *The Power of Promises: Rethinking Indian Treaties in the Pacific Northwest* (Seattle: University of Washington Press, 2008).

9   Mona K. Wright, "Transportation Systems on the Hanford Site Prior to 1943," paper delivered at forty-eighth annual Pacific Northwest History Conference, Richland, Mar. 25, 1995; D. W. Meinig, *The Great Columbia Plain: A Historical Geography, 1805–1910* (Seattle: University of Washington Press, 1968), chaps. 12, 14; Ted Van Arsdol, *Tri-Cities: The Mid-Columbia Hub* (Chatsworth, CA: Windsor Publications, Inc., 1990), chap. 3.

10   Natalie Cadoret, "Settlement in the Cold Creek Valley, Hanford Site, 1906–1943," paper delivered at forty-eighth annual Pacific Northwest History Conference, Richland, Mar. 25, 1995.

11   Evidence of people departing in the 1930s comes from Cadoret, "Settlement in the Cold Creek Valley"; "Family Histories from Hanford and White Bluffs, Washington," Summer 1982, MS, PRR Accession 9226 (replies to "pioneer questionnaires"). The Sever quotation comes from "Family Histories." The army's perception of decline was

disputed by those whose land was condemned and who sued to get a better price for it.

12 Robert E. Ficken, "Grand Coulee Dam, the Columbia River, and the Generation of Modern Washington," in *Politics in the Postwar American West,* ed. Richard Lowitt (Norman: University of Oklahoma Press, 1995), 277–94; Robert E. Ficken, "Rufus Woods, Wenatchee, and the Columbia Basin Reclamation Vision," *Pacific Northwest Quarterly* 87 (Spring 1996): 72–81; Robert E. Ficken, "Grand Coulee and Hanford: The Atomic Bomb and the Development of the Columbia River," in Hevly and Findlay, *Atomic West,* 21–38 (especially p. 27, for priority to Grand Coulee power); U.S. Army Corps of Engineers, "Manhattan District History, Book I—General, Volume 9—Priorities Program," in MEDPR, box 160, pp. 3.9–3.10; David E. Lilienthal to Julius P. Krug, draft letter, n.d. [Aug. 1947], MRCF-ACC, box 50.

13 *TCH,* Feb. 21, 1993; History Associates, Inc., *History of the Production Complex,* 79.

14 *TCH,* May 29, 1991.

15 Peter Bacon Hales, *Atomic Spaces: Living on the Manhattan Project* (Urbana: University of Illinois Press, 1997), see especially 12, 21, 185.

16 Hales (*Atomic Spaces,* 185) actually goes so far to suggest that the project came to regard the American people, and especially workers, as "the enemy," along with the Nazis and the Japanese.

17 Theodore M. Porter, *Trust in Numbers: The Pursuit of Objectivity in Science and Public Life* (Princeton, NJ: Princeton University Press, 1995), chap. 7.

18 For a helpful treatment of different groups and their perspectives on Columbia Basin dams in the postwar period, see Katrine Barber, *Death of Celilo Falls* (Seattle: University of Washington Press, 2005).

19 Information about dams and the Morse quotation come from Keith C. Petersen, *River of Life, Channel of Death: Fish and Dams on the Lower Snake* (Lewiston, ID: Confluence Press, 1995), 119–20.

20 *CBN,* Mar. 3, 21, Dec. 14, 1950, Feb. 8, 1951. The Korean emergency also could work against the interests of the AEC. Gordon Dean, worrying about getting priorities for the material demands of Hanford's largest production reactors in February 1952, complained that the joint chiefs were "spoiled. . . . The trouble is they always pull up the example: 'Do you want this pile built at the cost of bullets for Korea?'

This conflict, of course, never comes up." Gordon Dean office diary, Feb. 4, 1952, EHC.

21 *CBN*, Feb. 21, 1950, Feb. 28, 1951. A *News* editorial of Feb. 21, 1950, praised "the amazing changes wrought by man which have enabled him to mold the forces of nature to his use and comfort." In the same issue an advertisement taken out by local businesses described Pasco as a "City of Destiny" that had been "Founded Upon the Principles of the Land and the Forces of Nature, Guided by Human Hands."

22 Draft statement by Senator Warren G. Magnuson on Ben Franklin Dam, in Donald A. Pugnetti to Carl Downing, June 4, 1968, WGMP-4, 232:26; U.S. Army Corps of Engineers, "Manhattan District History, Book I—General, Volume 9—Priorities Program," pp. 3.9–3.10; David E. Lilienthal to Julius P. Krug, draft letter, n.d. [Aug. 1947], MRCF-ACC, box 50.

23 The AEC focused on "controlling or limiting the risk of *nuclear catastrophes*, placing far less emphasis on environmental hazards from *routine operation*." Rodney P. Carlisle, with Joan M. Zenzen, *Supplying the Nuclear Arsenal: American Production Reactors, 1942–1992* (Baltimore: Johns Hopkins University Press, 1996), 100.

24 H. H. Rockwell to Chief of Engineers, U.S. Army, Jan. 16, 1947, in MRCF-ACC, box 48; John F. Floberg to W. B. McCool, Apr. 7, 1958, 51–58Sec, 1280:4, appendix "A," 4.

25 E. F. Miller to files, June 14, 1949, "Reactor Safeguard Committee Recommendations for H. W. Pile Operations," production file box 4, EHC; AEC secretary's files, minutes, 709th AEC Meeting, June 11, 1952, 51–58Sec, 1282:10. On the conflicts between the Reactor Safeguard Committee, the General Advisory Committee, and the AEC concerning reactor safety at Hanford, see Brian Balogh, *Chain Reaction: Expert Debate and Public Participation in American Commercial Nuclear Power, 1945–1975* (Cambridge: Cambridge University Press, 1991), 128–31.

26 "AEC Releases 87,000 Acres of Land for New Hanford Production Plant," AEC press release, Jan. 8, 1953, 51–58Sec, 1280:5–6; McCullogh cited in AEC, "ACRS Report on Modifications to Hanford Reactors," Feb. 4, 1958, AEC 172/22, 51–58Sec, 1284:8.

27 Miller to files, June 14, 1949, p. 2. On the process of disciplining Oppenheimer for his advice on the question of development of a fusion weapon, among many accounts, see Richard Polenberg's

introduction to his edition of the Oppenheimer Personnel Security Board hearings, *In the Matter of J. Robert Oppenheimer: The Security Clearance Hearing* (Ithaca, NY: Cornell University Press, 2002).

28   "Testimony Taken at Public Hearing Concerning Wahluke Slope, Senator Henry M. Jackson, Presiding, Hearing held at Community House, Richland, Washington, On October 1, 1957, 10:00 A.M.," in Ted Van Arsdol, "The Wahluke Slope and Hanford," MS, 1980, UWL, 4, 6–9, 17.

29   Columbia Basin Commission to David G. Lilienthal, Dec. 30, 1946, LP6.

30   Van Arsdol, "The Wahluke Slope and Hanford," 15; "Hanford Operations Office: The Wahluke Slope—Secondary Zone Restrictions," (n.d. [Aug.–Oct. 1951]), GenMan 5440:7, pp. 30–31.

31   Van Arsdol, "The Wahluke Slope and Hanford," 1–2.

32   *TCH*, Dec. 30, 1958.

33   R. L. Dickman, J. W. Healy, and R. E. Tomlinson, "Report for Advisory Committee on Reactor Safeguards," Apr. 16, 1958, PRR, HW-55756.

34   Fred Clagget, conversation with John M. Findlay, Mar. 24, 1995, Richland, Washington; *CBN*, see especially issue of Feb. 21, 1950.

35   J. Gordon Turnbull, Inc., and Graham, Anderson, Probst, and White, Inc., *Master Plan for Richland, Washington* (Chicago: J. Gordon Turnbull, Inc., and Graham, Anderson, Probst, and White, Inc., 1948), 6–7; "Atomic Cities' Boom," *Business Week* (Dec. 18, 1948): 70.

36   Minutes of RCC, Nov. 16, 1953, Jan. 14, 1954, and Apr. 4, 1955.

37   *CBN*, Feb. 8, 1951.

38   Isochem, Inc., *Isochem at Hanford* (Richland, WA: Isochem, Inc., n.d.), located in "Richland—Businesses 1940–1970" file, RPL, 22.

39   "Town That Wouldn't Stay Down," *U.S. News and World Report* 59 (July 19, 1965): 96.

40   Ted Van Arsdol, "Richland Diversification (and Trends Toward Civilization)," MS of statement to Industrial Review Committee of the Atomic Energy Commission, Richland, Washington, June 11, 1962, UWL, 1–3, 5–7.

41   Apart from the matter of the physical setting, the economics of diversification were also difficult. Conditions at nuclear weapons sites around the country discouraged new investment for several reasons. The technology and skills required to manufacture bombs were unique and not directly transferable to other kinds of industry.

Moreover, the AEC had made its sites and technologies off-limits to outsiders, which prevented entrepreneurs from creating alliances. Nuclear weapons plants also tended to pay higher wages than non-defense employers could afford, and workers were either reluctant or unable to leave weapons work for other employment. Some prospective investors may have been scared off by questions about public health and pollution. See Weida, "Economic Implications of Nuclear Weapons," 524–26, 534–35.

42   Weida, "Economic Implications of Nuclear Weapons," 534; Battelle Memorial Institute, *Summary Report on Economic Analysis of the Tri-County Area: Benton, Franklin, and Walla Walla Washington to the Pacific Northwest Laboratories of Battelle Memorial Institute, Nov. 10, 1964* (Columbus, OH: Battelle Memorial Institute, 1964), 3.

43   On the lure of natural and cultural amenities in the American West after 1940, see John M. Findlay, *Magic Lands: Western Cityscapes and American Culture after 1940* (Berkeley: University of California Press, 1992), 272–76

44   C. D. Thimsen to Fred Clagett, Feb. 21, 1968, FCP3; Glenn C. Lee to Henry M. Jackson, Dec. 21, 1966, HMJP-4, 56:8.

45   Battelle–Pacific Northwest Laboratories, "Special Report: Hanford as a Site for Large Accelerators," Jan. 1965, WGMP-4, 232:23; Sam Volpentest to Warren G. Magnuson, May 25, 1965, WGMP-4, 241:38; Tri-City Nuclear Industrial Council to Paul McDaniel, June 9, 1965, WGMP-4, 232:23. The debate over the siting of the accelerators is covered nicely by Thomas P. Murphy, *Science, Geopolitics, and Federal Spending* (Lexington, MA: Lexington Books, 1971), chap. 10, and in Daniel S. Greenberg, *The Politics of Pure Science*, 2nd ed. (New York: New American Library, 1971), chap. 10. Greenberg reports that the Joint Committee on Atomic Energy was able to pry funding for Hanford's N reactor from the reluctant AEC by holding funding hostage for the Stanford Linear Accelerator, so in the end Washington benefited from the project; see pp. 235–36.

46   Christian Calmeyer Fleischer, "The Tri-City Nuclear Industrial Council and the Economic Diversification of the Tri-Cities, Washington, 1963–1974" (master's thesis, Washington State University, 1974), 93–110; clipping from *Seattle Argus*, n.d., in Glenn C. Lee to Warren G. Magnuson, Apr. 19, 1966, WGMP-4, 241:37; Lee to Magnuson, Apr. 19, 1966, WGMP-4, 241:37. When Magnu-

son made a Senate speech criticizing selection of Weston, Illinois (outside Chicago), as the site for one new accelerator, the *Tri-City Herald* editorialized approvingly. *TCH*, July 23, 1967, clipping in HMJP-4, 56:10. The AEC's designation of Hanford as a production center, rather than a national laboratory or research-and-development facility, contributed to the unfavorable perceptions of the Tri-Cities by ensuring there would be little institutional academic presence at Hanford.

47   *TCH*, Sept. 17, 1963; Floyd D. Robbins, "Atomic Energy—A Reality for the Pacific Northwest," *The Trend in Engineering* (College of Engineering, University of Washington) 14 (Oct. 1962): 29; *TCH*, Apr. 22, 1965. Atomic Energy Commission officials at Hanford predicted as early as 1965 that new nuclear power plants would produce as many as 30 million kilowatts of electricity (out of the Pacific Northwest's projected total of 71.5 million kilowatts) by the year 2000. See *TCH*, Apr. 15, 1965, clipping in "Hanford Works" file, Eells Northwest Room, Penrose Library, Whitman College, Walla Walla, Washington. On the expectations for eastern Washington surrounding Grand Coulee Dam, see Ficken, "Grand Coulee and Hanford," 21–38.

48   Chet Holifield, "Remarks . . . at Richland, Washington, Respecting Hanford and Its Possible Future Development," Apr. 23, 1970, RDF 6:7, pp. 2–3. See also Edward Bauser, "Remarks by Captain Edward Bauser at the Annual Meeting of the Tri-City Nuclear Industrial Council at Pasco, Washington," Jan. 15, 1975, RDF 11:2, p. 6.

49   Glenn C. Lee to Dixy Lee Ray, Apr. 25, 1974, RDF 6:10; Bauser, "Remarks," 1–2.

50   R. F. Philip to James R. Schlesinger, Nov. 10, 1971, RDF 4:6. See also Holifield, "Remarks," 2, 5–6.

51   "Richland—Hanford Nuclear Park Newsclippings—4/70 to 1/71" file, RDF 6:6; quotations from *TCH*, July 13, 3, 1969. The AEC seemed to share the *Tri-City Herald*'s response to the electoral defeat of nuclear power in Eugene. See G. J. Keto to John A. Erlewine, June 9, 1970, RDF 6:6.

52   *TCH*, Oct. 25, 1970.

53   See the *TCH* editorial of July 3, 1969, which quotes Oregon Governor Tom McCall on the region being "the last place in the country whose lakes and rivers are not inundated with slime and waste. Our cities are relatively smog-free."

54 See, for example, editorials in *TCH*, Mar. 4, July 6, 1970, Oct. 17, 1975, and Apr. 20, 1976.

55 Glenn T. Seaborg, "Large-Scale Alchemy—25th Anniversary at Hanford-Richland," AEC press release, June 7, 1968, FCP, pp. 11–12 (emphasis added).

56 See, for example, "New Frontier," *TCH*, Mar. 7, 1976.

57 *TCH*, Mar. 13, 1975 (quotation), Feb. 5, 1976.

58 Nancy Faller, "'Great! Let's Have It Here,'" *Progressive* 43 (Apr. 1979): 36; *TCH*, Jan. 28, 1982. On Dec. 26, 1975, *TCH* editorialized, "The alternative to nuclear power is stagnation and a non-growth economy." See also *TCH*, July 22, Oct. 17, 1975, Apr. 20, 1976.

59 *TCH*, Oct. 25, 1970; Glenn C. Lee to Dixy Lee Ray, Dec. 26, 1973, RDF 6:10; Philip to Schlesinger, Nov. 10, 1971, RDF 4:6.

60 *PI*, July 16, 1985.

61 *TCH*, Sept. 26, 1968.

62 Volpentest quoted in Paul Loeb, *Nuclear Culture: Living and Working in the World's Largest Atomic Complex* (New York: Coward, McCann & Geoghegan, 1982), 110–11.

63 K. M. Chrysler, "A City That Loves the Atom," *U.S. News and World Report* 89 (Aug. 25, 1980): 55–56; *ST*, Mar. 15, 1980.

64 On revision of forecasts of energy needs, see *TCH*, June 6, 1982. The collapse of WPPSS is covered by Daniel Jack Chasan, *The Fall of the House of WPPSS* (Seattle: Sasquatch, 1985); and Daniel Pope, *Nuclear Implosions: The Rise and Fall of the Washington Public Power Supply System* (New York: Cambridge University Press, 2008). On the Hanford workforce, see Loeb, *Nuclear Culture*.

65 *PI*, Sept. 13, 1979. See also *PI*, July 16, 1985.

66 Loeb, *Nuclear Culture*, 110.

67 For a contemporary history of the emerging breeder reactor technology, see Richard G. Hewlett, "Pioneering on Nuclear Frontiers: Two Early Landmarks in Reactor Technology," *Technology and Culture* 5 (Autumn 1964): 512–22. On the prospects for nuclear technology at the beginning of the 1980s, see Alvin Weinberg, *A Second Nuclear Era: A New Start for Nuclear Power* (New York: Praeger, 1985), and David E. Lilienthal, *Atomic Energy: A New Start* (New York: Harper and Row, 1980).

68 Initial consideration of breeder reactor designs began during the

Manhattan Project. See Ronald L. Kathren, Jerry B. Gough, and Gary T. Benefiel, eds., *The Plutonium Story: The Journals of Professor Glenn T. Seaborg, 1939–1946* (Columbus, OH: Battelle Press, 1994), 637, 701–2; Roy C. Hageman, "Hanford: Threshold of an Era" (1945), copy in Bill Mackey Papers, box 3, HMLMD, 119. C. H. Greenewalt to C. L. Wilson, Mar. 9, 1949, DuPont Atomic Energy Division Records, HMLMD, outlines DuPont's postwar interest: "We both agree that it would be most useful to have available an experimental high-flux reactor—first to take the load of test work away from your production areas and, second, for such obvious benefits as will be implied for your longer-term program."

69   George T. Mazuzan, "Atomic Power Safety: The Case of the Power Reactor Development Company Fast Breeder, 1955–1956," *Technology and Culture* 23 (July 1982): 341–71; Rebecca Lowen, "Entering the Atomic Power Race: Science, Industry and Government," *Political Science Quarterly* 102 (Autumn 1987): 459–79. Fears concerning the prospect of placing a commercially operated breeder near major cities in Michigan were compounded by an accident at the Idaho Experimental Breeder Reactor.

70   J. R. Wolcott, "The Fast Breeder Reactor: A Significant Trend in Reactor Technology," Apr. 8, 1954, HW-31404, DDRS.

71   For the administrative history and account of tasks undertaken with respect to breeder reactors at the beginning of the 1970s, see A. W. DeMershman, "HEDL Explanation of Historical MUF," Oct. 15, 1976, DDRS. On Volpentest's enthusiasm for the breeder reactor program as part of the Tri-Cities future, see Loeb, *Nuclear Culture*, 110.

72   C. F. Noonon and D. C. Stapp, "Research and Development," in *History of the Plutonium Production Facilities at the Hanford Site Historic District, 1943–1990*, ed. Rosalind E. Schrempf (Columbus, OH: Battelle Laboratories, 2002), 2-7.21 to 2-7.22; DOE-ROO, "Fast Flux Test Facility Facts," http://www.hanford.gov/rl/?page=305&parent=304 (accessed Jan. 23, 2009). On the fate of the FFTF, see Michele Stenehjem Gerber, *On the Home Front: The Cold War Legacy of the Hanford Nuclear Site,* 3rd ed. (Lincoln: University of Nebraska Press, 2007), 258–60.

73   On the product cycle for the FFTF, see C. M. Cox, "Cost and Schedule," HEDL-DP-810495, Mar. 20, 1986, DDRS; on breeder and anti-

nuclear sentiment, see Samuel McCracken, *The War Against the Atom* (New York: Basic Books, 1982), xi, 8, 33–36, 97.

74   Daniel Pope, "Anti-Nuclear Activism and the Pacific Northwest: WPPSS and its Enemies," in Hevly and Findlay, *The Atomic West* 235–54; *Wall Street Journal*, Jan. 17, 1980, June 29, 1981; *TCH*, June 26, 1981.

75   *NYT*, May 29, 1986; Barry S. Shanoff, "Tons of Nuclear Waste May Go West," *World Wastes* 28 (Jan. 1985): 21.

76   *TCH*, Oct. 8, 9, 1986. See also *TCH*, Feb. 25, 1985.

77   *ST*, Aug. 4, 1988; B. J. Williams, "The Decline of the Nuclear Family," *Pacific Northwest* 22 (Aug. 1988): 70. The role of Hanford in the Adams-Gorton race for the U.S. Senate seat was chronicled in *TCH* (see, e.g., issues of May 2, Aug. 17, Sept. 20, 29, and Oct. 8, 1986).

78   See Max S. Power, *America's Nuclear Wastelands: Politics, Accountability and Cleanup* (Pullman: Washington State University Press, 2008), chap. 6, especially pp. 97–102. Power notes that the amendment of the Nuclear Waste Policy Act in 1987 became known as the "screw Nevada bill" in that state.

79   *NYT*, Dec. 25, 1987; Sharon Begley, with Patricia King, "This Land Is Really Hot," *Time* (Feb. 8, 1993): 71.

80   Elouise Schumacher, "Weathering the Storm," *Pacific Magazine* (Sunday supplement to *ST*), June 21, 1987, pp. 26–27; *TCH*, Oct. 5, 1987.

81   Eliot Marshall, "Hanford's Radioactive Tumbleweed," *Science* 236 (June 26, 1987): 1616; *ST*, Feb. 4, 1990; *PI*, July 16, 1985.

82   For a comment on the sincerity of DOE's openness campaign of the late 1980s, see William J. Kinsella, "Nuclear Boundaries: Material and Discursive Containment at the Hanford Nuclear Reservation," *Science as Culture* 10 (June 2001): 163–94, especially pp. 178–81.

83   For a range of responses to the HTDS, see, for example, *Perspectives on the Hanford Thyroid Disease Study* (Hanford Health Information Network, Oct. 1999), http://www.doh.wa.gov/hanford/publications/overview/index.html, accessed Aug. 1, 2010. One critic noted that, "given the way in which the draft results of the study were communicated, the HTDS actually inflicted a good deal of harm on those whom the study was intended to serve. . . . If the scientific critics of the HTDS are found to be correct, it will come as little, if any, consolation for Hanford downwinders. Not only has the HTDS left them more confused than ever. It has also left them feeling more

betrayed than ever" (p. 5). In their introduction to a volume of essays on "landscapes of exposure," Gregg Mitman, Michelle Murphy, and Christopher Sellers note that "uncertainty is perhaps the single most pervasive characteristic of the history of exposure collectively sketched by our volume." Mitman, Murphy, and Sellers, "Introduction: A Cloud over History," *Osiris* n.s.19 (2004): 13. See also Scott Kirsch, "Harold Knapp and the Geography of Normal Controversy: Radioiodine in the Historical Environment," and Adriana Petryna, "Biological Citizenship: The Science and Politics of Chernobyl-Exposed Populations," in the same volume. For a critical and discerning analysis of the problems of applying epidemiological models in the modern West, see Linda Nash, *Inescapable Ecologies: A History of Environment, Disease, and Knowledge* (Berkeley: University of California Press, 2006), 192–202.

84   Barton C. Hacker, *Elements of Controversy: The Atomic Energy Commission and Radiation Safety in Nuclear Weapons Testing 1947–1974* (Berkeley: University of California Press, 1994), 119–30, 259–61.

85   Loeb, *Nuclear Culture*, 59; Loeb's informant went on to note that "if we'd told our kids more about how important the work was, they'd have been prepared when all those professors started criticizing it." On the California movement against nuclear power plants and the role of technical experts, see Thomas Raymond Wellock, *Critical Masses: Opposition to Nuclear Power in California 1958–1978* (Madison: University of Wisconsin Press, 1998), and Thomas R. Wellock, "Radical Initiative and Moderate Alternatives: California's 1976 Nuclear Safeguards Initiative," in Hevly and Findlay, *Atomic West*, 200–235.

86   On the 2000 legislation and the growing consensus about the fate of the river, see Gerber, *On the Home Front*, 270–72, 275–77. The publications of two writers who grew up in Richland exemplify the divergent responses of members of Hanford families. For example, see the angry poems of William Witherup in *Men at Work* (Boise: Ahsahta Press, 1989) and the more reflective poems of Debora Greger in *Desert Fathers, Uranium Daughters* (New York: Penguin Poets, 1996).

87   For another, related interpretation of the role of stakeholders and whistleblowers, see Kinsella, "Nuclear Boundaries."

88   *TCH*, May 15, 1963, in "Excerpts from Local Press Comments on AEC Efforts to Assist in Diversification and Encourage Industrial Development at Richland (Following Mar. 15, 1963)," 1964, RDF 4:1, p. 2.

89   *TCH*, July 26, 1965; Glenn C. Lee to Glenn T. Seaborg, Apr. 6, 1964; E. E. Miller to F. P Baranowski, July 13, 1964; and "Richland—Wahluke Slope" files; all in RDF 13:1–2. Booster cited in Fleischer, "Tri-City Nuclear Industrial Council and Economic Diversification," 111.

90   This and the following two paragraphs come from "Richland—Arid Lands Ecology (ALE)—1965–1968" file, RDF; "Richland—Arid Lands Ecology Reserve (ALE)," in "Briefing Data—Richland, for Chairman Seaborg's March 29, 1967, Visit," RDF 4:10; *TCH*, Mar. 29, 1967; William Rickard, telephone conversation with John Findlay, Sept. 1, 1999.

91   The McWhorter letter and the *Record-Bulletin* editorial may be found in the "Richland—Arid Lands Ecology (ALE)—1965–1968" Folder, RDF 4:10; and "Richland—Arid Lands Ecology Reserve (ALE)," in "Briefing Data—Richland, for Chairman Seaborg's March 29, 1967, Visit," RDF 4:10; *TCH*, Mar. 29, 1967.

92   Cited in *ST*, Feb. 14, 1993.

93   Alex G. Fremling to G. J. Keto, Dec. 31, 1973, RDF 5:5; Richland Operations Office, "A Proposed Study of the Hanford Complex as a Site for a National Environmental Pollution Study Center (NEPSC)," May 1969, HJMP-4, 85:3. Even today ALE is described in somewhat confusing terms. The reserve is part of the Hanford National Environmental Research Park, a site that is "largely undisturbed by human activity." The ALE reserve is represented as a remnant of "the Washington shrub-steppe landscape that is still in pristine condition" (http://www.pnl.gov/nerp, accessed July 27, 2010). Could the site have been affected by radioactive emissions, and at the same time be "undisturbed" and "pristine"?

94   Forest Western, "Problems of Radioactive Waste Disposal," *Nucleonics* 3 (Aug. 1948): 48.

95   *Nucleonics* 4 (Mar. 1949): 19.

96   "Report of the Safety and Industrial Health Advisory Board," AEC1232, "Health and Safety Program," 47–51Sec (quotation from p. 70). For a detailed history and assessment of the dilution strategy for dealing with wastes from Hanford, see Gerber, *On the Home Front*.

97   "Proposed visit of Reactor Safeguard Committee to the United Kingdom," Aug. 23, 1949, AEC 43/167, 337/Reactor Safeguard Committee file, 47–51Sec.

98   Clarence C. Ohlke to R. F. Philip, Oct. 23, 1964, RDF; G. J. Keto to R. E.

Hollingsworth, June 23, 1971, RDF 3:5; draft statement by Warren G. Magnuson, in Donald A. Pugnetti to Carl Downing, June 4, 1968, WGMP-4, 232:26; Harold Harty, "The Effects of the Ben Franklin Dam on the Hanford Site," Richland, Washington, Apr. 1979, PRR, PNL-2821.

99 Columbia River Conservation League, "A Proposal for the Hanford National Recreation Area," Richland, June 1972, RDF 3:6, pp. 1–4.

100 R. F. Philip to James R. Schlesinger, Aug. 21, 1972, RDF 3:6; Philip to Schlesinger, Sept. 12, 1972, RDF 3:6; Glenn C. Lee to Schlesinger, Aug. 31, 1972, RDF 3:6.

101 Alex G. Fremling to G. J. Keto, Dec. 31, 1973, RDF 5:5.

102 G. J. Keto to Mike McCormack, May 15, 1973, copy, RDF 3:6.

103 Carl Patzwaldt et al., "Ad Hoc Committee Draft of a Proposal for Management of Public Use of Portions of Hanford AEC Reservation," Oct. 23, 1973, in Fremling to Keto, Dec. 31, 1973; Fremling to Keto, Dec. 31, 1973.

104 Patzwaldt et al., "Ad Hoc Committee Draft of A Proposal for Management of Portions of Hanford"; F. P. Baranowski to J. L. Liverman, Jan. 24, 1974, RDF 5:5; G. J. Keto to John C. Ryan, Feb. 19, 1974, RDF 6:1; G. J. Keto to Alex F. Fremling, May 29, 1974, RDF 3:6; TCH, May 23, 1993.

105 Gerber, On the Home Front, 210–11.

106 This shift in DOE policy is described by Michael D'Antonio, Atomic Harvest: Hanford and the Lethal Toll of America's Nuclear Arsenal (New York: Crown, 1993), chap. 10. See also ST, Feb. 4, 1990, Aug. 23, 1992.

107 ST, June 1, 1992, Nov. 3, 1993, July 5, 1999; SSR, Nov. 13–17, 1994.

108 SSR, Nov. 13, 1994; Bill Richards, "Nuclear Site Learns to Stop Worrying and Love the Boom," Wall Street Journal, Aug. 28, 1992; TCH, Dec. 20, 1992. There remained much concern locally that the job of cleanup required skills that the local population did not possess, and that established residents would be displaced by newcomers with different skills. ST, Aug. 23, 1992.

109 On "locking up" resources, referring to Wahluke Slope, see quotation from congressman Doc Hastings in TCH, Apr. 11, 1999.

110 NYT, May 3, 1989; TCH, May 26, 1974. On Tri-Cities supporters of river development, see Portland Oregonian, Sept. 28, 1994. Booster interests were explained in a column in TCH, Feb 12, 1995. For a late-

1970s study of the feasibility of damming the Hanford Reach, see Harty, "Effects of the Ben Franklin Dam on the Hanford Site."

111   *TCH*, Nov. 6, 1987.

112   Portland *Oregonian*, Sept. 28, 1994; *TCH*, Oct. 9, 1994.

113   *TCH*, Feb. 12, 1995.

114   *NYT*, May 3, 1989; *ST*, July 2, 1992; Portland *Oregonian*, Sept. 28, 1994; National Park Service, Pacific Northwest Regional Office, "The Hanford Reach of the Columbia River: Comprehensive River Conservation Study and Environmental Impact Statement," draft, Seattle, June 1992, in authors' possession.

115   *TCH*, Apr. 11, 1999; *ST*, Nov. 6, 1999, Feb. 8, 2002.

116   *TCH*, Feb 12, 1995. At local meetings in 1999 on the disposition of the Wahluke Slope, the majority of speakers advocated *federal* control and minimal development of the slope. See *TCH*, May 21, June 4, 1999.

117   Portland *Oregonian*, Sept. 28, 1994.

118   C. E. Cushing, D. G. Watson, A. J. Scott, and J. M. Gurtiseu, "Decrease of Radionuclides in Columbia River Biota Following Closure of Hanford Reactors," *Health Physics* 41 (July 1981): 66; *TCH*, Apr. 24, June 28, 29, 1999.

119   Moses Lake *Columbia Basin Herald*, Dec. 21, 2005; bottle label, Desert Wind Winery, 2004 Wahluke Slope Semillon.

120   Richards, "Nuclear Site."

Notes to Epilogue

1   *The Phantom Empire*, Mascot Pictures, 1935. "Murania" appears to be a near-anagram of the word "uranium." On popular conceptions of mad scientists and their connection to the power of radioactivity in the interwar period, see Spencer R. Weart, "The Physicist as Mad Scientist," *Physics Today* 41 (June 1988): 28–37; Weart, *Nuclear Fear: A History of Images* (Cambridge, MA: Harvard University Press, 1988), part 1, and especially chap. 4. See also Lawrence Badash, "Radium, Radioactivity and the Popularity of Scientific Discovery," *Proceedings of the American Philosophical Society* 122 (1978): 145–54; Badash, *Radioactivity in America: Growth and Decay of a Science* (Baltimore: Johns Hopkins University Press, 1979). Autry settled in southern California in real life, and in the 1960s built up a big-league

baseball franchise, with its home in Anaheim. Disneyland, Anaheim's major attraction, resembled Radio Ranch and Richland in its
mixture of frontier and futuristic motifs. See John M. Findlay, *Magic
Lands: Western Cityscapes and American Culture After 1940* (Berkley:
University of California Press, 1992), chap. 2. Roy Rogers had similar
problems on his own ranch.

2   Merritt Roe Smith, *Harpers Ferry Armory and the New Technology:
The Challenge of Change* (Ithaca, NY: Cornell University Press, 1977).

3   The nuclear historian Richard G. Hewlett suggested the need to take
such an approach decades ago. See Hewlett, "Pioneering on Nuclear
Frontiers: Two Early Landmarks in Reactor Technology," *Technology
and Culture* 5 (Autumn 1964): 512–22.

4   Anthony F. C. Wallace, *St. Clair: A Nineteenth-Century Coal Town's
Experience with a Disaster-Prone Industry* (Ithaca, N.Y.: Cornell University Press, 1988), chap. 5; death figures for 1858 are given on p. 253.

5   The argument has been developed at greater length, with reference
to the entire American West, in John M. Findlay, "The Nuclear West:
National Programs and Regional Continuity Since 1942," *Journal of
Land, Resources, and Environmental Law* 24, no. 1 (2004): 1–15.

6   Barton C. Hacker, "Radiation Safety, the AEC and Nuclear Weapons
Testing: Writing the History of a Controversial Program," *Public Historian* 14 (Winter 1992): 31–53.

Bibliographic Essay

The endnotes to each chapter contain detailed references to the works consulted for this book. This essay offers starting points for historical research and summarizes the nature of selected types of materials. We begin with an assessment of the pertinent general literature, focusing particularly on secondary works, and then discuss the key primary sources relied upon for our four chapters. Regarding secondary materials, preference is given to books rather than articles or chapters, and, unless otherwise noted, all references are to the first edition of a publication.

The historical literature on U.S. nuclear weapons programs is quite extensive, with most of it framed as a national story. Three titles by Richard Rhodes—*The Making of the Atomic Bomb* (New York: Simon & Schuster, 1986); *Dark Sun: The Making of the Hydrogen Bomb* (New York: Simon & Schuster, 1995); and *Arsenals of Folly: The Making of the Nuclear Arms Race* (New York: Knopf, 2007)—offer a general introduction. A more institutional perspective is provided by the three volumes of *A History of the United States Atomic Energy Commission*: Richard G. Hewlett and Oscar E. Anderson, Jr., *The New World, 1939/1946* (University Park: Pennsylvania State University Press, 1962); Richard G. Hewlett and Francis Duncan, *Atomic Shield, 1947/1952* (University Park: Pennsylvania State University Press, 1969); and Richard G. Hewlett and Jack M. Holl, *Atoms for Peace and War, 1953-1961: Eisenhower and the Atomic Energy Commission* (Berkeley: University of California Press, 1989). Useful insights into the deliberations of the Atomic Energy Commission (AEC) may be found in Roger M. Anders, ed., *Forging the Atomic Shield: Excerpts from the Office Diary of Gordon E. Dean* (Chapel Hill: University of North Carolina Press, 1987). For a very useful survey of the toll of nuclear weapons programs on U.S. politics and society, consult Stephen I. Schwartz, ed., *Atomic Audit: The Costs and Consequences of U.S. Nuclear Weapons since 1940* (Washington, D.C.: Brookings Institution Press, 1998).

Nuclear weapons deserve to be treated as a national story because, beginning with the Manhattan Project during World War II, the development, manufacture, testing, transport, storage, and environmental fallout of those weapons were dispersed across the United States and beyond. Yet a number of specific sites played particularly large and lasting roles in this American story, and they have been well documented by scholars. Peter Bacon Hales, *Atomic Spaces: Living on the Manhattan Project* (Urbana: University of Illinois Press, 1997), interprets the first generation of bomb building through close study of society and culture at Oak Ridge, Tennessee; Los Alamos, New Mexico; and Hanford, Washington. During World War II, Oak Ridge was in many ways Hanford's counterpart in the Manhattan Project. Its story has been told by Charles W. Johnson and Charles O. Jackson, *City Behind a Fence: Oak Ridge, Tennessee, 1942–1946* (Knoxville: University of Tennessee Press, 1981). The postwar history of another institution that, like Hanford, arose from the research program at the Chicago Metallurgical Laboratory during World War II is given in Jack M. Holl, *Argonne National Laboratory, 1946–1996* (Urbana: University of Illinois Press, 1997).

While Oak Ridge was hardly the only "eastern" site to play a large role in America's nuclear weapons programs, the states and towns of the American West played host to a disproportionately large number of federal weapons programs in the years after 1942. Los Alamos, with its large complement of prominent scientists, has attracted more historical attention than any other site. For aspects of its story, see James W. Kunetka, *City of Fire: Los Alamos and the Atomic Age, 1943–1945* (Englewood Cliffs, NJ: Prentice-Hall, 1978); two works by Ferenc Morton Szasz, *The Day the Sun Rose Twice: The Story of the Trinity Site Nuclear Explosion, July 16, 1945* (Albuquerque: University of New Mexico Press, 1984), and *British Scientists and the Manhattan Project: The Los Alamos Years* (New York: St. Martins Press, 1992); and John Hunner, *Inventing Los Alamos: The Growth of an Atomic Community* (Norman: University of Oklahoma Press, 2004), which follows the town's development into the 1950s. On the engineering facility spun off from Los Alamos and operated by the Bell Telephone Company for the AEC, see Necah Stewart Furman, *Sandia National Laboratories: The Postwar Decade* (Albuquerque: University of New Mexico Press, 1990). In 1952 the University of California opened Lawrence Livermore Laboratory, a California counterpart to Los Alamos. There is as yet no book-length history of the Livermore site, but Hugh Gusterson offers a valuable anthropological study in *Nuclear Rites: A Weapons Laboratory at the End of the Cold War* (Berkeley: University of California Press, 1996). Also in 1952, the Atomic Energy Commission opened a facility at Rocky Flats, Colorado, for manufacturing plutonium "triggers" for nuclear weapons. With its industrial orientation and related pollution problems, Rocky Flats may have been the Cold War site that most resembled Hanford. Len Ackland narrates its history engagingly in *Making a Real Killing: Rocky Flats and the Nuclear West* (Albuquerque: University of New Mexico Press, 1999).

Considering its large and enduring role as well as its widespread impact on the environment, Hanford has not received enough historical attention — no doubt in large part because it was devoted to manufacturing rather than research and development, and because its leading employees were engineers rather than physicists. Michele Stenehjem Gerber, *On the Home Front: The Cold War Legacy of the Hanford Nuclear Site* (Lincoln: University of Nebraska Press, 1992), offers the most thorough account, emphasizing the plant's contamination of the surrounding environment. Revised editions in 2002 and 2007 have updated the story. Developments during World War II have been documented by S. L. Sanger, with Robert W. Mull, *Hanford and the Bomb: An Oral History of World War II* (Seattle: Living History Press, 1989). The title was reissued in expanded form as S. L. Sanger, *Working on the Bomb: An Oral History of WWII Hanford*, ed. Craig Wollner (Portland: Continuing Education Press, Portland State University, 1995). Other important sources for the wartime period at Hanford include Ronald L. Kathren, Jerry B. Gough, and Gary T. Benefiel, eds., *The Plutonium Story: The Journals of Professor Glenn T. Seaborg, 1939-1946* (Columbus, OH: Battelle Press, 1994), and David A. Hounshell, "Du Pont and the Management of Large-Scale Research and Development," in *Big Science: The Growth of Large-Scale Research,* ed. Peter Galison and Bruce Hevly (Stanford, CA: Stanford University Press, 1992), 236–61. In recent years, as the federal government overturned many of the secrets surrounding Hanford, the history of the site has been revisited by journalists, activists, scientists, social scientists, and litigants. Some of the products of their efforts are mentioned below.

In short, the history of U.S. nuclear weapons programs tends to have been written at either the national or the local level. We have surveys of the "entire" Manhattan Project and the Atomic Energy Commission, and we have studies of such outposts as Oak Ridge, Los Alamos, and Hanford that do not routinely look beyond site boundaries to understand each site's relationship to its surroundings. This book attempts a different focus, one that might be called "regional" history. We explore how Hanford interacted with an assortment of neighbors, and how that interaction affected or mediated relations between the local and the national. For us, "region" suggests more than one scale: as the individual towns that together became known as the Tri-Cities; as the Columbia Basin; as the state of Washington; as the Pacific Northwest; and as the American West. Each of these regions, of course, to a greater or lesser extent represents a social and cultural construction that came to be imposed on the "natural" setting. The idea of the Tri-Cities was very much the creation of local boosters and journalists, for example, while the construct of the Columbia Basin is based on a natural feature—a watershed—yet took on greater significance once the U.S. government redefined the place through planning, damming, and irrigating programs during the mid-twentieth century.

A few truly *regional* studies of U.S. nuclear weapons programs focus primarily on region at the scale of the American West. Patricia Nelson Limerick stood among the very first historians to sketch the outlines of the atomic West in *The Legacy of Conquest: The Unbroken Past of the American West* (New York: Norton, 1987), and "The Significance of Hanford in American History," in *Washington Comes of Age: The State in the National Experience,* ed. David H. Stratton (Pullman: Washington State University Press, 1992), 153–71. After she drew broad connections between nuclear weapons sites and regional history, others began to fill in considerable detail. On the history of the Columbia Basin in the twentieth century, including Hanford's place in it, see Richard White, *The Organic Machine: The Remaking of the Columbia River* (New York: Hill and Wang, 1995). For general studies, see Bruce Hevly and John M. Findlay, eds., *The Atomic West* (Seattle: University of Washington Press, 1998); John M. Findlay, "The Nuclear West: National Programs and Regional Continuity Since 1942," *Journal of Land, Resources, and Environmental Law* 24, no. 1 (2004): 1–15; and Bruce Hevly, "Hanford's Postwar Voices," in Sanger, *Working on the Bomb,* 227–42 (cited above). More specific accounts of regional atomic development include Tad Bartimus and Scott McCartney, *Trinity's Children: Living Along America's Nuclear Highway* (New York: Harcourt Brace Jovanovich, 1991); Raye Ringholz, *Uranium Frenzy: Boom and Bust on the Colorado Plateau* (New York: Norton, 1989); and Michael A. Amundson, *Yellowcake Towns: Uranium Mining Communities in the American West* (Boulder: University Press of Colorado, 2002). Thomas R. Wellock, *Critical Masses: Opposition to Nuclear Power in California, 1958–1978* (Madison: University of Wisconsin Press, 1998), sets anti-nuclear power activism in the context of California's political culture and, particularly, in the context of the emerging human potential movement in the Bay Area.

Our study draws as much on the historiography of science and technology as it does on the historiography of the American West. Although most nuclear history has been written from a political or institutional perspective, a number of works by historians of science and technology (in addition to those cited above) are particularly useful. Jeff Hughes, *The Manhattan Project: Big Science and the Atomic Bomb* (New York: Columbia University Press, 2002), puts the wartime Anglo-American nuclear weapons project into a broader context of the interactions between science and industry that is reminiscent of some of the industrial accounts produced in the 1950s, and provides important background for the approach we take in this volume. Mark Walker, *German National Socialism and the Quest for Nuclear Power, 1939-1949* (Cambridge: Cambridge University Press, 1989), is a landmark work for its nuanced and reasoned argument for decentering academic physicists and reconsidering the relationships between science and the state. Similarly, amid a large literature on science advisors in the United States since World War II, S. S. Schweber, *In the Shadow of the Bomb: Oppenheimer, Bethe and the Moral Responsibility of the Scientist* (Princeton, NJ: Princeton University Press, 2000), points out the ways in which moral autonomy was negotiated by important actors during the Cold War, and the extent to which the responses of individuals were conditioned by their experiences of science and society.

On the development of reactor technology in the United States, the AEC histories cited above, along with the history of Argonne Laboratory, are useful starting points. Brian Balogh, *Chain Reaction: Expert Debate and Public Participation in American Commercial Nuclear Power, 1945-1975* (Cambridge: Cambridge University Press, 1991), should be read in tandem with the earlier work by Steven L. Del Sesto, *Science, Politics and Controversy: Civilian Nuclear Power in the United States, 1946-1974* (Boulder, CO: Westview Press, 1979), for insights into how reactor designs for civilian power emerged within the political economy of the Cold War. For details on the influence of the U.S. Navy on reactor design, see Richard G. Hewlett and Francis Duncan, *Nuclear Navy 1946-1962* (Chicago: University of Chicago Press, 1974).

Damage to surrounding populations from nuclear production and, especially, weapons testing has been dealt with from a variety of perspectives. Attempts to formulate the basic questions concerning nuclear weapons effects, including the impacts of radioactive contamination, are recounted in M. Susan Lindee, *Suffering Made Real: American Science and the Survivors at Hiroshima* (Chicago: University of Chicago Press, 1994). On the cultural construction of risk from radioactive substances in the twentieth century, see Spencer Weart, *Nuclear Fear: A History of Images* (Cambridge, MA: Harvard University Press, 1988). The history of attempts to monitor and interpret worker and regional safety is chronicled in two volumes by Barton C. Hacker, *The Dragon's Tail: Radiation Safety in the Manhattan Project, 1942-1946* (Berkeley: University of California Press, 1987), and *Elements of Controversy: The Atomic Energy Commission and Radiation Safety in Nuclear Weapons Testing, 1947-1974* (Berkeley: University of California Press, 1994). Hacker has also written an insightful essay on the problems of writing history at the intersection of nuclear technology, federal bureaucracy, and public controversy entitled "Radiation Safety, the AEC and Nuclear Weapons Testing: Writing the History of a Controversial Program," in *The Public Historian* 14 (Winter 1992): 31–53. For the evolution of concepts of radiological safety, see J. Samuel Walker, *Permissible Dose: A History of Radiation Protection in the Twentieth Century* (Berkeley: University of California Press, 2000). Max S. Power discusses the political framework for cleaning up nuclear industrial sites, with special reference to his own experience in Washington State, in his book *America's Nuclear Wastelands: Politics, Accountability, and Cleanup* (Pullman: Washington State University Press, 2008).

While responding to issues raised in the secondary literature, this study depends heavily on investigations in primary sources. Those sources were generated in the six decades after 1940 by a number of entities, ranging from the federal government to individual women and men, and they may be found in all corners of the continental United States. As creator, supervisor, and chief beneficiary of nuclear weapons programs, regulator of nuclear power plants, and manager of nuclear waste projects, the national government was responsible for producing most of the primary sources available for the study of Hanford. Many government documents of course remain classified. Nonetheless, like virtually any other twentieth-century state endeavor, Hanford became the focus of an enormous number of primary sources generated by federal agencies. Between 1942 and 1946, the U.S. Army oversaw the Hanford Site, along with the rest of the Manhattan Project. Between 1946 and 1974, the U.S. Atomic Energy Commission presided over America's nuclear weapons programs, including Hanford. After the period between 1974 and 1977, when the Energy Research and Development Administration succeeded the AEC, the U.S. Department of

Energy (DOE) assumed responsibility for Hanford, and remains the lead agency at the site today.

The documents created on behalf of these successive federal entities may be found in several locations. In Richland, Washington, the U.S. Department of Energy maintains the Public Reading Room (PRR) of the Richland Operations Office. Its holdings include more than 45,000 items created by the army, the AEC, and the DOE, along with their contractors, since 1943. The collection, while not comprehensive, ranks first among collections of primary sources for historical studies of Hanford. Items in the PRR are customarily identified by document number (for example, "HAN-10970"). Of particular importance are the notes and diary of colonel Franklin T. Matthias, 1942–46. Also in Richland, a treasure trove of historical photographs is maintained on behalf of the DOE by the Photography Division at Rockwell Hanford Operations. Many historical documents and photographs may also be found online via the online Hanford Declassified Document Retrieval System (Hanford DDRS) at http://www2.hanford.gov/declass/.

Numerous federally generated records concerning Hanford remain in or near the nation's capital. The DOE Office of History and Heritage Resources, in Germantown, Maryland, maintains the Historical Research Center, which preserves and makes available records on U.S. nuclear weapons programs. We found those holdings, especially the Richland Diversification Files and the Energy History Collection, particularly helpful for understanding the AEC period at Hanford (1947–74). Of course, many army, AEC, and DOE records have been transferred to the U.S. National Archives and Records Administration (NARA). The collections pertinent to Hanford predominantly reside at the NARA branch (National Archives II) at College Park, Maryland (for example, the David Lilienthal Papers within the records of the AEC). At the end of World War II, moreover, a number of documents pertaining to Hanford's role in the Manhattan Project were combined with materials from the Oak Ridge, Tennessee, site and therefore came to be housed at the NARA Southeast Region branch in Morrow, Georgia (outside of Atlanta). We especially benefited from consulting the letters to and from Manhattan Project officials associated with Hanford in this collection.

The archival holdings hosted by federal agencies contain not only documents generated by government officials but also materials produced by the numerous contractors who operated nuclear weapons facilities under army, AEC, and DOE supervision. Employees of DuPont, General Electric (GE), and Battelle–Pacific Northwest Laboratories in particular played sizeable roles at Hanford, and holdings at PRR, DOE, and NARA amply document their efforts. While we did not visit the company archives of all contractors, we did consult the Hanford-related records of E. I. du Pont de Nemours and Company in the manuscripts collections of the Hagley Museum and Library, Wilmington, Delaware.

Our understanding of the towns surrounding Hanford begins with the information gathered by the U.S. Bureau of the Census and with the reports and plans of the army, DuPont, the AEC, and GE. While American government agencies and their corporate contractors were crucial founts of information concerning the communities and rural areas near Hanford, many other sources were consulted as well. Newspapers proved to be crucial. Federal officials and private contractors helped to start two newspapers during World War II: the *Sage Sentinel*, which served the Hanford Camp, and the *Richland Villager*, which served the wartime townspeople and which continued publication after the war. Hanford and the surrounding area were covered by Pasco's *Columbia Basin News* from 1950 to 1963, and by the *Tri-City Herald* after 1949. Pasco's publication was oriented more toward farmers, while the *Herald* focused more on the city readers in Pasco, Kennewick, and Richland. Hanford and the Tri-Cities received useful coverage as well from the *Spokane Spokesman-Review*, Portland's *Oregonian*, the *Seattle Times*, and the *Seattle Post-Intelligencer*. All four big-city papers at times undertook serious investigative reports on environmental issues and on the economics of the Washington Public Power Supply System (WPPPS) in the 1970s and 1980s. The Richland Public Library maintains a clippings file drawn from local papers, as well as files for a host of miscellaneous materials concerning the town.

The Special Collections Division of the University of Washington Libraries has several resources pertaining to community development in and around the Tri-Cities. See especially the Fred Clagett Papers (FCP), which contain not only the personal papers of a Hanford chemist who became a prominent civic official in Richland but also a comprehensive run of the minutes and records of the Richland Community Council (RCC), the advisory board consulted by AEC and GE between 1947 and 1958 regarding town matters; the Seattle Urban League Papers, which document conditions faced by African Americans in the Cold War Tri-Cities; the Nell Lewis McGregor Papers, especially her 1969 memoir "I Was at Hanford" concerning World War II experiences; and the William Witherup Papers, donated by a writer who grew up in the Tri-Cities during the 1950s.

The Special Collections Division of the University of Washington Libraries is an important starting point for understanding the political history of Hanford as well. It contains the voluminous papers of Henry M. Jackson and Warren G. Magnuson, two Democrats who served for decades as congressmen and senators from Washington. Both men took a close interest in Hanford and eventually came to toil on behalf of Tri-Cities interests. Their papers include extensive letters from constituents regarding Hanford; however, guidelines for the use of the papers require that the names of selected individual correspondents remain confidential. For information about Dixy Lee Ray, we are indebted to Erik Ellis, "Dixy Lee Ray, Marine Biology and the Public Understanding of Science in the United States, 1930–1970" (Ph.D. diss., Oregon State University, 2006). In discussing diversification at Hanford, we have drawn heavily from *Nucleonics*, a postwar journal devoted to the nuclear industry. Rather than investigate the WPPSS firsthand, we have relied on Daniel Pope, *Nuclear Implosions: The Rise and Fall of the Washington Public Power Supply System* (New York: Cambridge University Press, 2008).

Since the mid-1980s, a new generation of historical materials has appeared as people have begun to assess Hanford's impact on the environment and public health, and to develop programs to clean up the site. For instance, the DOE and the U.S. Centers for Disease Control and Prevention (CDC) sponsored the Hanford Environmental Dose Reconstruction Project (HEDR), which between 1987 and 1995 produced numerous reports touching on many aspects of how people became exposed to radiation from the site. The Government Publications Division, University of Washington Libraries, Seattle, is one repository for HEDR reports. Between 1988 and 2002, the U.S. government sponsored the Hanford Thyroid Disease Study (HTSD), which assessed radiation releases between 1944 and 1957. Conducted by the CDC and the Fred Hutchinson Cancer Research Center, Seattle, the study is online at http://www.cdc.gov/nceh/radiation/hanford/htdsweb/index.htm. Both HEDR and HTSD reflect the recent trend of scientists writing the history of Hanford. Their efforts have complemented the work of the attorneys whose lawsuits regarding the damage caused to downwinders from exposure to Hanford radiation are also revisiting Hanford's past. And by keeping abreast of the story of the troubled cleanup at Hanford, newspapers from around the Pacific Northwest continue to cover the history of the locale.

Index

dust storms. *See* Hanford Site, and winds

E. I. du Pont de Nemours and Company. *See* DuPont
East Pasco. *See* Pasco
economic diversification. *See* diversification, economic
Einstein, Albert, 4
Eisenhower, Dwight D., 162–63, 239
energy crisis, 66, 188–89, 230–31, 236
Energy Employees Occupational Illness Compensation Act of 2000, U.S., 248
Energy Research and Development Administration, U.S. (ERDA), 65, 187
engineers, 50–51, 53, 269. *See also* production mission
Environmental Protection Agency, U.S. (EPA), 72, 195, 188, 201, 257
environmentalism and environmentalists, 10–11, 193, 196–99, 204–5, 230, 232, 235–36, 240, 244–51, 255, 257–58, 269–70. *See also* Atomic Energy Commission; Columbia River Conservation League; Sierra Club
Eugene, Oregon, 242
Evans, Daniel J., 143, 153, 178, 186, 191, 192, 243

F reactor, 31, 32, 34–35, 91, 273
farmers and farming, 144, 176, 202, 204, 249, 269; before advent of Manhattan Project, 12, 209–10; as form of economic diversification, 249, 261–63; as influenced by Hanford, 158–59, 207, 211, 213–14, 218–22, 250. *See also* Bureau of Reclamation; Columbia Basin Project; Columbia Valley Authority; Wahluke Slope

Fast Flux Test Facility (FFTF), 11, 68–69, 186, 236–37, 240–42, 244, 245
Federal Facility Agreement and Consent Order. *See* Tri-Party Agreement
federal government, 4–5, 8–10, 73, 141–43, 196–98, 205, 211; and Tri-Cities identity, 77–78. *See also names of specific agencies*
Federal Housing Administration, U.S. (FHA), 113, 116
Fermi, Enrico, 17, 23, 32, 33, 210
Fremling, Alex, 256
frontier, 96–98, 100–102, 135, 166–67, 226–27. *See also* American West; pioneers and pioneering
fuel fabrication, 32, 240
Fyall, Adam J., 198

General Electric Company (GE), 46–47, 49, 54, 55, 90, 147; and breeder reactors, 239–40; and economic diversification, 161, 176–77, 181, 182; and management of Richland, 92–94, 105–7, 111–12, 117, 150–52
Gilstrap, Robert W., 184
Gore, Albert, 163
Gorton, Slade, 194, 244
Grand Coulee, 18
Grand Coulee Dam, 10, 12, 157, 211, 230
graphite creep, 91–92
Green Run, 57–58
Groves, Leslie, 16–19, 22–25, 28, 33, 35–38, 79, 146; and design of Richland, 85–88

H reactor, 42, 48
Hales, Peter Bacon, 212
Hall, A.E.S., 18–19
Hanford (town), 18, 22, 209